Engineering mechanics

Oxford Engineering Science Texts

Editors: A. M. Howatson, F. A. Leckie

J. M. Prentis: *Engineering mechanics*

A. C. Palmer: *Structural mechanics*

Engineering mechanics

J. M. Prentis
University of Cambridge

Clarendon Press · Oxford
1979

Oxford University Press, Walton Street, Oxford OX2 6DP
OXFORD LONDON GLASGOW
NEW YORK TORONTO MELBOURNE WELLINGTON
KUALA LUMPUR SINGAPORE JAKARTA HONG KONG TOKYO
DELHI BOMBAY CALCUTTA MADRAS KARACHI
NAIROBI DAR ES SALAAM CAPE TOWN

Published in the United States by Oxford University Press, New York

British Library Cataloguing in Publication Data

Prentis, James Martin
Engineering mechanics.—(Oxford engineering science texts).
1. Mechanics 2. Engineering
I. Title II. Series
531′.02′462 TA350 79-41137

ISBN 0-19-85620-5-5
ISBN 0-19-85620-6-3 Pbk

Typeset in Northern Ireland at The Universities Press (Belfast) Ltd.
and printed by The Pitman Press, Bath.

Preface

THE purpose of this book is to relate the motion of bodies to the forces that are applied to them. The body under consideration may be an electron in an oscilloscope, a car moving along a motorway, or one of the components of the engine that drives it; it may be a spacecraft under power from its rocket motors, or a satellite subject only to gravitational forces. In every case the question to be answered is basically the same: given the forces acting on it, what is the motion of a particle, or a body, or a system of particles and bodies?

The question can be posed the other way round: what forces are required to cause a body to move in a specific way?

Before these questions can be answered it is necessary to pay separate attention to the two entities 'force' and 'motion'. Both must be capable of being described and measured. We have then to understand the effect of combining two or more forces, or two or more simultaneous motions. The separate studies of force and motion are referred to respectively as statics and kinematics, and are taken together under the title dynamics.

The word statics suggests anything but motion, and is often taken to imply only a consideration of the equilibrium of stationary bodies, with the term kinetics being applied to forces involving motion. The distinction between statics and kinetics is, however, fine enough for us to abandon the latter term in favour of the former. We shall take mechanics to mean the application of dynamics in engineering, particularly in relation to machinery.

The members of a first-year class of engineering students at a university or polytechnic will usually differ widely from each other in the extent to which they have previously studied mechanics and mathematics. Some students will already be familiar with all the aspects of mathematics that are necessary to support the study of mechanics at the first-year level. Others will be much less well equipped. There is likely to be a similar variation in the students' previous experience of basic mechanics. Bearing this in mind, I attempt to provide a complete account of the subject for the absolute beginner while maintaining the interest of the reader with previous experience by presenting the material at a lively pace and in a context that is likely to be different from that which will have been met earlier.

I assume that the reader already has sufficient knowledge of elementary calculus, in particular that he can differentiate and integrate simple functions and can solve elementary differential equations. It is in relation to vector algebra that there may be considerable variation in the previous experience of readers. This aspect of mathematics is taken from scratch and is developed as it is required in what is intended to be a self-contained account of the mechanics. The coverage in mechanics is limited to what is likely to be generally accepted as being suitable for first-year undergraduate study in engineering.

Students are often impatient with wordy introductions, and perhaps for them what I have already written is a sufficient statement of my intensions. My colleagues will each have their own approach to our subject and may be interested to know why I have chosen to present the material in one way rather than another, and some further comments may be of interest to them.

The first two chapters attempt a full exploitation of the basic vectorial nature of forces and displacements: a displacement has direction and magnitude and addition is in accordance with the parallelogram rule. The same is true for forces. The use of vector products is circumvented in Chapters 1 and 2 to enable the student to whom the very idea of a vector is new to digest that idea. On the other hand, by treating kinematics before statics, practically the whole class is immediately offered something new. The second chapter, on statics, includes a brief introduction to material which might already have been more fully covered in Theory of Structures, and can in that case be omitted. Experience shows, however, that it is advisable to point out that the statics of a mechanism is not the same as for a conventional pin-jointed structure.

The concept of vector multiplication is likely to be new to a large proportion of the class unless it has already been covered in an undergraduate course. As well as providing a prelude to further work Chapter 3 introduces the important twin concepts of the instantaneous screw and the wrench.

Differentiation of vectors is introduced later and is used to reinforce the rigid-body kinematics in Chapter 1. Having assembled the necessary mathematical equipment, the mechanics is developed in the conventional order through particle dynamics and on to the elementary treatment of rigid-body dynamics. Separate chapters on virtual work and particle streams are included.

The book concludes with a chapter on power transmission. This looks like, and indeed is, something of a postscript. There are really two reasons for its inclusion. First, it is to introduce engineering applications of the basic mechanics presented earlier in the book that are too important to be treated in passing by way of one or two examples, but whose full treatment at an earlier stage would interrupt the general flow too much. Secondly, if the general content is thought of as providing a one-year course, the final chapter provides a useful revision of some of the principles developed earlier in a context where the emphasis is on the hardware rather than the mathematics.

Many of the set problems in this book have either been taken directly from examinations set in the Engineering Department of the University of Cambridge, or are based on questions set in these examinations. The author is grateful to the Syndics of the Cambridge University Press for permission to use these questions, and to the many colleagues who composed them.

Cambridge
September 1978

J. M. P.

Contents

1. Planar kinematics—finite displacements and instantaneous motion

1.1. Scalars and vectors

FORCE and motion are both said to be vector quantities. Vector quantities are distinguished by their nature from many other physical quantities such as temperature, mass, volume, density, work, energy, electric charge, and so on, primarily by the way in which they are measured and by the ways they may combine together. The latter group, temperature, mass, etc., are said to be scalar, and their common distinguishing feature is that, in calculations involving them, all the normal operations of simple arithmetic—addition, subtraction, multiplication, and division—apply. We take it as self-evident that masses may be added and subtracted, and similarly with volumes; density is taken by definition as mass divided by volume.

To describe the displacement of a particle from one point O to another point P it necessary to specify the direction of OP as well as its length. Likewise to describe a force through O, it is necessary to specify its direction as well as its magnitude. A vector is required, by definition, to have both a magnitude and a direction associated with it. This distinguishes it from a scalar, which is fully determined by its magnitude.

A second distinction between scalars and vectors comes from the way in which vectors combine together, and for which the rules of simple arithmetic are insufficient.

Consider the displacement of a particle, Fig. 1.1, first from O to P and thence to Q. The resulting movement is clearly from O to Q, with the distance and direction of OQ depending on the directions of the two component displacements as well as their magnitudes. It is natural to refer to the total displacement as the sum of the two component displacements and to write

$$\overrightarrow{OQ} = \overrightarrow{OP} + \overrightarrow{PQ},$$

where the superimposed arrows denote the vectorial nature of the displacements and emphasize that something more than the simple arithmetic addition of lengths is implied. We shall find it more convenient to label the vector itself, rather than its end-points, and to write

$$\mathbf{c} = \mathbf{a} + \mathbf{b},$$

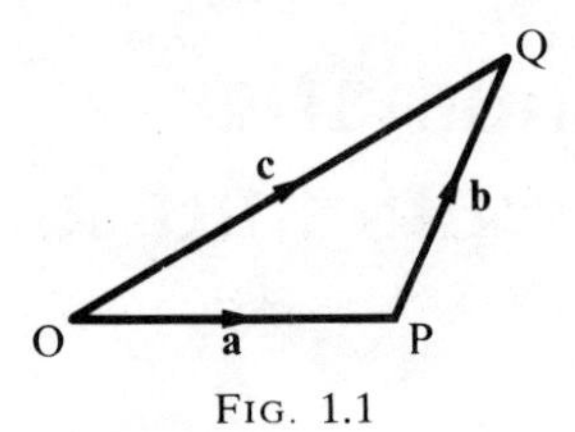

FIG. 1.1

using bold-face type to distinguish vectors from scalars, which will be written in italics, e.g. m. In manuscript it is usual to denote vectors by underlining, e.g.

$$\underline{c} = \underline{a} + \underline{b}.$$

1.2. The parallelogram rule: definition of a vector

If, instead of considering displacements from O to P and P to Q, attention is turned to the two component displacements **a** and **b**, it can be seen that the order in which they take place is immaterial. Indeed, they can happen simultaneously. It follows that **c** can be regarded as the resultant of **a** and **b**, and be represented by the diagonal between two adjacent sides **a** and **b** of a parallelogram, as in Fig. 1.2. Vector addition can therefore be regarded as being in accordance with either a triangle rule, Fig. 1.1, or a parallelogram rule, Fig. 1.2.

A vector quantity is defined as being one which is fully represented by a directed line segment and obeys the parallelogram (*or triangle*) *rule of addition.* For displacements, the length of the directed line segment equals the length of the displacement; in all other cases a suitable scale factor must be employed.

One consequence of this rule is that the resultant of two concurrent vectors must be coplanar with its components. It may be noted that, for the moment, we leave on one side consideration of the addition of two non-concurrent vectors.

As well as the displacement of a particle, its velocity and acceleration are vectors, as are force, momentum, angular velocity, angular momentum, electric field intensity, and so on. To justify this statement it is necessary to study each quantity separately. The need to specify direction as well as magnitude is in each case self-evident. Obedience to the parallelogram rule is less obvious. Sometimes this can be proved

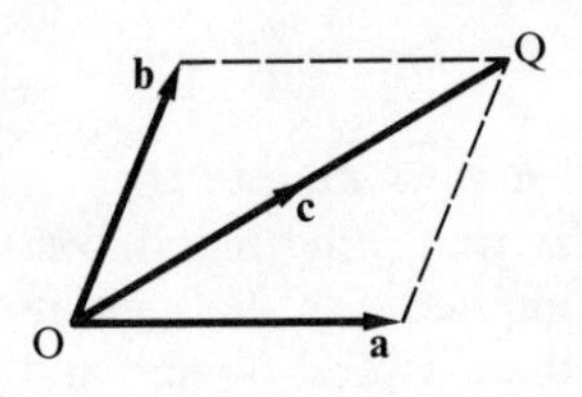

FIG. 1.2

mathematically, but in other cases it must be regarded simply as an observed or experimental fact. There is hardly any question of proving that particle displacements accord with the parallelogram rule; the evidence of our own eyes and common sense is sufficient.

If it is accepted, for the moment, that we can extend our study of particle displacements to demonstrate that particle velocity and acceleration are vectors too, and we also accept Newton's Second Law in the form that the force on a particle equals its mass multiplied by its acceleration as an experimental fact, we can deduce that force is a vector. It is, however, equally valid to regard satisfaction of the parallelogram rule by forces as a fact proven by direct experiment, using pieces of string and spring balances.

It must not be assumed that all quantities that are fully specified by a directed line-segment are thereby vectors. The rotation of a rigid body through a finite angle about a fixed axis can be represented by a line segment along that axis and with its length proportional to the angle of rotation. Verification that it is a vector would require the demonstration that a further rotation about another axis, assumed for simplicity to be concurrent with the first, leads to a resultant rotation which accords with the parallelogram rule. A simple experiment shows that this does not happen, and hence that a finite rotation cannot be regarded as a vector. Figure 1.3 shows a die on which, in accordance with standard practice, the number of spots on opposite faces always adds up to seven. In Fig. 1.3(a) the die is in position in relation to a set of cartesian axes $Oxyz$ where Oy comes vertically out of the page towards the reader. The 3-spot face is on top with the 1, 2, 5, and 6-spot faces vertical as indicated. In Fig. 1.3(b) the die has been turned through 90° about Oz, and in Fig. 1.3(c) has suffered a further rotation through 90° about Oy. The assumption that these rotations may be represented by vectors along Oz and Oy leads to the conclusion from the parallelogram rule that the resultant rotation is about a line in the Oyz plane. As it is clearly not possible to

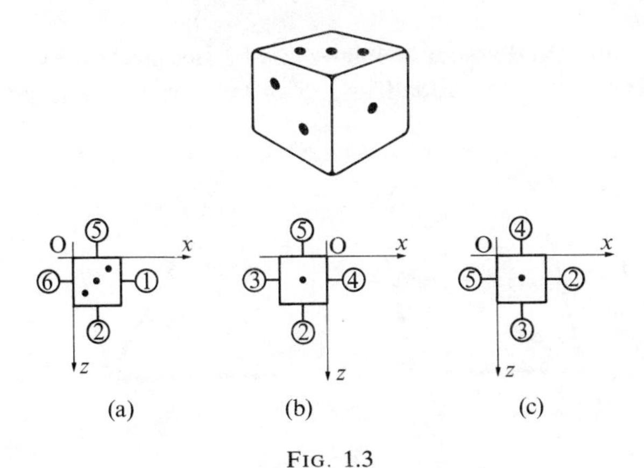

FIG. 1.3

return the die to its original orientation, with the 3-spot on the top, by any such rotation we must conclude that the parallelogram rule does not apply. In fact, the resultant rotation is about an axis through the origin O and the corner at the intersection of the 1, 2, and 3-spot faces.

1.2.1. The commutative law

The commutative law for vector addition is an immediate consequence of the parallelogram rule and states that vectors may be added in any order. There is no reason to assume that either **a** or **b** has precedence in Fig. 1.2, so that

$$\mathbf{a}+\mathbf{b}=\mathbf{c}=\mathbf{b}+\mathbf{a}.$$

It is of interest to note that finite rotations, which as we have seen may not be regarded as vectors, do not obey the commutative law. If the order of the rotations depicted in Fig. 1.3 is reversed, it is found that the final orientation is different, and so is the resultant axis of rotation.

1.2.2. The associative law

The associative law for vector addition states that in a repeated application of the parallelogram rule to add three or more vectors, these vectors may be grouped together arbitrarily. In Fig. 1.4

$$\begin{aligned}(\mathbf{a}+\mathbf{b})+\mathbf{c}&=\mathbf{e}+\mathbf{c}\\&=\mathbf{d}\\&=\mathbf{a}+\mathbf{f}\\&=\mathbf{a}+(\mathbf{b}+\mathbf{c}).\end{aligned}$$

Applying the commutative law,

$$\begin{aligned}\mathbf{a}+(\mathbf{b}+\mathbf{c})&=(\mathbf{b}+\mathbf{c})+\mathbf{a}\\&=\mathbf{b}+\mathbf{c}+\mathbf{a}.\end{aligned}$$

The commutative and associative laws can be summed up by stating that a string of vectors may be added in any order and with any grouping.

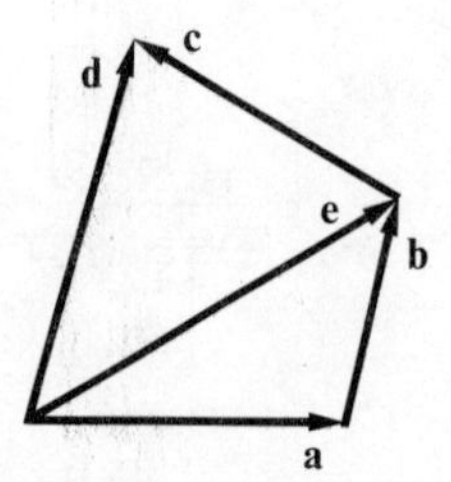

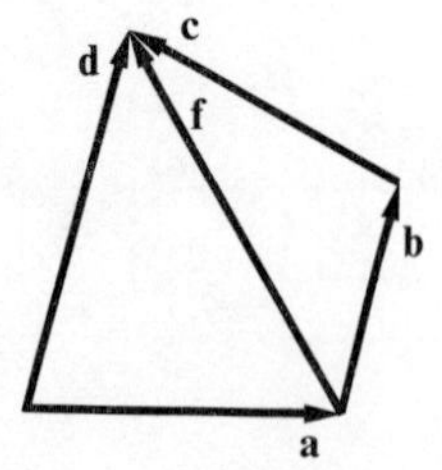

FIG. 1.4

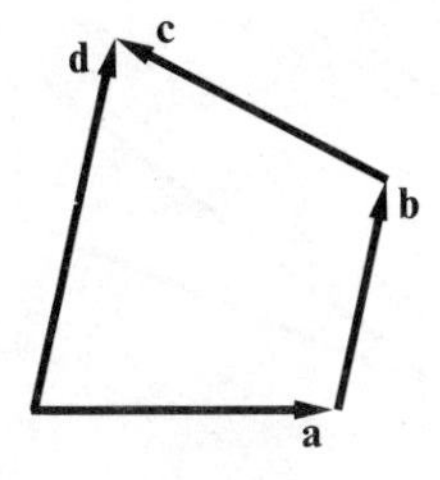

FIG. 1.5

1.2.3. The polygon rule

Further reference to Fig. 1.4 leads to the conclusion that it is not necessary to introduce either of the vectors **e** and **f** to arrive at **d**. If a number of vectors are represented by the properly directed sides of a polygon, the sum is represented by the closing side, and is in the direction from the starting point of the first vector in the summation to the terminal point of the last, Fig. 1.5.

It is to be noted that the vectors **a**, **b**, and **c** in Fig. 1.5 do not need to be coplanar, and that consequently the polygon is not necessarily a plane figure.

1.2.4. Subtraction

The operation of subtraction is defined by ruling that

$$\mathbf{a}-\mathbf{a}=0.$$

Writing this equation as

$$\mathbf{a}+(-\mathbf{a})=0$$

and applying the triangle rule we see that $(-\mathbf{a})$ is equal in length to **a** but points in the opposite direction, Fig. 1.6(a). To evaluate $(\mathbf{a}-\mathbf{b})$ we add $(-\mathbf{b})$ to **a**, Fig. 1.6(b).

1.3. The velocity of a particle

We have observed that the position of a particle can be defined by a vector drawn from a fixed point, or origin, to the particle, and that displacements of the particle are represented by vectors drawn between the successive locations.

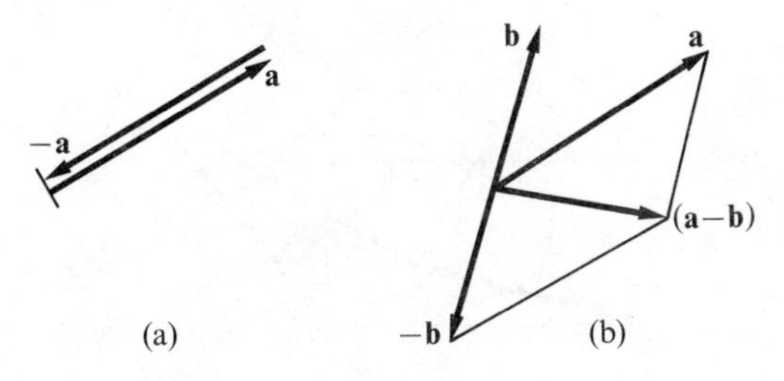

FIG. 1.6

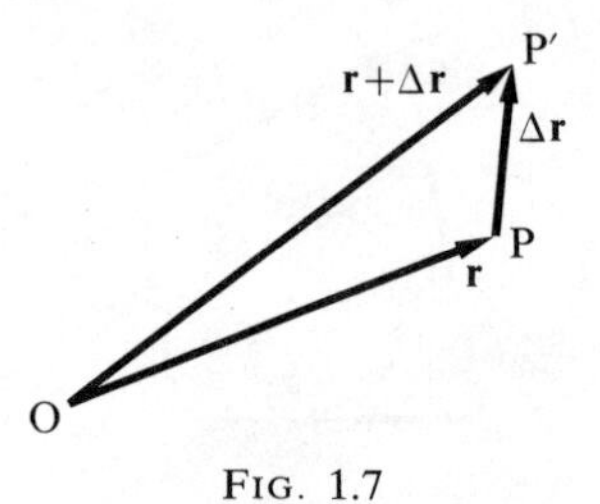

FIG. 1.7

The velocity of a particle is defined as its instantaneous rate of change of position. Thus, if in a small time interval Δt a particle moves from P to P′, Fig. 1.7, with its positions defined by $\mathbf{r}$ and $\mathbf{r}+\Delta\mathbf{r}$ respectively, the change of position is $\Delta\mathbf{r}$ and its velocity is

$$\mathbf{v} = \lim_{\Delta t \to 0} \frac{\Delta \mathbf{r}}{\Delta t} = \frac{\mathrm{d}\mathbf{r}}{\mathrm{d}t}. \tag{1.1}$$

It is to be noted that $\mathrm{d}\mathbf{r}/\mathrm{d}t$ is not, in general, in the same direction as $\mathbf{r}$. Clearly $\mathbf{v}$ has both magnitude and direction, and it may be considered self-evident that it is a vector. The matter is settled beyond all doubt by demonstrating adherence to the parallelogram rule. For this it is sufficient to refer back to Figs. 1.1 and 1.2, and to replace the finite vectors $\mathbf{a}$, $\mathbf{b}$, and $\mathbf{c}$ by infinitesimal vectors $\Delta\mathbf{a}$, $\Delta\mathbf{b}$, and $\Delta\mathbf{c}$.

1.4. Finite displacements of a lamina

More information is needed to describe the displacement of a lamina in its own plane than can be contained in a single vector. The most natural way to describe the displacement of a lamina from position 1 to position 2, Fig. 1.8, is in terms of the displacement of a point such as A and an accompanying rotation. Clearly there are alternative ways of specifying the displacement, e.g.

$$(A_1 \to A_2) + \text{rotation } \theta$$

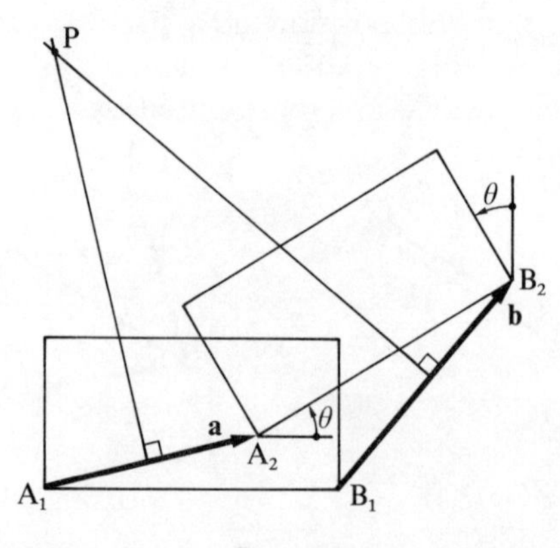

FIG. 1.8

or

$$(B_1 \rightarrow B_2) + \text{rotation } \theta$$

The figure makes it clear that, whilst $\mathbf{a} \neq \mathbf{b}$, the rotation is θ in both cases. It may be concluded that: *any displacement of a lamina in its own plane is equivalent to a translation of the lamina accompanied by a rotation.* The amount of the translation depends on the location of the reference point (e.g. A or B) in the moving plane; the angle of rotation is the same for all such reference points.

This conclusion applies also to spatial (i.e. three-dimensional) displacements of a rigid body. It may be noted that the direction of the axis of rotation is the same for all reference points as well as the angle of rotation.

If the amount of translation depends on the choice of reference point, can this point be so chosen that the translation is zero? For plane motion the answer is yes. The displacement in Fig. 1.8 can clearly be achieved by a rotation θ about P, which lies at the intersection of the perpendicular bisectors of A_1A_2 and B_1B_2. We conclude that: *a general displacement of a lamina in its own plane is equivalent to a rotation about a unique point (the pole) in the plane.* In the special case when the pole is at infinity the displacement is a pure translation.

Except in special cases, a spatial displacement is not equivalent to a rotation about a pole. In Fig. 1.9, ABCD is the plan view of a rectangular body. It suffers a translation so that A_2D_2 is collinear with A_1D_1 and a rotation of 180° about AD. It can be seen by inspection that it is not possible to achieve the same displacement by a rotation about a single point. In fact this particular displacement has already been described in the simplest possible terms. Of all the points in the body those along AD move the least amount, and in a direction which is parallel to the axis of rotation. The displacement can be visualized as the result of a screwing motion about AD. Spatial motion is clearly more complex than plane motion, and it must not be assumed that conclusions reached regarding plane motion may be easily extended to spatial motion. The rest of this chapter is devoted to plane motion.

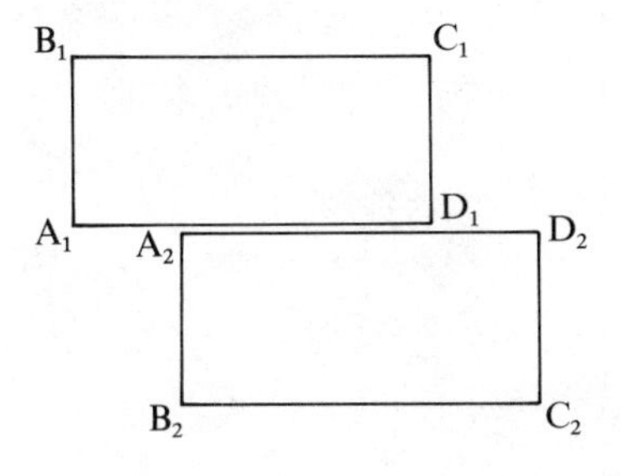

FIG. 1.9

1.5. The velocity field of a lamina

1.5.1. The instantaneous centre

The way in which the velocity varies throughout a body in plane motion, that is its *velocity field*, can be deduced by allowing the body to suffer an infinitesimal displacement over a time interval Δt and proceeding to the limit as Δt tends to zero. Referring to Fig. 1.8, it can be seen that there are alternative ways of describing the instantaneous motion of the body, thought of as a lamina for convenience. The immediate alternatives are:

$$\left.\begin{matrix}\mathbf{v}_A\ (=\mathrm{d}\mathbf{a}/\mathrm{d}t)\\ \mathbf{v}_B\ (=\mathrm{d}\mathbf{b}/\mathrm{d}t)\end{matrix}\right\}\text{together with a rotation } \theta.$$

If the lamina is extended to encompass the pole P the velocity there is zero, as there is no displacement at that point in the time interval Δt. In the context of instantaneous motion the pole is usually referred to as the instantaneous centre, and is labelled I. The instantaneous motion is fully specified, as far as the first derivative with respect to time, by the location of I and the value of $\mathrm{d}\theta/\mathrm{d}t$. The anticlockwise direction is taken as positive by convention.

Figure 1.10 shows Fig. 1.8 redrawn to leave A_1B_1 and A_2B_2 to represent the two positions of the lamina, and with A_1, A_2, B_1, and B_2 joined to P ($\equiv$I).

The displacement from position 1 to position 2 is achieved by a rotation of triangle AIB about I through the angle $\Delta\theta$. As I lies on the perpendicular bisectors of A_1A_2 and B_1B_2, triangles A_1IA_2 and B_1IB_2 are similar. It follows that the lengths A_1A_2 and B_1B_2 are in the same ratio as IA_1 and IB_1. Also, as $\Delta\theta \to 0$, the angle $PA_1A_2 \to 90°$. Hence: *at any instant the velocity of any point A in a lamina moving in its own plane is in a direction perpendicular to IA, where I is the instantaneous centre, and of magnitude* $(\mathrm{d}\theta/\mathrm{d}t)\times IA$.

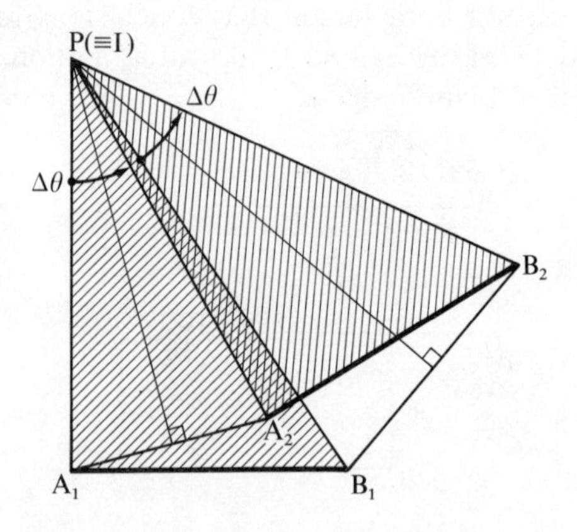

FIG. 1.10

For a point whose position with respect to the instantaneous centre is $\mathbf{r}$, the velocity $\mathbf{v}$ will be written $\boldsymbol{\omega}\times\mathbf{r}$ and spoken of as $\boldsymbol{\omega}$ *cross* $\mathbf{r}$. This will be taken to denote a vector perpendicular to $\mathbf{r}$ of magnitude $\omega\mathbf{r}$, where $\omega=|\boldsymbol{\omega}|=\mathrm{d}\theta/\mathrm{d}t$. To write $\mathbf{v}=\omega\mathbf{r}$ would be incorrect as it would imply that $\mathbf{v}$ is parallel to $\mathbf{r}$, which it clearly is not. We have yet to prove that the angular velocity $\boldsymbol{\omega}$ is a vector quantity, and hence that it is in accordance with our convention to print it in bold type. However, as long as we are concerned only with planar motion, the point is immaterial. The relationship now proposed, $\mathbf{v}=\boldsymbol{\omega}\times\mathbf{r}$, will be found to be consistent with what follows later on spatial motion. The velocities at a series of points A, B, C in a lamina are therefore as depicted in Fig. 1.11, the velocity distribution along typical vectors $\mathbf{r}_A$, $\mathbf{r}_B$, and $\mathbf{r}_C$ being as shown.

The velocity field shown in Fig. 1.11 is the same as that for a lamina rotating with angular velocity $\boldsymbol{\omega}$ about a fixed point I. It is, however, misleading to refer to I as the instantaneous centre 'of rotation', and to $\boldsymbol{\omega}$ as the angular velocity about I, as this may be taken, incorrectly, to imply that I is the centre of curvature of the paths of A, B, and C. The reason why I is not the centre of curvature of the path at A, say, is that in locating I and the velocity of A we have considered only two infinitesimally separated positions of the lamina, and hence only two locations of A in its path. To locate the centre of curvature of the path of A it is necessary to allow a second infinitesimal displacement to obtain a third point on the path of A. It will be found that the pole for the second displacement is different from the first, albeit that the difference is infinitesimal. This matter can be most simply illustrated by considering the motion of a circular disc as it rolls without slip along a straight line, Fig. 1.12. The velocity is zero, and the instantaneous centre of the disc is located, at the point of contact between the disc and the line. The velocity of the centre C of the disc is perpendicular to IC, as in Fig. 1.11. However, as the radius of curvature of the path of C is infinite, I is clearly not the centre of curvature of that path.

Figures 1.11 and 1.12 both tend to give the impression that $\boldsymbol{\omega}$ is the angular velocity about I. But this is a needlessly restrictive way of

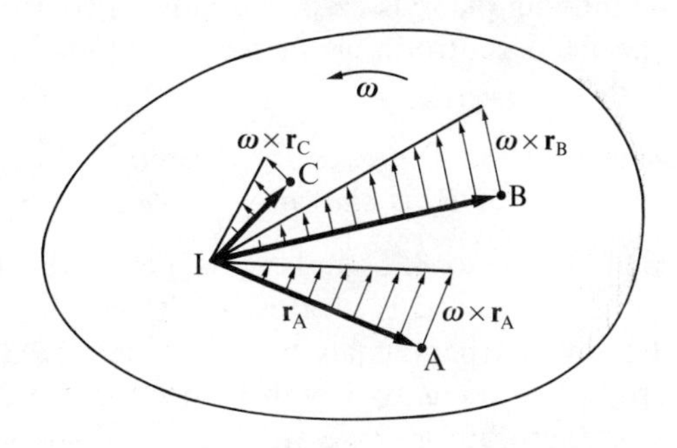

FIG. 1.11

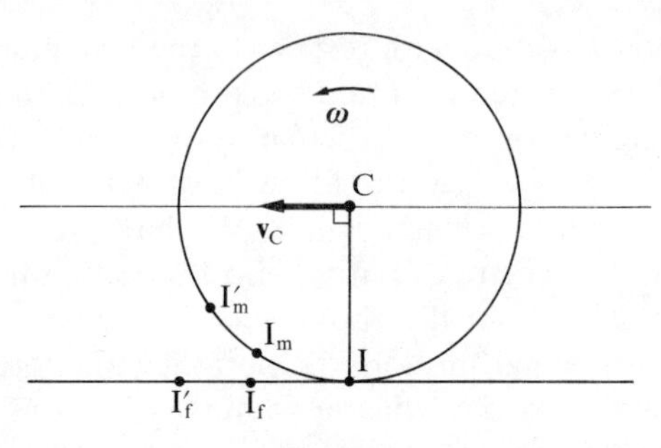

FIG. 1.12

regarding the angular velocity. In Fig. 1.10 the triangle APB turns bodily through an angle $\Delta\theta$. This means, for example, that the angle between A_1B_1 and A_2B_2 is also $\Delta\theta$, and more generally that every line drawn in the plane is turned through the same angle. Hence to determine the rotation of a plane it is sufficient to measure the angle between the two positions of any line in the plane. The angular velocity is the rate of change of orientation of any such line.

1.5.2. Centrodes

Figure 1.12 shows that although the instantaneous centre is the point of zero velocity, its location changes continuously. After a small time interval, point I_m on the moving disc will coincide with point I_f on the fixed line, and will be the instantaneous centre at that instant. Later I'_m will coincide with I'_f, and so on. Hence, notwithstanding that I is the point of zero velocity in the moving plane, it is continuously changing its position both in that plane and in the fixed plane with respect to which the motion is being observed. The loci traced in the moving and fixed planes are called the moving centrode and fixed centrode respectively. Any planar motion can be described in terms of the centrodes, though they are not usually as obvious as in Fig. 1.12, where the original circle and the straight line are themselves the moving and fixed centrodes.

As the point of contact between the two centrodes is always at I, where the velocity of the moving plane is zero, the moving centrode always rolls without slip on the fixed centrode, as in this example. Two other examples will be studied.

Example 1.1: Determine the centrodes for the rod AB in Fig. 1.13 constrained by guides at A and B as shown.

As $\mathbf{v}_A$ is perpendicular to IA, and $\mathbf{v}_B$ is perpendicular to IB, the instantaneous centre of the rod is located as shown. As the axes of the guides at A and B are perpendicular to each other $\angle AIB = 90°$ for all positions of the rod. The locus of I with respect to the rod, that is the moving centrode, is therefore a circle with AB as diameter.

To determine the fixed centrode we note that $OI = AB$, and hence that

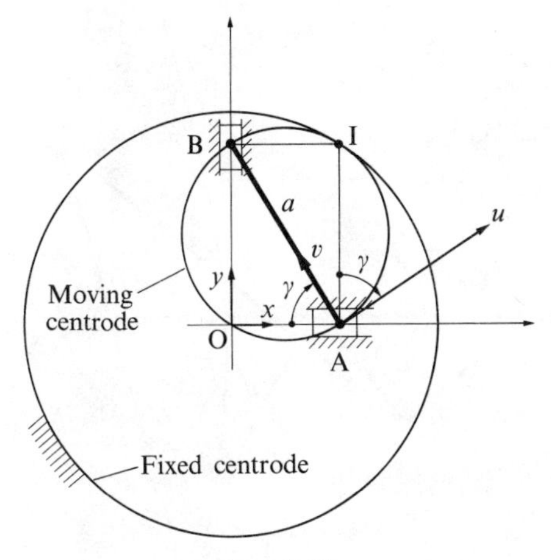

FIG. 1.13

it is of fixed length. The fixed centrode is therefore also a circle and of radius equal to AB with its centre at O.

Equations for the moving and fixed centrodes are derived by expressing the coordinates of I with respect to two sets of Cartesian axes, one set (u, v) being attached to the moving member, the other (x, y) being fixed. The two origins A and O are chosen for convenience. In the fixed frame of references the coordinates of I are

$$x = a \cos \gamma \quad \text{and} \quad y = a \sin \gamma.$$

On eliminating γ the equation of the fixed centrode is found to be

$$x^2 + y^2 = a^2.$$

In the moving frame of reference the coordinates of I are

$$u = a \cos \gamma \sin \gamma \quad \text{and} \quad v = a \sin^2 \gamma.$$

On eliminating γ the equation of the moving centrode is found to be

$$u^2 = v(a - v)$$

$$\Rightarrow \qquad u^2 + (v - a/2)^2 = (a/2)^2.$$

Example 1.2: In Fig. 1.14, ABCD is a four-bar mechanism; AD is a fixed link and carries at its ends cranks AB and DC, the joints at A and D being hinges whose axes are perpendicular to the plane of the paper. The two cranks are connected through similar hinges to the coupler BC. In this particular case AD = BC = a, AB = CD = b, and the coupler and fixed link cross as shown. Determine the centrodes for BC.

At the given instant AB is being turned at an unspecified rate. The velocities $\mathbf{v}_B$ and $\mathbf{v}_C$ are in directions that are perpendicular to cranks AB and CD respectively, so that the instantaneous centre I of BC is at the intersection of AB and CD produced.

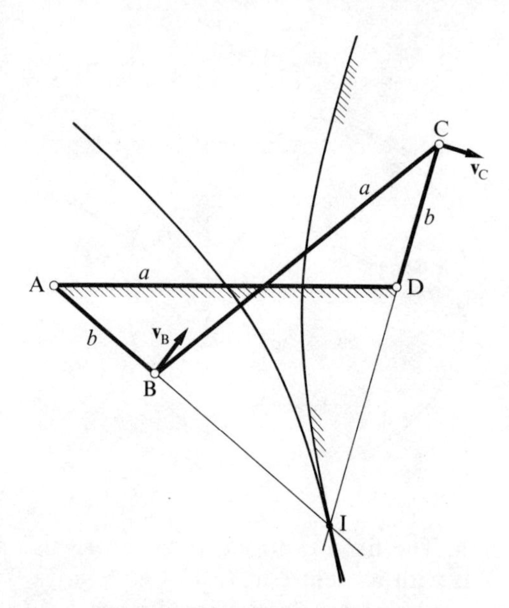

FIG. 1.14

It follows from the symmetry of the figure that $AI - ID = b$. Hence the locus of I with respect to the fixed link is such that the differences of its distances from the two fixed points A and D is a constant ($= b$). The fixed centrode is therefore a hyperbola with foci at A and D.

The moving centrode is a similar hyperbola with foci at B and C.

1.5.3. Velocity image

Figure 1.15 shows three points A, B, and C of a lamina moving in its own plane, its instantaneous centre at I, and with angular velocity $\boldsymbol{\omega}$. The velocities of A, B, and C are $\boldsymbol{\omega} \times \mathbf{r}_A$, $\boldsymbol{\omega} \times \mathbf{r}_B$, and $\boldsymbol{\omega} \times \mathbf{r}_C$, and are represented by vectors *oa*, *ob*, and *oc* respectively, where *o* is an arbitrary fixed

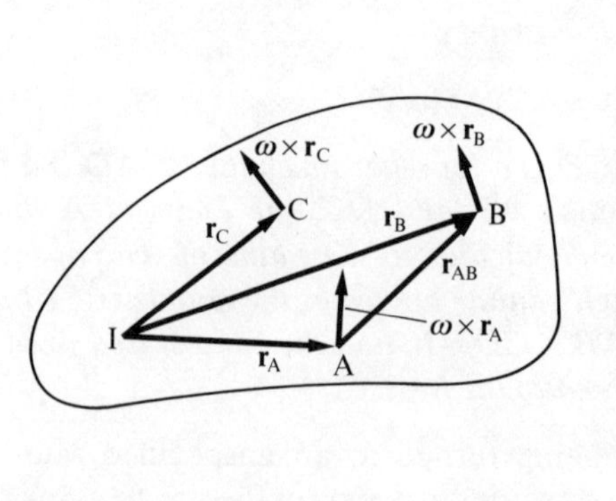

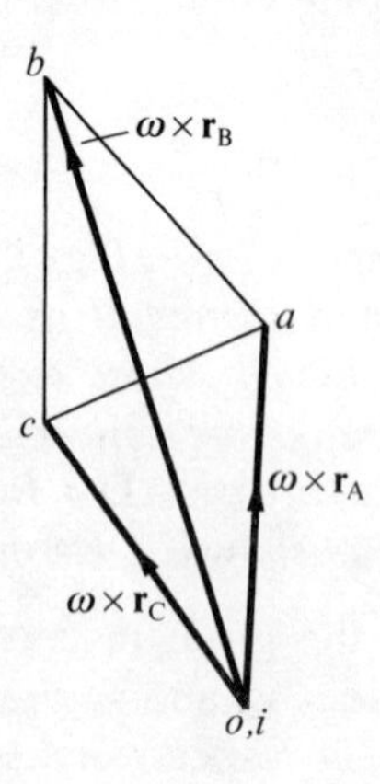

FIG. 1.15

origin. To distinguish between the two diagrams in Fig. 1.15 we will refer to them respectively as the space diagram and the velocity diagram.

As I has zero velocity i in the velocity diagram must be concurrent with o, as shown. The vector ac represents the difference between the velocities of points A and C. When drawn from a to c it represents $(\mathbf{v}_C - \mathbf{v}_A)$, and when drawn from c to a it represents $(\mathbf{v}_A - \mathbf{v}_C)$. Now $oa = |\boldsymbol{\omega} \times \mathbf{r}_A| = \omega \mathrm{IA}$. Similarly $oc = \omega \mathrm{IC}$. It follows that $ac = \omega \mathrm{AC}$ and that $cb = \omega \mathrm{CB}$. We conclude that triangle abc is similar to triangle ABC: it is referred to as the *velocity image* of ABC.

This argument can be extended to apply to any such group of points of the moving plane, and the general conclusion is embodied in the *image theorem: if A, B, C ⋯ are the vertices of a polygon drawn on a lamina moving in its own plane, the associated points a, b, c ⋯ in the velocity diagram are at the vertices of a geometrically similar polygon.* As ia is perpendicular to IA, it can be concluded further that the polygon in the velocity diagram is turned through 90°, in the direction of $\boldsymbol{\omega}$, relative to the polygon in the space diagram.

The velocity field is fully determined if the velocities of two points A and B are known, because the image theorem can then be used to determine the velocity of any other point such as C. It is to be noted that the velocities of A and B can not be chosen completely arbitrarily, for it is essential that ab in the velocity diagram should be perpendicular to AB in the space diagram. The relationship between $\mathbf{v}_A$ and $\mathbf{v}_B$ is very important.

1.5.4. Relationship between the velocities of points of a lamina in plane motion

Referring to Fig. 1.15, as triangles IAB and iab are similar

$$\vec{ab} = \boldsymbol{\omega} \times \mathbf{r}_B - \boldsymbol{\omega} \times \mathbf{r}_A = \boldsymbol{\omega} \times \mathbf{r}_{AB}$$

but

$$\vec{ab} = \mathbf{v}_B - \mathbf{v}_A,$$

hence

$$\mathbf{v}_B = \mathbf{v}_A + \boldsymbol{\omega} \times \mathbf{r}_{AB}. \tag{1.2}$$

This equation is fundamental; it shows that, insofar as B may be any point in the lamina, the velocity field is fully determined if the velocity of one point such as A is known together with the angular velocity ω of the plane.

1.6. Relative motion

The foregoing work assumes that the motion of the moving body is specified relative to a fixed frame of reference. Such a specification describes the motion as it would be seen by an observer who is stationary in the fixed frame of reference. The motion perceived by the observer would be different if he were himself moving. His description of the motion would be with reference to a frame attached to the vehicle that carried him. There are really two aspects to the matter: if the motions of two bodies are specified relative to a fixed frame of reference, it is of interest to determine the motion of one body relative to the other, that is the motion which would be seen by an observer sitting on one of the bodies. If the motion of one body is specified relative to the fixed frame of reference, and the motion of the second body is specified relative to the first, it is of interest to determine the motion of the second body relative to the fixed frame of reference.

1.6.1. Relative motion between a particle and a moving lamina

Fig. 1.16(a) shows a lamina for which the velocity $\mathbf{v}_A$ and the angular velocity ω are known. A particle P is situated at B on the lamina, though for convenience the figure shows P and B slightly apart. Taking A as the origin for the plane, the position of P is defined by $\boldsymbol{\rho}$. The velocity of the particle relative to the plane is the rate of change of $\boldsymbol{\rho}$. To emphasize that this is only part of the total velocity of P it is helpful to write this relative velocity as $\partial\boldsymbol{\rho}/\partial t$, where ∂ is referred to as *curly d*. The absolute displacement of P in an infinitesimal time interval δt is clearly the vector sum of the displacement of B and the displacement of the particle relative to the lamina. It follows that

$$\begin{aligned}\mathbf{v}_P &= \mathbf{v}_B + \partial\boldsymbol{\rho}/\partial t \\ &= \mathbf{v}_A + \boldsymbol{\omega}\times\mathbf{r}_{AB} + \partial\boldsymbol{\rho}/\partial t.\end{aligned} \tag{1.3}$$

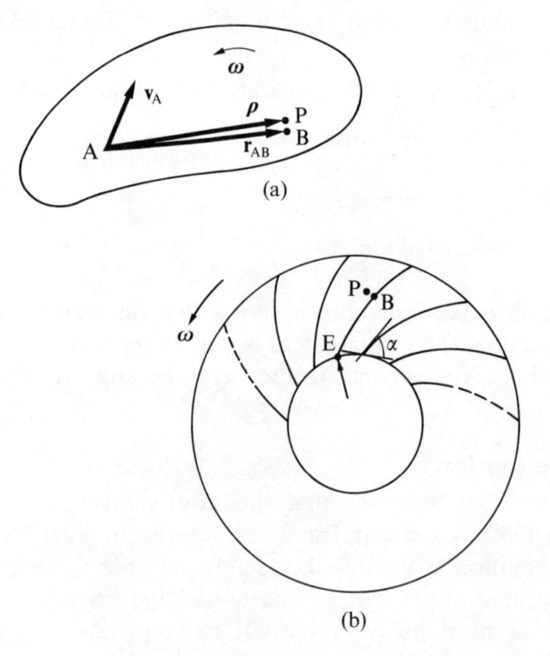

FIG. 1.16

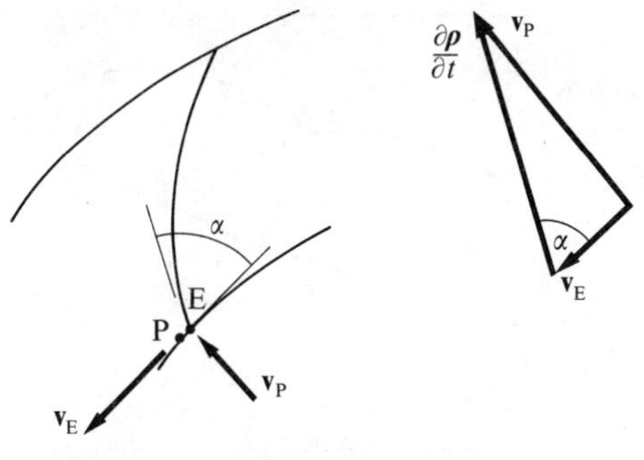

FIG. 1.17

At the given instant $\mathbf{r}_{AB} = \boldsymbol{\rho}$ so

$$\mathbf{v}_P = \mathbf{v}_A + \boldsymbol{\omega} \times \boldsymbol{\rho} + \partial\boldsymbol{\rho}/\partial t. \tag{1.4}$$

Example 1.3: The velocity of particles of fluid in the impeller of a centrifugal pump.

In Fig. 1.16(b) water flows outward through the impeller of a centrifugal pump. The velocity of the impeller blade at a typical point B is determined by the drive velocity $\boldsymbol{\omega}$. The velocity of a particle of water that is in contact with the impeller at B is given by eqn (1.3). The direction of $\partial\boldsymbol{\rho}/\partial t$ at B must be tangential to the blade of the impeller. Its magnitude depends on the geometry of the impeller: usually the axial width of the impeller decreases with radius so as to keep the radial component of the fluid velocity constant. Conditions at entry to the impeller, point E in Fig. 1.17, are particularly interesting. The velocity of the water particles $\mathbf{v}_P$ is often constrained to be in the radial direction. The velocity of the impeller $\mathbf{v}_E$ is then normal to the fluid velocity. If these two velocities are given then $\partial\boldsymbol{\rho}/\partial t$ is fully determined by the equation

$$\mathbf{v}_P = \mathbf{v}_E + \partial\boldsymbol{\rho}/\partial t.$$

However, the instant that a particle of water actually enters the rotor, the relative velocity $\partial\boldsymbol{\rho}/\partial t$ must be tangential to the blade. To avoid a sudden change in the velocity of the water particles, and consequential shock losses, the angle α between the blade and the circumferential direction is chosen so as to equal $\tan^{-1}(v_P/v_E)$.

1.6.2. Relative motiom of two laminae

In Fig. 1.18 lamina 1 is fixed and laminae 2 and 3 are moving in their own plane but independently of each other. An observer is riding on plane 2. In a finite time interval Δt he observes that lamina 2 has turned through an angle $\Delta\theta_{12}$ relative to lamina 1. Likewise he observes that lamina 3 has turned through an angle $\Delta\theta_{23}$ relative to lamina 2. The total angle

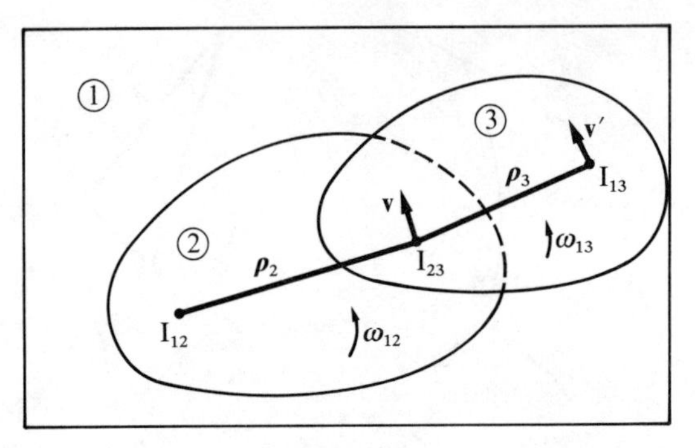

FIG. 1.18

through which lamina 3 has turned is clearly the sum of these two angles so

$$\Delta\theta_{13}=\Delta\theta_{12}+\Delta\theta_{23}.$$

The corresponding relationship between angular velocities is

$$\boldsymbol{\omega}_{13}=\boldsymbol{\omega}_{12}+\boldsymbol{\omega}_{23}. \tag{1.5a}$$

$\boldsymbol{\omega}_{13}$ is the angular velocity of plane 3 perceived by an observer on the fixed plane. An observer riding on plane 3 will have the impression that plane 1 has an angular velocity $\boldsymbol{\omega}_{31}$ which is equal and opposite to $\boldsymbol{\omega}_{13}$ so $\boldsymbol{\omega}_{31}=-\boldsymbol{\omega}_{13}$. An alternative form for eqn (1.5a) is therefore

$$\boldsymbol{\omega}_{12}+\boldsymbol{\omega}_{23}+\boldsymbol{\omega}_{31}=0. \tag{1.5b}$$

The instantaneous centre for lamina 2 is at I_{12}. An observer sited there, whether on plane 1 or plane 2 would perceive no relative motion apart from the angular velocity. Lamina 3 moves independently of laminae 1 and 2 and its instantaneous centre is at I_{13}. An observer on plane 2 will not readily locate I_{13}: he will be aware that at I_{12} the velocity relative to the ground is zero, and that the velocity of 3 relative to 2 is zero at I_{23}, which is the instantaneous centre for the relative motion between 2 and 3. To locate I_{13} we note first that at I_{23}

$$\mathbf{v}=\boldsymbol{\omega}_{12}\times\boldsymbol{\rho}_2$$

and that this velocity is common to laminae 2 and 3. At I_{13}

$$\begin{aligned}\mathbf{v}'&=\mathbf{v}+\boldsymbol{\omega}_{13}\times\boldsymbol{\rho}_3\\&=\boldsymbol{\omega}_{12}\times\boldsymbol{\rho}_2+\boldsymbol{\omega}_{13}\times\boldsymbol{\rho}_3.\end{aligned}$$

But at I_{13} the velocity is zero, so

$$\boldsymbol{\omega}_{12}\times\boldsymbol{\rho}_2+\boldsymbol{\omega}_{13}\times\boldsymbol{\rho}_3=0.$$

If the sum of two vectors is zero they must be equal in magnitude but opposite in direction. But $\boldsymbol{\omega}_{12}\times\boldsymbol{\rho}_2$ is perpendicular to $\boldsymbol{\rho}_2$, and $\boldsymbol{\omega}_{13}\times\boldsymbol{\rho}_3$ is perpendicular to $\boldsymbol{\rho}_3$. It follows that $\boldsymbol{\rho}_2$ and $\boldsymbol{\rho}_3$ are collinear, and hence that I_{12}, I_{23}, and I_{13} are collinear.

Although, to simplify the argument, we have assumed lamina 1 to be fixed, this condition is not essential. The conclusion is summed up in the *three-centre (Kennedy–Aronhold) theorem: if three laminae are in co-planar motion their three relative instantaneous centres are collinear.*

The three-centre theorem can be used to locate all the relative instantaneous centres for the members of a planar linkage mechanism, for instance the four-bar mechanism in Fig. 1.19. The four links are numbered 1 to 4. If link 1 is fixed it can be argued that, as the velocity of B is perpendicular to AB and the velocity of C is perpendicular to CD, the instantaneous centre for BC is at the intersection of AB and CD as in Fig. 1.14. Alternatively the three-centre theorem can be used: I_{12} coincides with the axis of hinge A which connects links 1 and 2; I_{23} lies similarly at B. Hence, by the three-centre theorem, I_{13} lies on the line joining I_{12} and I_{23}. Likewise I_{13} lies on the line that contains I_{34} and I_{41}. This reasoning does not depend on link 1 being fixed and it can be similarly concluded that the instantaneous centre for the relative motion between links AB and CD lies at the intersection of AD and BC.

Two applications of the three-centre theorem that are of particular interest relate to cam mechanisms. In Fig. 1.20 a shaped disc 2 (the cam) turns about a fixed centre A and makes edge contact at C with a follower 3 which turns about a fixed centre. Cam mechanisms are widely used as a means of generating a variety of periodic motions. Usually the cam rotates with constant speed and the angular velocity of the follower varies. Usually also, the relative motion between the cam and follower at their point of contact C is a combination of rolling and sliding motions. By suitable choice of cam profile it is possible to ensure, for an arbitrary shape of follower profile, either that the velocity ratio $\boldsymbol{\omega}_{12}/\boldsymbol{\omega}_{13}$ is constant, or that the relative motion at C is always one of pure rolling, that is without slip.

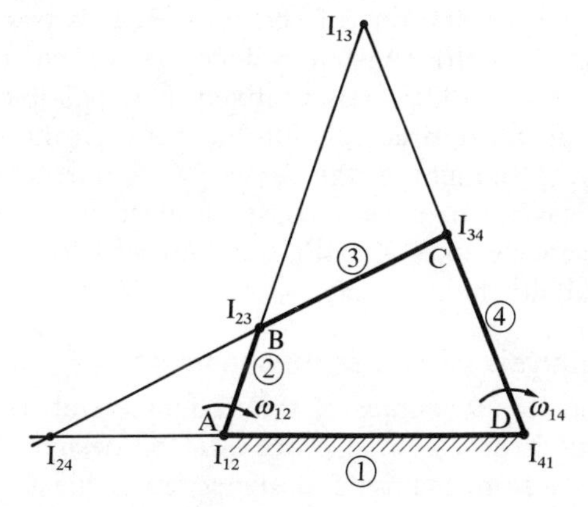

FIG. 1.19

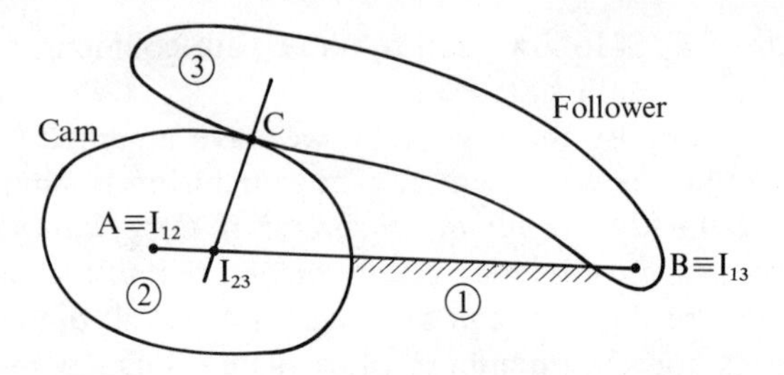

FIG. 1.20

1.6.3. Constant speed ratio cam drive

Provided that contact is maintained, the only possible relative motion at C, Fig. 1.20, is one of sliding along the common tangent. It follows that the instantaneous centre I_{23} for the relative motion lies on the common normal to the two profiles at C. But, by the three-centre theorem, I_{23} is collinear with I_{12} (≡A) and I_{13} (≡B). Hence I_{23} is located as shown in the figure. At I_{23} the velocity of the cam is $\boldsymbol{\omega}_{12} \times \overrightarrow{I_{12}I_{23}}$ and the velocity of the follower is, or would be if it extended so far, $\boldsymbol{\omega}_{13} \times \overrightarrow{I_{23}I_{13}}$. For zero relative velocity at this point $\boldsymbol{\omega}_{13}/\boldsymbol{\omega}_{12} = I_{12}I_{23}/I_{23}I_{13}$. If this speed ratio is to be constant, I_{23} must clearly be a fixed point. This condition is summed up as follows: *for constant speed ratio the common normal at the point of contact between a cam and follower must pass through a fixed point on the line of centres.* This requirement is of particular relevance to the geometry of gear-tooth profiles, for which it is normally a basic requirement that they should give a constant speed ratio. In achieving this condition an arbitrary profile may be chosen for either the cam or the follower, but not both.

1.6.4. Rolling contact

The relative velocity at C, Fig. 1.20, is zero if C coincides with I_{23}. There will, therefore, be pure rolling motion between the two profiles if the point of contact lies on the line of centres AB. It is possible to achieve this condition, as with the constant velocity condition, for an arbitrary choice of one of the profiles. Alternatively it is possible to choose the profiles so as to give a desired relationship between the rotations of the cam and follower, though in this case, of course, neither profile is arbitrary. It is possible to have a constant speed ratio and pure rolling contact simultaneously only if both the cam and follower profiles are circles rotating about their centres.

1.7. Velocity analysis of planar mechanisms

The calculation of the velocities of the various members of linkage and other mechanisms is an essential process in the design of such mechanisms. Not only are mechanisms designed to achieve certain motion characteristics, but the analysis of velocities, and later accelerations, is closely bound up with the forces within, and transmitted by, them. Two allied but distinct techniques emerge from the topics considered so far.

1.7.1. The instantaneous-centre method

It has been shown that the instantaneous centre I of a body in planar motion is the point where the velocity is zero, and that the velocity at any point P in that body is $\boldsymbol{\omega} \times \mathrm{IP}$, that it is perpendicular to IP, and has a magnitude $\omega\mathrm{IP}$, where $\boldsymbol{\omega}$ is the angular velocity of the body.

Example 1.4: Four-bar mechanism.

Referring to Fig. 1.19, we have seen that I_{13}, the instantaneous centre of BC, is located at the intersection of AB and CD. The velocity of B is of magnitude $v_B = \omega_{12} \times \mathrm{AB}$. The angular velocity of BC is $\omega_{13} = v_B/I_{13}B$, and the velocity of C is $v_C = \omega_{13} \times I_{13}C$. Finally, the angular velocity of DC is $\omega_{14} = v_C/\mathrm{CD}$, thus giving the speed ratio between the input and output cranks as $\omega_{14} = \mathrm{AB} \times I_{13}C/(I_{13}B \times \mathrm{CD})$.

Example 1.5: Slider-crank mechanism.

The basic mechanism of reciprocating engines, such as are used in the majority of automobiles, is known as the slider-crank mechanism. It consists of a slider A (e.g. a piston and cylinder combination), a coupler (or connecting rod) AB, and a crank BC, shown schematically in Fig. 1.21(a). The velocity of A is directed along the axis of the slider so that the instantaneous centre I_{12} of AB lies on the perpendicular to the axis at A. Following the argument used for the four-bar mechanism, the instantaneous centre for AB also lies on BC produced. Hence I_{12} is located as

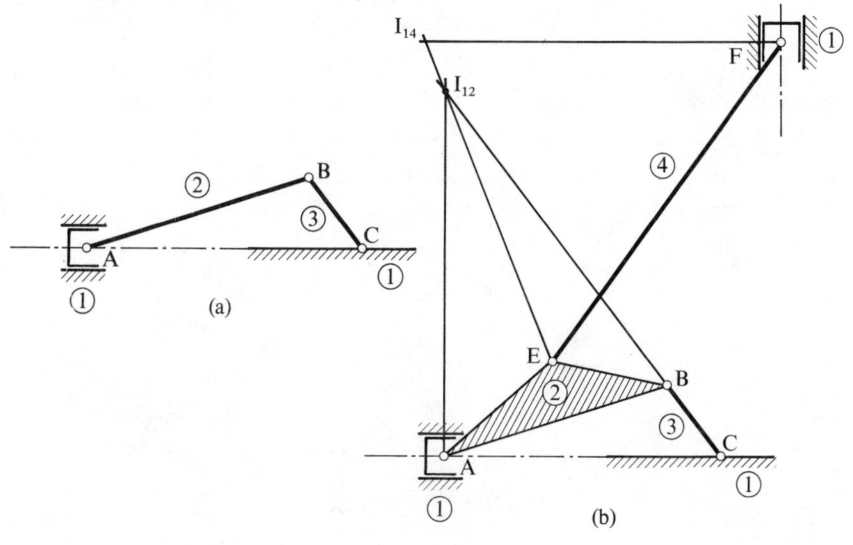

FIG. 1.21

shown. Calculations similar to those for the four-bar mechanism enable the angular velocity of the coupler AB, and the velocity of the slider A to be determined in terms of the angular velocity of the crank BC.

To determine the velocity of the secondary slider F, which is connected by coupler EF to AB, Fig. 1.21(b) we start by noting that the velocity of E is perpendicular to $I_{12}E$, and hence that I_{14} lies on this line. But I_{14} also lies on the perpendicular to the axis of the slider at F, and so I_{14} is located. With ω_{12} already determined, $v_E = \omega_{12} \times I_{12}E$; the angular velocity ω_{14} of EF has magnitude $v_E/I_{14}E$; and $v_F = \omega_{14} \times I_{14}F$.

Although the three-centre theorem is implicit in this analysis, no direct use has been made of it and, indeed, it is not even necessary to be aware of it. With some mechanisms the argument is very much facilitated by a direct appeal to the three-centre theorem.

Example 1.6: Quick-return mechanism.

This mechanism consists of a swinging link CD, and a crank AB connected together by a slider at B, Fig. 1.22. In normal operation the crank turns with constant angular velocity, and the swinging link oscillates. The particular feature which is utilized in various applications is that the times taken for the total movement of the swinging link in either direction are different. I_{12} is at A, and I_{13} is at C. Hence I_{32} lies on AC. To locate I_{32} we note that at B the relative velocity between the crank and the swinging link is the sliding velocity along CD. It follows that I_{32} lies on the perpendicular to CD at B. AB and CD have the same velocity at I_{32}, hence $\omega_{12} \times I_{12}I_{32} = \omega_{13} \times I_{13}I_{32}$. It may be noted that this equation gives $\omega_{13} = 0$ when AB is perpendicular to CD, as would be expected from a study of the geometry.

Example 1.7: Slotted-plate mechanism.

In Fig. 1.23 coupler ABC is connected to the plate 2, which rotates about D, through blocks B and C sliding in slots in the plate. Point A slides along a fixed axis thereby determining one line on which I_{31} must

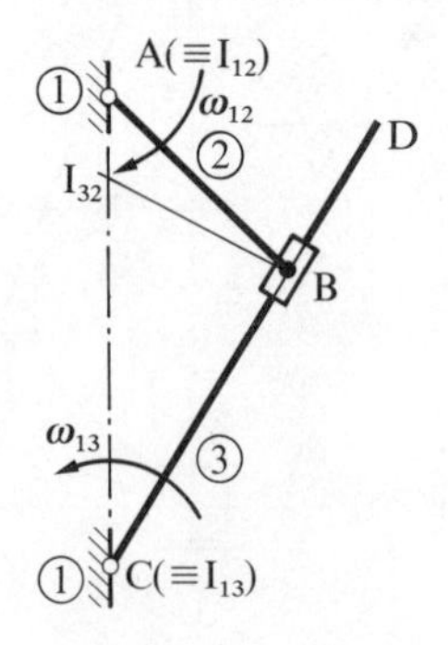

FIG. 1.22

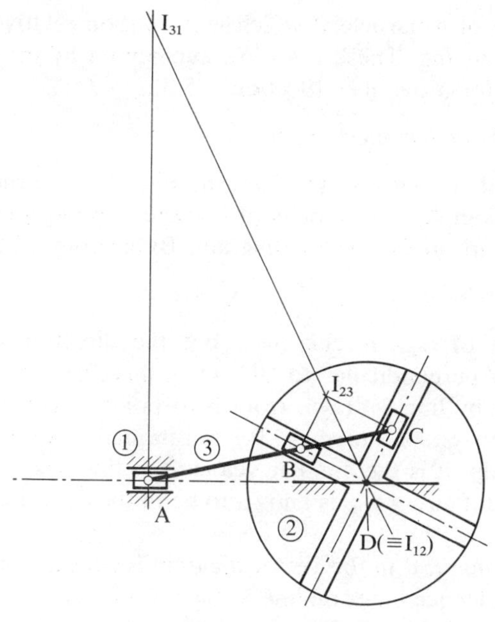

FIG. 1.23

lie, namely the perpendicular through A to the axis of the fixed slider. There is no obvious second point on the coupler where the direction of the velocity is known, so that a direct determination of the location of I_{31} is not possible. However, I_{12} is known to be at D and I_{23} is readily determined: the relative velocities between the coupler and the disc are along the sliders at B and C respectively, so the intersection of the normals to the axes of the two slots at B and C fixes I_{23}. By the three-centre theorem I_{12}, I_{23}, and I_{31} are collinear, thus fixing I_{31}.

The instantaneous-centre method is adequate and easy to apply for relatively simple mechanisms, provided that none of the relevant instantaneous centres lie off the edge of the piece of paper on which the mechanism is drawn. It does, however, become cumbersome when more than three or four members are involved, and it does not extend readily to the analysis of accelerations.

1.7.2. Velocity-diagram method

The velocity diagram for a mechanism is essentially a set of velocity images for the members of the mechanism combined into a single diagram. The separate images are constructed by successive applications of the polygon rule. The basic vector equations are

$$\mathbf{v}_B = \mathbf{v}_A + \boldsymbol{\omega}_{AB} \times \mathbf{r}_{AB} \tag{1.2}$$

which relates the velocities of points A and B of the member AB, and

$$\mathbf{v}_P = \mathbf{v}_A + \boldsymbol{\omega} \times \boldsymbol{\rho} + \partial \boldsymbol{\rho}/\partial t \tag{1.4}$$

for the velocity of a particle P which is in motion relative to a member which is itself moving. These rules are augmented by the concept of the velocity image for a member (Section 1.5.3).

Example 1.8: Four-bar mechanism.

Crank AB of a four-bar mechanism, Fig. 1.24, turns with angular velocity $\boldsymbol{\omega}_{AB}$ about A. The velocity of B is therefore $\mathbf{v}_B = \boldsymbol{\omega}_{AB} \times \mathbf{r}_{AB}$ and is represented by *ab* in the vector diagram. By analogy with eqn (1.2)

$$\mathbf{v}_C = \mathbf{v}_B + \boldsymbol{\omega}_{BC} \times \mathbf{r}_{BC}.$$

The magnitude of $\boldsymbol{\omega}_{BC}$ is unknown, but the direction of $\boldsymbol{\omega}_{BC} \times \mathbf{r}_{BC}$ is known, namely perpendicular to BC. $\mathbf{v}_C$ is therefore represented in the vector diagram by line 3 drawn from *b* to point *c*, as yet undetermined (the velocity image of member 3). Additionally, we know that $\mathbf{v}_C = \boldsymbol{\omega}_{DC} \times \mathbf{r}_{DC}$. Again, it is the angular velocity, in this case $\boldsymbol{\omega}_{DC}$, which is as yet unknown. But $\boldsymbol{\omega}_{DC} \times \mathbf{r}_{DC}$ is known to be in the direction perpendicular to DC.

As D is stationary, *d* in the vector diagram is concurrent with *a*, as A is also stationary. Hence *c* lies on line 4 (the velocity image of member 4) at its intersection with line 3. The complete vector diagram, known as the velocity diagram for the mechanism, is normally drawn to scale. Hence to determine the magnitude of $\boldsymbol{\omega}_{DC}$ it merely remains to measure *dc* and to evaluate $\omega_{DC} = dc/DC$. Directions are important. We note that the direction of $(\mathbf{v}_C - \mathbf{v}_D)$ is to the left, and hence infer that the direction of $\boldsymbol{\omega}_{DC}$ is anticlockwise. Visualization of how the mechanism will move confirms this conclusion.

$(\mathbf{v}_C - \mathbf{v}_B)$ is downwards (*bc* in the diagram), indicating a clockwise direction for ω_{BC}.

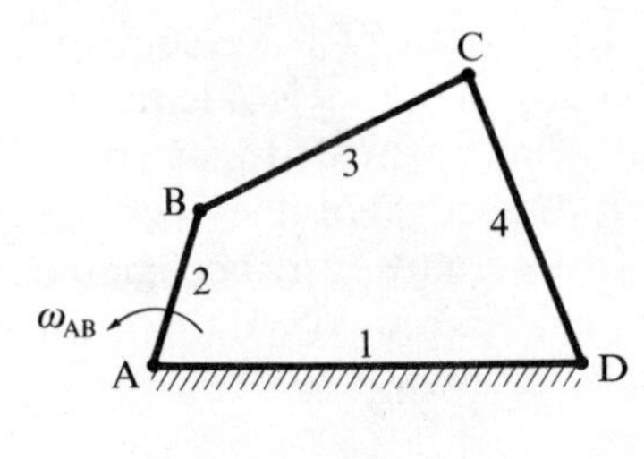

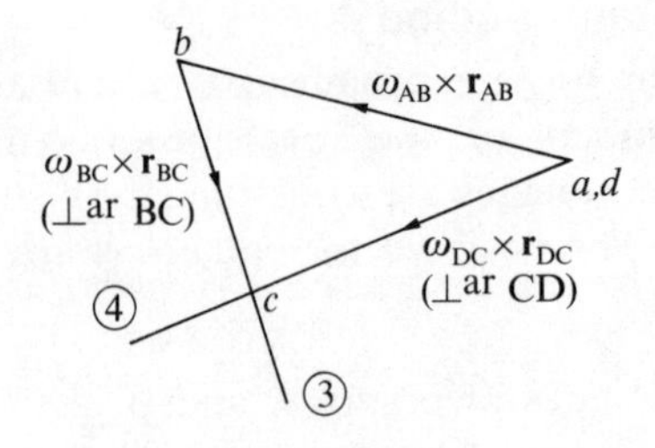

FIG. 1.24

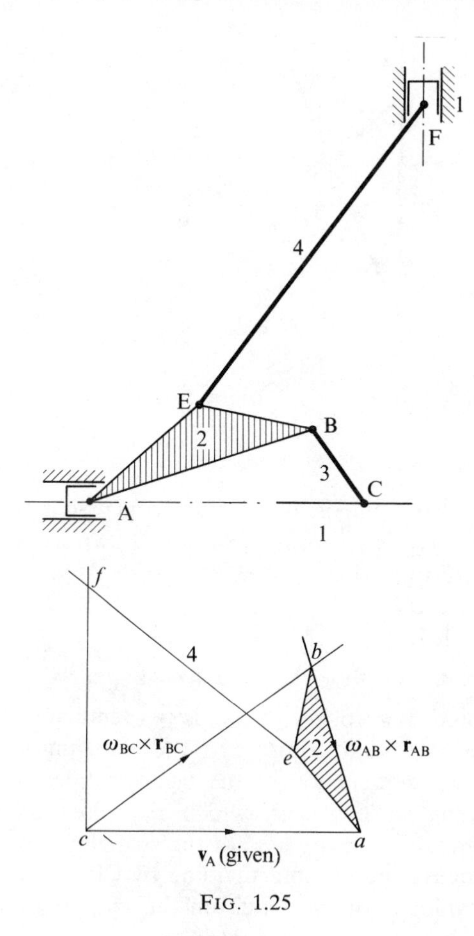

FIG. 1.25

Example 1.9: Slider-crank mechanism.

Assumed $\mathbf{v}_A$ to be specified, the velocity diagram *abc* for the slider-crank mechanism ABC in Fig. 1.25 is drawn following the method of the last example, *c* being the origin for the diagram. Using the velocity-image theorem the triangle *eba* is drawn similar to triangle EBA (alternatively *eb* and *ea* are drawn perpendicular to EB and EA respectively).

$\boldsymbol{\omega}_{EF} \times \mathbf{r}_{EF}$ is perpendicular to EF, so line 4 is drawn in the velocity diagram. But $\mathbf{v}_F$ is parallel to the slider at F, and hence *cf* is drawn in that direction, remembering that C is a fixed point. The magnitude of $\mathbf{v}_F$ is now measured directly from the completed diagram.

Example 1.10: Quick-return mechanism.

This mechanism, Fig. 1.26, differs from the two just considered in that there is now a block B which slides on a moving link, and eqn (1.4) applies.

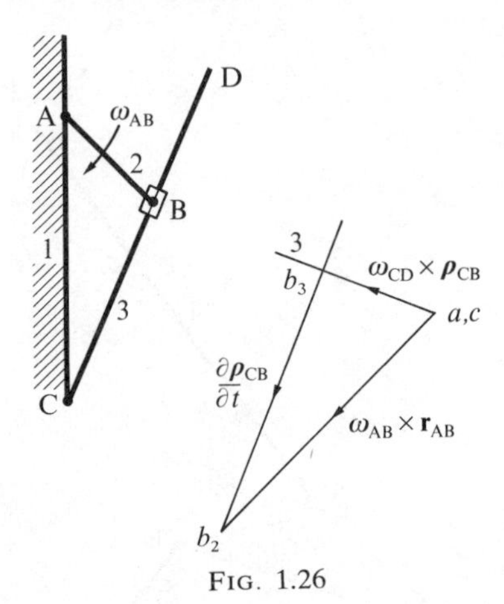

FIG. 1.26

As A and C are fixed hinges, a and c are concurrent at the origin of the vector diagram. If ω_{AB} is given, $\boldsymbol{\omega}_{AB} \times \mathbf{r}_{AB}$ is drawn as ab_2 in the velocity diagram. The subscript 2 denotes that the point B under consideration is on number 2, the crank.

Following eqn (1.4),

$$\mathbf{v}_B = \mathbf{v}_C + \boldsymbol{\omega}_{CD} \times \boldsymbol{\rho}_{CB} + \partial \boldsymbol{\rho}_{CB}/\partial t.$$

$\mathbf{v}_C$, as already noted, is zero. $\boldsymbol{\omega}_{CD} \times \boldsymbol{\rho}_{CB}$ is perpendicular to CD, line 3 in the velocity diagram. $\partial \boldsymbol{\rho}_{CB}/\partial t$ represents the changing length CB, and is parallel to CB. The intersection of the last two lines is labelled b_3 and refers to the point on CD with which the block B is momentarily coincident. The angular velocity ω_{CD} of the swinging link can now readily be obtained by measuring cb_3 and dividing by CB. Measurement of b_2b_3 gives the sliding velocity of the block on the swinging link.

Comparison between the vector solution of these problems and their solution by the instantaneous-centre method is invited. The choice between the two methods is largely a matter of taste. The velocity-diagram method will be seen to be advantageous when the transmission of forces is considered (Chapter 7). Also, ideas developed in the construction of velocity diagrams extend to accelerations, but the use of the instantaneous centre is not readily extended. However, it will be noted that the final example using instantaneous centres, Fig. 1.23, has not been solved by means of a velocity diagram. If the attempt is made to do this, it will be found that the solution is not as straightforward as it has been in the other examples. The problem is soluble using a velocity diagram, but it requires the development of peripheral techniques that will not be considered here.

1.8. Mobility

Almost without exception, the mechanisms which have been analysed quite obviously function. Relative motion between the members is possi-

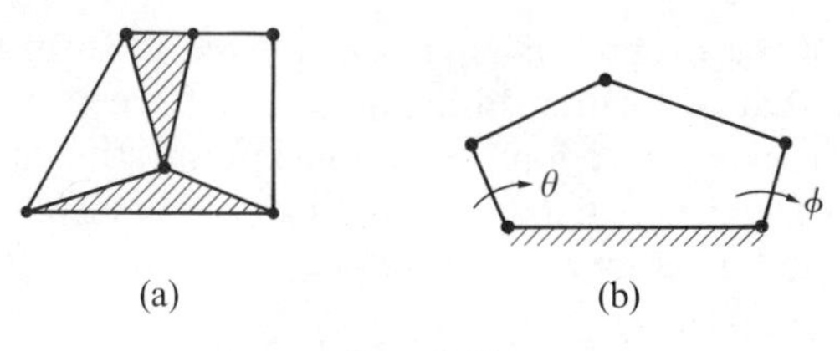

FIG. 1.27

ble, and for every position of the input member there is a definite position for every other member. Such mechanisms are said to have unit mobility. The examples in Fig. 1.27(a) and (b) have zero mobility and mobility of degree two respectively. The first is a just rigid structure, the second requires two members to be positioned before the configuration is determined and is a differential mechanism. The degree of mobility of a mechanism depends on the number of members and the way in which they are connected together. The examples so far presented have been very simple, and formal analysis is hardly necessary to show that they will 'work', but with more members and complex interconnections the degree of mobility is less obvious and a simple test is of interest.

A single body moving in a plane freely is said to have three degrees of freedom. It requires three co-ordinates to specify its location in relation to a set of fixed axes: two co-ordinates to locate a chosen point in the body and a further co-ordinate to orientate the body. The total number of co-ordinates needed to specify the configuration of n unconnected bodies is $3n$. If the bodies are connected together by hinges, sliders, and so forth, degrees of freedom are lost. For example, the two separate bars in Fig. 1.28(a) have a total of six degrees of freedom, but when hinged together, Fig. 1.28(b) they have only four degrees of freedom. This is because only one co-ordinate ϕ is now required to determine the relative positions of the two bars; the hinge destroys two degrees of freedom. A slider allows two degrees of freedom s and ϕ, Fig. 1.28(c). If there are j_1 connections which allow one degree of freedom and j_2 which allow two degrees of freedom, the total number of degrees of freedom which have been destroyed is $(2j_1+j_2)$.

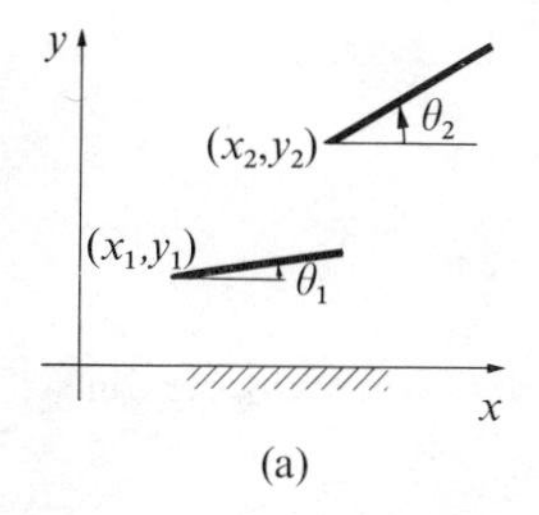

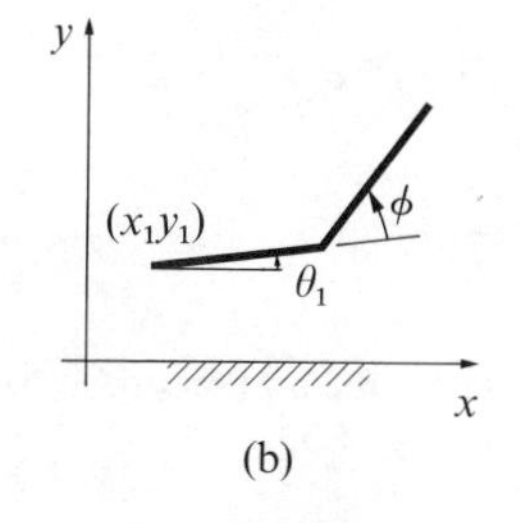

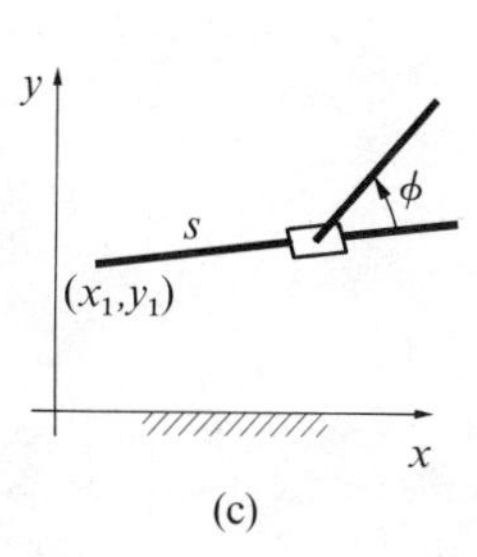

FIG. 1.28

Usually one of the original n members is fixed to form the base of the mechanism, so that a further three degrees of freedom are lost. The number of degrees of freedom of a mechanism (its mobility) with n members, one of which is fixed, j_1 hinges or similar connections, and j_2 sliders or equivalents, is given by Grübler's formula

$$F = 3(n-1) - (2j_1 + j_2).$$

It can readily be verified for the two examples in Fig. 1.27 that $F = 0$ and 2 respectively. The quick-return mechanism in Fig. 1.22 has $n = 3$, $j_1 = 2$, and $j_2 = 1$, giving $F = 1$ as expected. The mechanism in Fig. 1.23 also has $n = 3$ but $j_1 = 1$ and $j_2 = 3$, again yielding $F = 1$.

1.9. Problems

1. A particle is initially at the origin of a set of cartesian axes Oxy. It suffers three successive displacements: 2 units of length at 30° to the positive direction of Ox, and 60° to the positive direction of Oy; one unit in the positive direction of Oy; and finally, 1·5 units at 45° to both the negative Ox and negative Oy directions. What is the final position of the particle?

Confirm that the result is the same when the same set of displacements is made in a different order.

2. Two ships leave the same harbour simultaneously. Ship A sails a straight course due north at 22·5 km h^{-1}. Ship B sails a straight course NE at 16 km h^{-1}.

What appears to be the course and speed of A as seen from B? By considering the positions of the two ships after, say, 30 minutes and 60 minutes, confirm that the bearing from B to A is constant, and that the distance between them increases at a rate equal to the relative speed.

Assume that the initial separation of the two vessels is negligible.

3. Determine the location of the pole and the rotation for the movement of AB from position 1 to position 2 for the two cases shown in Fig. 1.29.

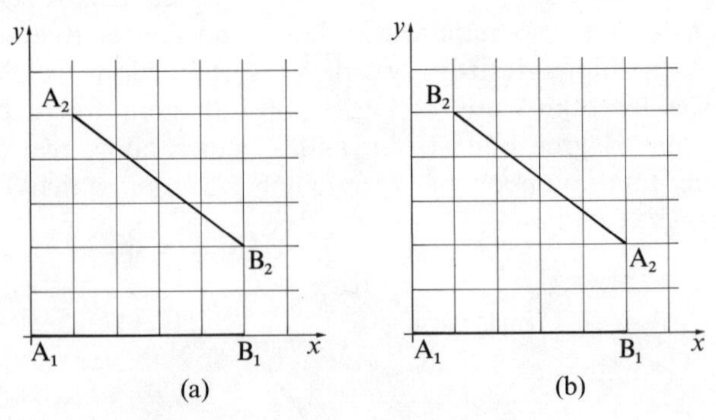

FIG. 1.29

4. Locate the instantaneous centre for the motion of AB for each case as it slides in contact with the fixed surfaces s shown in Fig. 1.30.

5. Show for the mechanism in Fig. 1.14 that if $a < b$ the centrodes are ellipses.

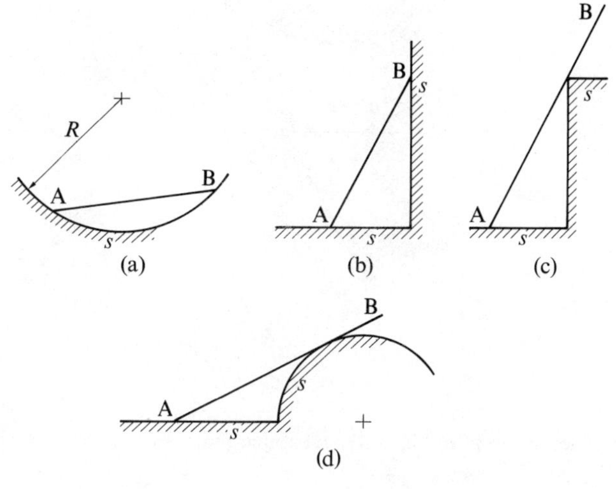

FIG. 1.30

6. A cam rotates about a fixed point A driving a straight rocker-follower BC pivoted at B. Determine the polar equation to the profile of the cam if the cam and follower are to rotate at equal speeds in opposite directions. Let AB = a.

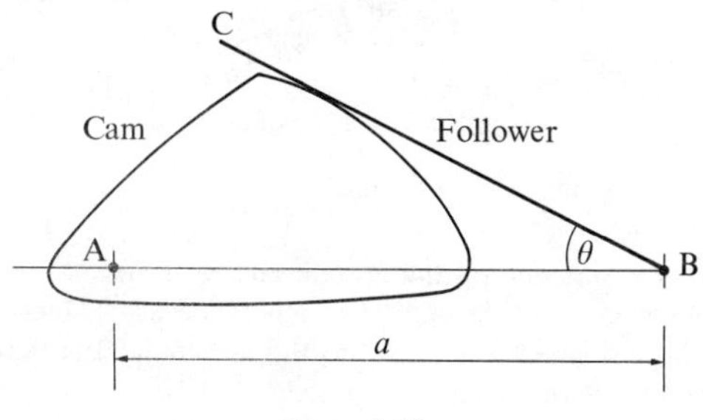

FIG. 1.31

7. Two standard properties of an ellipse are:

(i) The sum of the distances from the foci S and S′ to any point P on the ellipse equals the length $2a$ of the major axis.
(ii) SP and S′P make equal angles with the tangent to the ellipse at P.

A cam and a follower both have the same elliptical profile, and each profile pivots about one of its foci. The two fixed foci are distance $2a$ apart. Use the given geometrical properties of an ellipse to show that the cam and the follower rotate without slip at their point of contact.

8. The mechanism shown in Fig. 1.32 has two links AB and CBD pinned together at B. Points C and D are constrained to move along perpendicular lines as shown. The link AB passes through a swivel at S and rotates clockwise at a constant angular velocity of 50 rad s^{-1}.

Find, for the configuration shown in the figure,
(a) the instantaneous centres of CBD and AB, and
(b) the velocity of the point in AB which is instantaneously at S.

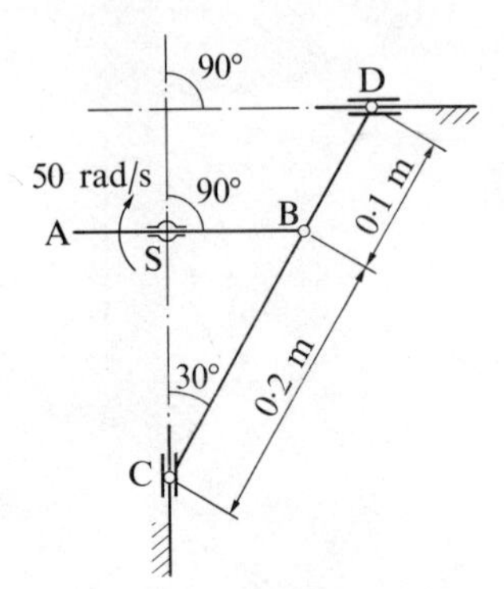

FIG. 1.32

9. For the mechanism shown in Fig. 1.33, determine the angular velocity of crank DB and the velocity of point C,
   (a) by drawing a velocity diagram, and
   (b) by use of the instantaneous centre.

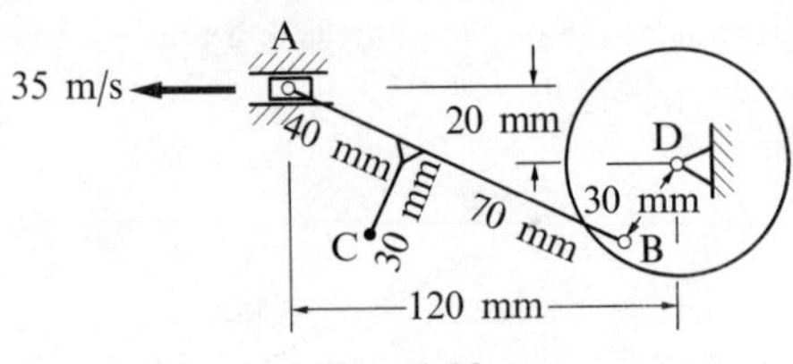

FIG. 1.33

10. A plank 4 m long has one end on the ground and rests against a wall 2·5 m high. The bottom end is slipping away from the wall at a rate of 1·5 m s^{-1}. Draw a velocity diagram for the plank at the instant when it is at 50° to the horizontal and determine,
   (a) the angular velocity of the plank,
   (b) the velocity of the top end, and
   (c) the point on the plank that has the smallest velocity.

11. In the mechanism depicted in Fig. 1.34, BCE is a rigid link. Determine for the configuration shown,

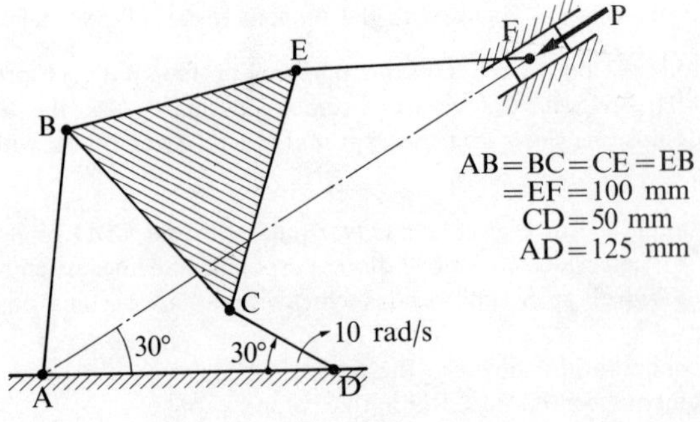

FIG. 1.34

(a) the instantaneous centres of links BCE and EF,
(b) the angular velocities of links BCE and EF,
(c) the velocity of F, and
(d) the rubbing speed on the pin at E if its diameter is 10 mm.

12. Fig. 1.35 shows a cam lifting a flat-footed follower. Find the velocity of the follower in the position shown, and the sliding speed between the cam and the follower.

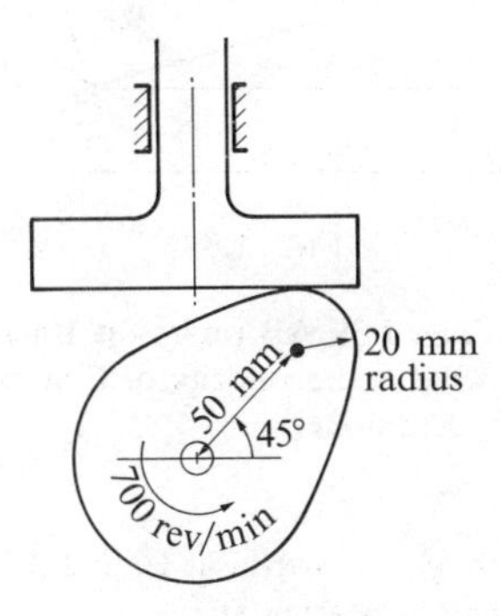

FIG. 1.35

13. Fig. 1.36 shows a cam OA which rotates about O at 1800 rev min^{-1}. The rocker arm CB is pivoted at C and carries a roller at B.

For the position shown, find the angular velocity of CB. Note that for a limited range of motion on either side of the given position AB is of constant length, so that the cam and follower can be regarded as a four-bar mechanism OABC.

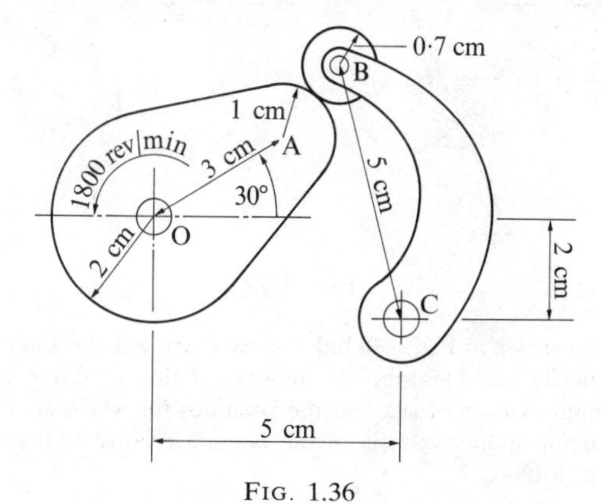

FIG. 1.36

14. In the linkage shown in Fig. 1.37 AB = 300 mm, AD = 600 mm, and BC = 1200 mm. AB rotates with an angular velocity of 10 rad s^{-1}. Determine the velocity of C at the instant when $\theta = 120°$.

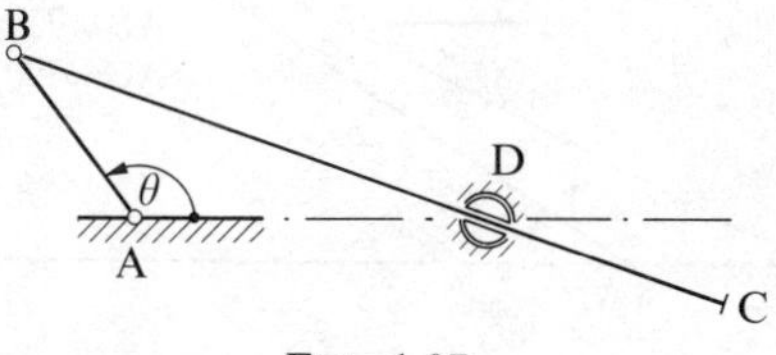

FIG. 1.37

15. In the mechanism shown in Fig. 1.38 crank AB rotates at 120 rad s^{-1}. Determine the velocity of E.

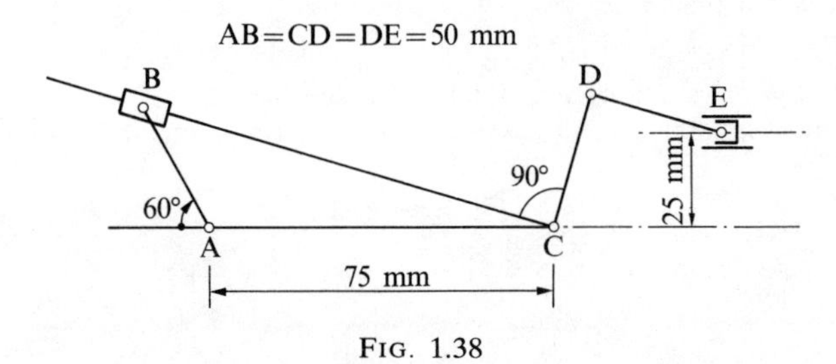

FIG. 1.38

16. In the mechanism shown in Fig. 1.39 AB rotates at 100 rad s^{-1}. At what values of θ is D moving parallel to AC, and what is the velocity of F at each of these instants?

For the instant when $\theta = 45°$, determine,

(a) the velocity of C,
(b) the velocity of F,
(c) the velocity of sliding through the swivel at E, and
(d) the relative angular velocity at bearing D.

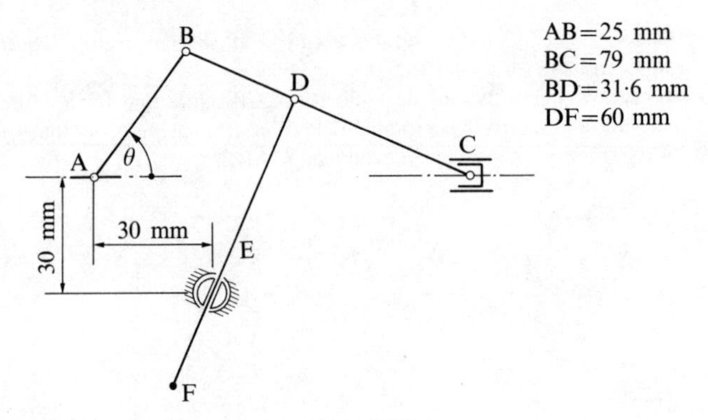

FIG. 1.39

17. In the mechanism shown in Fig. 1.40 link 1 is fixed and link 2 rotates anticlockwise with uniform angular velocity $\boldsymbol{\omega}_2$. Describe the motion of link 4 during a cycle, giving the maximum and minimum values of $\boldsymbol{\omega}_4$, and the positions for which $\boldsymbol{\omega}_4 = \boldsymbol{\omega}_2$.

What is the maximum sliding velocity of the block 3 relative to the link 4? Locate the instantaneous centre of block 3.

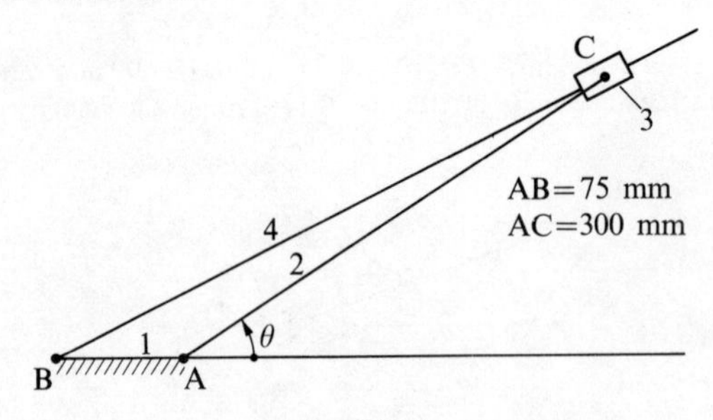

FIG. 1.40

18. ABCD is a four-bar mechanism made up of straight links pinned at their ends. AD is fixed. In any position, M is the point on BC (or BC produced) such that AM is parallel to DC. Prove that the angular velocities of the moving links are related by

$$\omega_{CD}=\frac{AM}{DC}\omega_{AB} \quad \text{and} \quad \omega_{BC}=\frac{BM}{BC}\omega_{AB}.$$

In a particular case AB = 300 mm, BC = 750 mm, CD = 600 mm, and DA = 900 mm. AB rotates continuously in one direction causing CD to oscillate. Find the angles that CD makes with the fixed link in its extreme positions.

19. Determine for each of the systems shown in Fig. 1.41 whether it is a structure for which $F=0$, a mechanism with $F=1$, or a differential mechanism with $F=2$.

Note. When three bars are hinged together to form a rigid triangle, as in case (c), it is permissible to regard them as a single body. This makes it easier to study cases such as (c) where the shaded triangle can be treated as one body, and so can the ground.

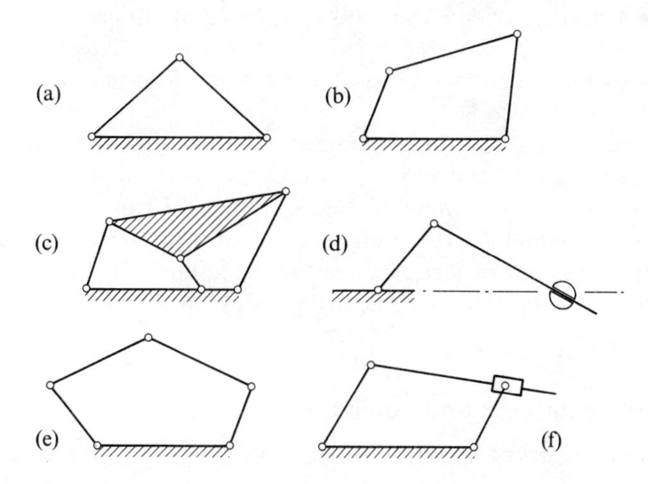

FIG. 1.41

2. Mainly planar statics

2.1. Force as a vector

THE vector nature of force has already been mentioned. The addition of force vectors in accordance with the parallelogram rule can be verified directly by experiment, or deduced by reference to Newton's Second Law of Motion expressed in the form $\mathbf{P} = m\mathbf{a}$. In this equation $\mathbf{P}$ is the force acting on a particle of mass m, and $\mathbf{a}$ is the resulting acceleration of the particle. Although Newton's Second Law is not directly used in the study of statics, its introduction is necessary for it provides the means whereby the basic unit of force is defined.

The standard unit of force is the newton and is defined as the force that will impart an acceleration of 1 metre per second squared to a mass of one kilogram. In more homely terms, it can be thought of as being approximately equal to the weight of a moderately sized apple. The definition of the unit of force assumes that the units of length, mass, and time have also been defined. These definitions will be referred to in due course, in Chapter 6.

2.2. Force resultants and equilibrium

2.2.1. Equilibrium of coaxial forces: bound and sliding vectors

The study of kinematics in Chapter 1 was concerned always with the motion at particular points in a moving plane, the velocity having been shown, in general, to vary from point to point. Consequently vectors were always associated with, or tied to, particular points. Addition of velocities or displacements always referred to what was happening at particular points. Such vectors are referred to as *bound vectors*. In one sense, forces likewise are often associated with particular points of a body or a structure, for they are applied at particular points, and the forces and stresses induced within a structure depend on the precise manner of its loading. In considering the over-all equilibrium of the loads their actual points of application become less important.

Consider for example a bar acted upon by forces at either end, Fig. 2.1. The forces must be coaxial for otherwise the bar would spin. Given that they are coaxial, it does not matter for over-all equilibrium whether the forces push or pull on the ends of the bar. Either way the equilibrium equation is $\mathbf{P}_1 + \mathbf{P}_2 = 0$, or $\mathbf{P}_2 = -\mathbf{P}_1$. The bar will feel the difference: in one case the force is tending to stretch it, in the other the force is compressive. For equilibrium, what matters is simply that $\mathbf{P}_1$ and $\mathbf{P}_2$ must

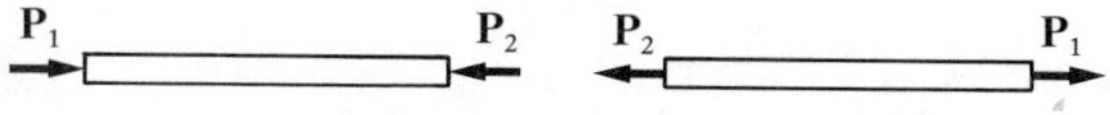

$\mathbf{P}_1$ $\mathbf{P}_2$ $\mathbf{P}_2$ $\mathbf{P}_1$

FIG. 2.1

be equal in magnitude and opposite in direction. The same pair of forces would still be in equilibrium if the bar were twice as long. Insofar as one is concerned only with the equilibrium of forces, only their magnitudes, directions, and lines of action matter. They can be thought of as free to slide along their lines of action, and consequently are referred to as *sliding vectors.*

2.2.2. Addition of concurrent forces

The addition of two concurrent forces is effected by means of the triangle rule, Fig. 2.2(a), where **P** and **Q** combine to give the resultant force **R**. The addition of three concurrent forces is achieved by successive application of the triangle rule or by using the polygon rule, Figs 2.2(b) and (c).

If **P**, **Q**, and **S** form a triangle their resultant **R**′ is zero and the three forces are said to be in equilibrium. The third forces **S** must be equal and opposite to the resultant **R** of the other two forces **P** and **Q**. So

$$\mathbf{P}+\mathbf{Q}=\mathbf{R}=-\mathbf{S}$$

giving

$$\mathbf{P}+\mathbf{Q}+\mathbf{S}=0.$$

Likewise, if four or more force vectors form a closed polygon, the forces, assuming them to be concurrent, are in equilibrium. In accordance with the commutative rule the order of addition of vectors does not alter their resultant, so that closure of a force polygon is not affected by the order in which the vectors are taken, Fig. 2.3.

It should be noted that although three forces in equilibrium must be coplanar, there is no such restriction when four or more forces are involved: the polygons in Fig. 2.3 do not have to be plane figures.

2.2.3. Resolution into components

Reference to the polygon of forces as a means of adding a number of concurrent forces carries an implication that the addition is performed

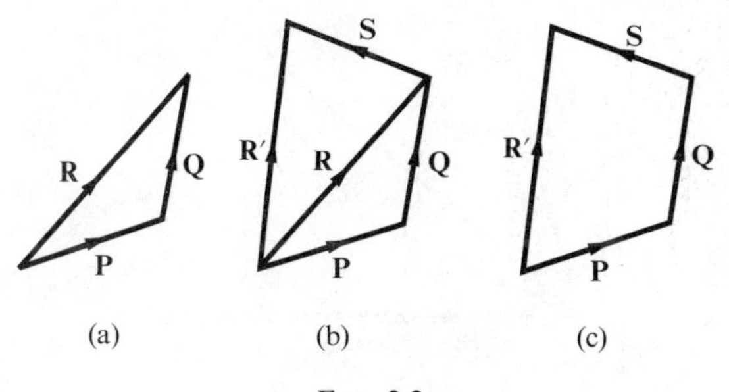

FIG. 2.2

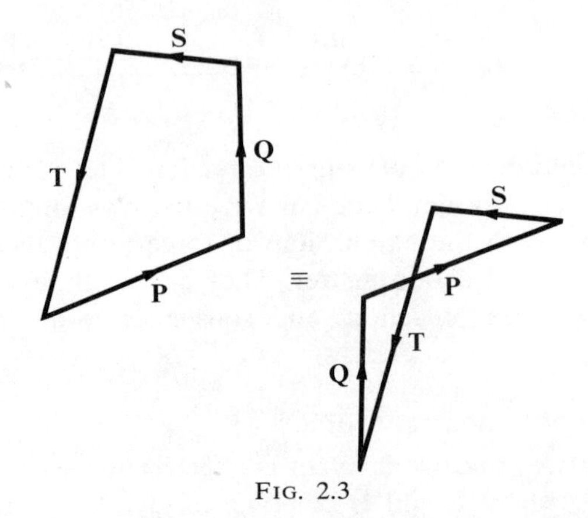

FIG. 2.3

graphically. In the solution of many problems graphical construction does indeed prove to be convenient, but equally, calculation will often be preferred. In this case the procedure is to use the parallelogram rule to decompose the vectors to be added into components in two convenient directions. In Fig. 2.4 the vectors **P** and **Q** are shown split up into components in the u and v directions so that

$$\begin{aligned}\mathbf{P}+\mathbf{Q} &= (\mathbf{P}_u+\mathbf{P}_v)+\mathbf{Q}_u+\mathbf{Q}_v) \\ &= (\mathbf{P}_u+\mathbf{Q}_u)+(\mathbf{P}_v+\mathbf{Q}_v) \\ &= \mathbf{R}_u+\mathbf{R}_v=\mathbf{R}.\end{aligned}$$

As Fig. 2.4 makes clear, it is not necessary for the two chosen directions to be normal to each other, though in practice they almost invariably will be. The resolution of vectors into normal components and subsequent addition is a calculation which can readily be performed on a small electronic calculator which has trigonometric functions. The operation is effectively carried out automatically on a calculator that has direct conversions from polar to Cartesian co-ordinates.

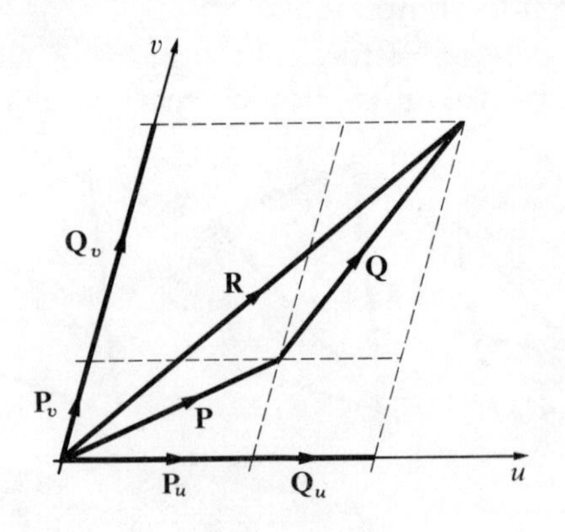

FIG. 2.4

As it is impractical to draw force polygons in three dimensions, the addition of three or more concurrent forces which do not lie in a single plane usually necessitates the resolution of those forces into components.

One way of regarding the procedure for resolving a general force **R** into components in the directions Oxyz, Fig. 2.5(a), is to take it in stages whereby the parallelogram rule is applied successively. **R** is first resolved into components $\mathbf{P}_z$ along Oz and $\mathbf{R}_{xy}$ in the plane Oxy. Then $\mathbf{R}_{xy}$ is resolved into $\mathbf{P}_x$ and $\mathbf{P}_y$.

Clearly, it makes no difference if a different order of resolution is chosen, Fig. 2.5(b). By resolving **R** first into $\mathbf{P}_y$ and $\mathbf{R}_{xz}$, and then decomposing $\mathbf{R}_{xz}$ into $\mathbf{P}_x$ and $\mathbf{P}_z$ the same set of components $\mathbf{P}_x$, $\mathbf{P}_y$, and $\mathbf{P}_z$ is obtained as before.

Any number of forces can be resolved into components in the same way. Their resultant $\sum \mathbf{R}$ is found as

$$\sum \mathbf{R} = \sum \mathbf{P}_x + \sum \mathbf{P}_y + \sum \mathbf{P}_z.$$

Very often it is not necessary to split each of the original set of forces into three components, and it is sufficient to proceed only as far as the first resolution, and to represent the resultant as

$$\sum \mathbf{R} = \sum \mathbf{P}_z + \sum \mathbf{R}_{xy}.$$

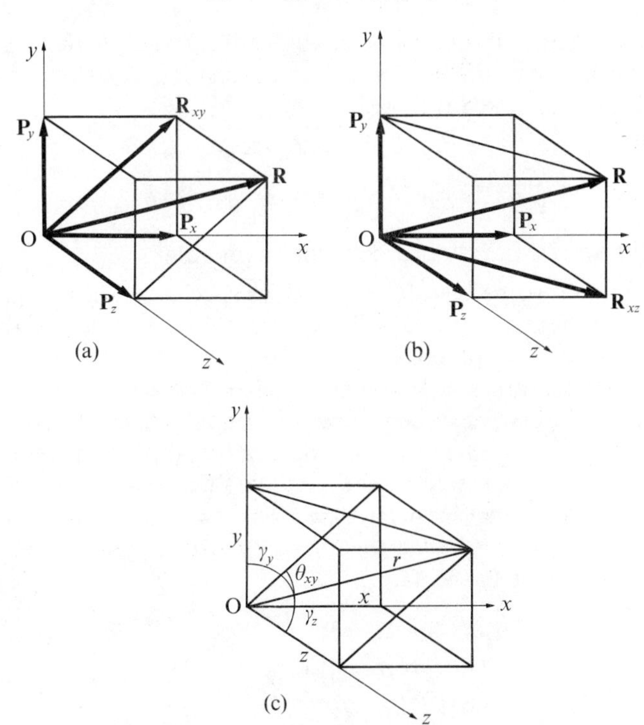

FIG. 2.5

Here each of the given forces is considered to be made up of two components, one in the direction Oz, and the other in the plane Oxy. The over-all component in the plane Oxy might conveniently be obtained by application of the polygon rule. The best procedure for one problem is not always the best for another.

The actual calculation of the magnitudes of the components $\mathbf{P}_x$, $\mathbf{P}_y$, and $\mathbf{P}_z$ is easier than might be supposed. Considering Figs 2.5(a) and (c) together, we see that

$$R_{xy} = R \cos \theta_{xy}$$

and

$$P_z = R \cos \gamma_z$$

where θ_{xy} is the angle between $\mathbf{R}$ and the plane Oxy, and γ_z is the angle between $\mathbf{R}$ and the axis Oz.

On comparing Figs 2.5(b) and (c), we see that

$$P_y = R \cos \gamma_y$$

and deduce that

$$P_x = R \cos \gamma_x.$$

If the terminal point of $\mathbf{R}$ is at the point (x, y, z) it follows that

$$P_x = Rx/r, \qquad P_y = Ry/r, \quad \text{and} \quad P_z = Rz/r.$$

The quantity $\cos \gamma$ is known as a direction cosine of the force. Two direction cosines are sufficient to define a direction. The third direction is related to the other two through

$$x^2 + y^2 + z^2 = r^2$$

$$\Rightarrow \qquad \cos^2\gamma_x + \cos^2\gamma_y + \cos^2\gamma_z = 1.$$

2.2.4. Addition of parallel forces: the lever rule

The addition of two parallel forces requires special consideration for they are not concurrent except at infinity, and the use of the triangle rule might appear to be impracticable.

Let the parallel forces be **P** and **Q**, Fig. 2.6. Add a pair of coaxial equal and opposite forces $\pm\mathbf{F}$ whose line of action crosses **P** and **Q**. The resultant force will be unaffected by the addition of the extra forces. Now combine **P** with one of the forces **F**, and combine **Q** with the other using the parallelogram rule in each case. The two resultants $\mathbf{R}_1$ and $\mathbf{R}_2$ intersect and are in turn added together to give their resultant **R**, which is also the resultant of **P** and **Q**:

$$\begin{aligned}\mathbf{R} &= \mathbf{R}_1 + \mathbf{R}_2 \\ &= (\mathbf{P} + \mathbf{F}) + (\mathbf{Q} - \mathbf{F}) \\ &= \mathbf{P} + \mathbf{Q}.\end{aligned}$$

R must perforce be parallel to **P** and **Q**.

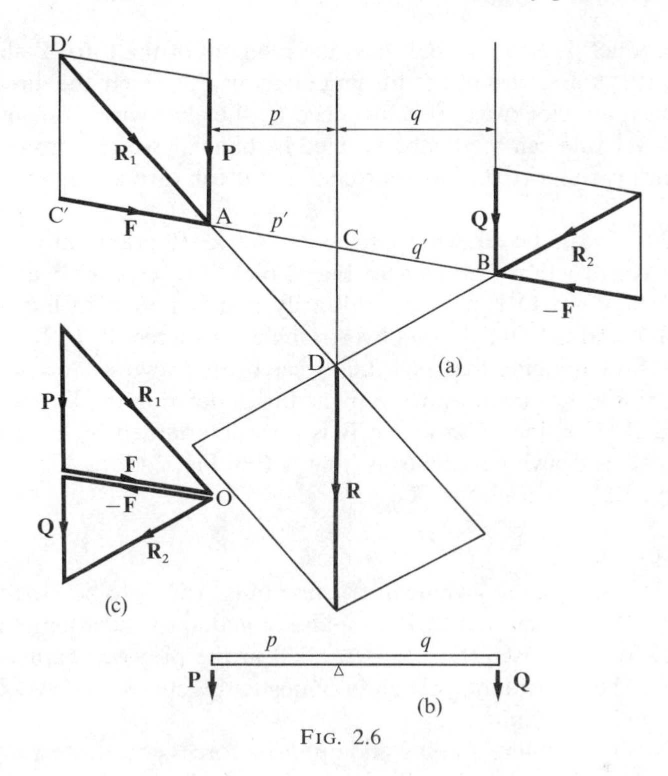

FIG. 2.6

To calculate the line of action of **R** we note first that triangles ACD and A′C′D′ in Fig. 2.6 are similar. Hence

$$C'D'/C'A = CD/CA$$

so

$$P/F = CD/p'$$

giving

$$p' . P = CD . F$$

Similarly

$$q' . Q = CD . F.$$

Hence

$$p' . P = q' . Q.$$

It follows that

$$p . P = q . Q$$

where p and q are the perpendicular distances between the resultant and **P** and **Q** respectively. This is known as the *lever rule.* For if a lever is supported on a fulcrum, Fig. 2.6(b), and subjected to parallel forces on either side, the lever is in balance if this condition holds.

The product $p\,.\,P$ is referred to as the moment of the force **P** about the point C. It is a measure of the turning effect of **P**. Clearly the direction of this effect is anticlockwise. It is balanced by the clockwise moment due to **Q**. The lever rule can readily be verified by hanging weights from a metre stick which is supported at its centre so that it can turn about a horizontal axis.

Fig. 2.6(a) can be drawn to scale to locate R graphically. A more compact construction is shown in Fig. 2.6(c). The vectors **P** and **Q** are drawn first. Point O is chosen arbitrarily and is joined to the terminal points of **P** and **Q**. This defines two triangles of forces ($\mathbf{P}$, $\mathbf{F}$, $\mathbf{R}_1$) and ($\mathbf{Q}$, $-\mathbf{F}$, $\mathbf{R}_2$). Now imagine that Fig. 2.6(c) has been drawn as described, and that Fig. 2.6(a) has been drawn only as far as defining the lines of action of **P** and **Q**. The line of action of **R** is now determined by locating D as follows: AB is drawn parallel to **F** (that is **F** in Fig. 2.6(c)), AD is parallel to $\mathbf{R}_1$ and BD is parallel to $\mathbf{R}_2$.

2.2.5. Couples

If $\mathbf{Q}=-\mathbf{P}$, Fig. 2.6, the given construction fails. This is to be expected, for as $\mathbf{R}=\mathbf{P}+\mathbf{Q}=0$ it can hardly have a line of action. If the attempt is made to complete the construction as described in the previous paragraph, $\mathbf{R}_1$ and $\mathbf{R}_2$ will be coincident, though in opposite directions, and AD and BD parallel, with D at infinity.

A pair of non-collinear equal and opposite forces constitute a couple. If the forces are of magnitude P and the perpendicular distance between them is p the couple is said to be of magnitude $p \times P$. The couple can also be said to be of moment $p \times P$, measured in newton metres (N m), for the net moment of the two forces is $p \times P$ about any point in the plane, Fig. 2.7

$$\begin{aligned} M_O &= (h+p)\times P - h\times P \\ &= p\times P \text{ clockwise.} \end{aligned}$$

2.2.6. Equilibrium of a general set of coplanar forces

A general set of coplanar forces reduces either to a single resultant force, or to a couple. It is easy to see that this must be so, for, having chosen from the original set any pair of forces which do not by themselves constitute a couple, we can determine their resultant. That resultant can

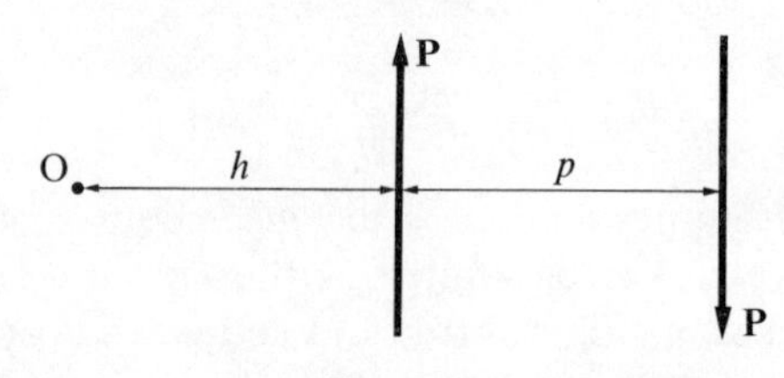

FIG. 2.7

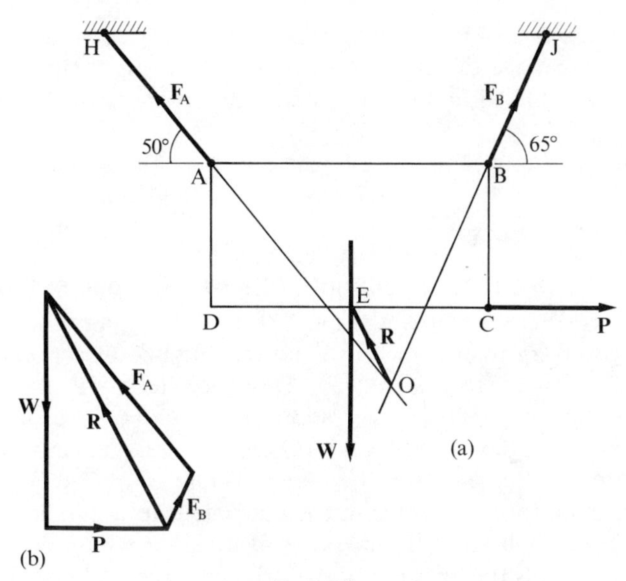

FIG. 2.8

be combined with a third force of the original set, and so on. Eventually the system must reduce to a single force or couple as stated. In the particular circumstances that the resultant force is zero, and there is also no couple, the set of forces are in equilibrium.

Example 2.1: A board ABCD, Fig. 2.8(a) hangs in a vertical plane supported by two strings AH and BJ. To maintain AB and DC horizontal, a horizontal force **P** *is applied at C. Determine the magnitude of* **P**, *and the tensions in the strings in terms of the weight* **W**, *which acts through the geometric centre of the board.*

We cannot proceed exactly as just described because the magnitudes of three forces are unknown, but the same general line of reasoning applies.

$\mathbf{F}_A$ and $\mathbf{F}_B$ intersect at O, and hence their resultant **R** passes through O.

P and **W** intersect at E and hence their resultant passes through E. For over-all equilibrium this force must be in equilibrium with **R**. It follows that the line of action of the two resultants, **R** of $\mathbf{F}_A$ and $\mathbf{F}_B$, and $-\mathbf{R}$ of **W** and **P** is along OE.

Knowing the direction of **R**, the two triangles of forces $\mathbf{F}_A$, $\mathbf{F}_B$, **R** and **W**, **P**, $-\mathbf{R}$ can be drawn as shown and the magnitudes of all the forces determined in terms of **W**, Fig. 2.8(b).

This method of solution is the most convenient in the given circumstances, and the most revealing. If the directions of $\mathbf{F}_A$ and $\mathbf{F}_B$ had been such that O was to the left of **W**, it would immediately have been clear that the direction of **P** needed to be reversed. It can be seen that if **P** is rotated slightly anticlockwise about C, then point E will drop. If E now lies on the line of $\mathbf{F}_A$, then $\mathbf{F}_b$ is zero. A further rotation of **P** would require $\mathbf{F}_B$ to change direction. This requires a compressive force in BJ and is incompatible with one of the stated conditions, namely that BJ is a string.

In spite of the wealth of information that the graphical solution provides, it will often be found more convenient, perhaps quicker, to proceed by calculation.

2.2.7. Equations for the equilibrium of a set of coplanar forces

A single force **P** acts through a point A. Introduce a pair of forces **Q** and $-\mathbf{Q}$ at an arbitrary point O which is not on the line of action of **P**, Fig. 2.9, and let $\mathbf{Q}=\mathbf{P}$ (that is to say, **Q** is parallel to **P** and of equal magnitude). The over-all effect of adding these extra forces is zero.

Taken together, the forces **P** and $-\mathbf{Q}$ constitute a couple of magnitude $p \times \mathbf{P}$, where p is the length of the perpendicular from O to **P**. It follows that the original force **P** through A is equivalent to a parallel and equal force $\mathbf{Q}=\mathbf{P}$ through an arbitrary point O together with a couple whose moment $\mathbf{M}_O$ equals the moment of the original force **P** about O. We shall see later that this moment is a vector whose line of action is perpendicular to the plane of the forces **P** and **Q**, and it will conform with that work if we write $\mathbf{M}_O=\mathbf{p}\times\mathbf{P}$, although for the time being nothing more that simple arithmetic multiplication is implied. If **P** acts on a body which is pivoted at O, the body will tend to turn about O. The direction of that rotation is the direction associated with the moment $\mathbf{M}_O$. This direction is anticlockwise in Fig. 2.9 and may be conveniently depicted as shown. To sum up: *any force* **P** *is equivalent to an equal parallel force through an arbitrary point* O *together with a couple equal to the moment of* **P** *about* O.

The moment of a force **P** about a point O has been defined as the product of **P** and the perpendicular distance from O to the line of action of **P**. An alternative view is obtained from Fig. 2.10, where **P** is regarded as being applied at a particular point A on its line of action. The parallelogram rule is used to resolve **P** into two components, $\mathbf{P}_n$ in the direction of OA and $\mathbf{P}_t$ perpendicular to OA. As these two components are perpendicular to each other triangles ONA and MGA are similar.

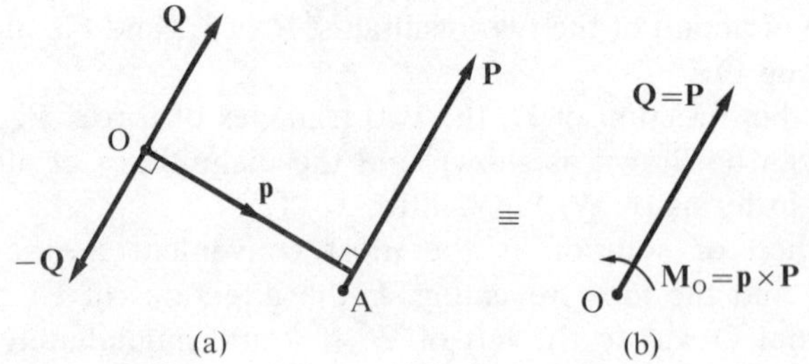

(a) (b)

FIG. 2.9

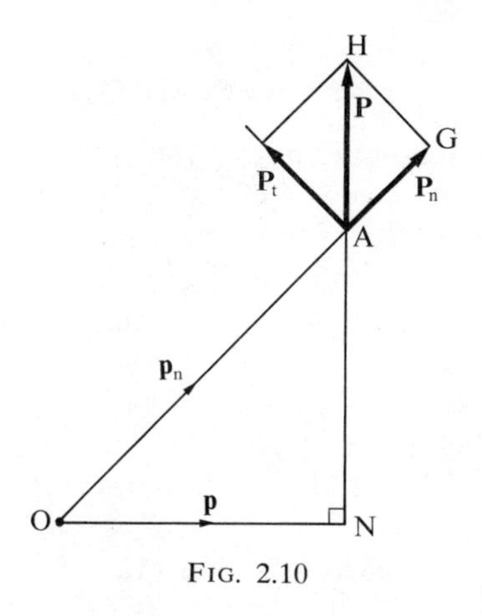

FIG. 2.10

Hence

$$p/p_n = P_t/P$$

and

$$\mathbf{p} \times \mathbf{P} = \mathbf{p}_n \times \mathbf{P}_t.$$

The moment about a point O of a force **P** acting at A equals the moment of the component of **P** perpendicular to OA.

Now consider two forces **P** and **Q** which intersect at A, Fig. 2.11. Their resultant $\mathbf{R} = \mathbf{P} + \mathbf{Q}$ and, from Section 2.2.3, $\mathbf{R}_t = \mathbf{P}_t + \mathbf{Q}_t$, where $\mathbf{R}_t$, $\mathbf{P}_t$, and $\mathbf{Q}_t$ are the components of **R**, **P**, and **Q** perpendicular to OA. Multiplying through this equation by $\mathbf{r}_n$, $\mathbf{p}_n$, and $\mathbf{q}_n$ as appropriate (they are identically equal to $\overrightarrow{OA}$) we have

$$\mathbf{r}_n \times \mathbf{R}_t = \mathbf{p}_n \times \mathbf{P}_t + \mathbf{q}_n \times \mathbf{Q}_t,$$

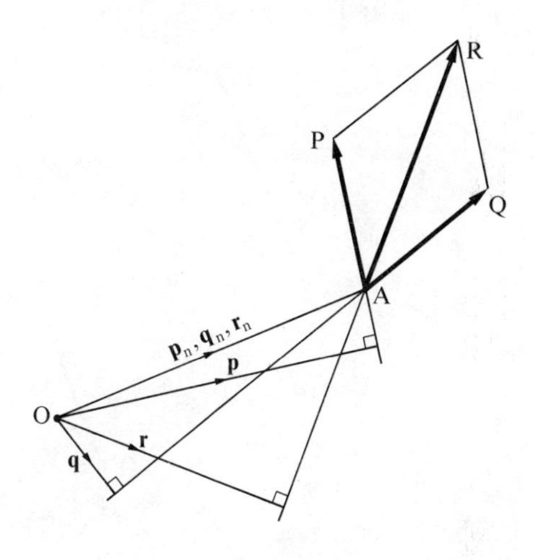

FIG. 2.11

which in turn gives

$$\mathbf{r}\times\mathbf{R}=\mathbf{p}\times\mathbf{P}+\mathbf{q}\times\mathbf{Q}.$$

The moment of the resultant $\mathbf{R}$ *of two forces* $\mathbf{P}$ *and* $\mathbf{Q}$ *about a point* O *equals the sum of the moments of* $\mathbf{P}$ *and* $\mathbf{Q}$ *about* O.

It follows immediately that any set of coplanar forces is equivalent to their resultant acting through an arbitrary point O and a couple whose moment $\mathbf{M}_O$ equals the sum of the moments about O of the separate forces.

The moment of a couple is the same about any point in a plane. So if we now refer to another arbitrary point O′, the moment of the original set of forces is the couple $\mathbf{M}_O$ added to the moment about O′ of $\mathbf{R}$ acting through O. In Fig. 2.12

$$\mathbf{M}_{O'}=\mathbf{M}_O+\mathbf{e}\times\mathbf{R}.$$

If $\mathbf{M}_O=0$ and $\mathbf{R}=0$, $\mathbf{M}_{O'}=0$.

If the resultant of a set of coplanar forces is zero and the resultant moment about an arbitrary point is zero, the moment about any other point is also zero, and the set of forces is in equilibrium.

This statement of the condition for equilibrium is summed up by the equations

$$\sum\mathbf{P}=\mathbf{R}=0 \quad\text{and}\quad \sum\mathbf{M}_O=0. \tag{6.1}$$

The algebraic equivalent of the force equation is obtained by resolving the forces $\mathbf{P}$ into two convenient directions, u and v say, to give

$$\sum P_u=0, \qquad \sum P_v=0, \quad\text{and with}\quad \sum M_O=0 \text{ as before.} \tag{6.2}$$

If more than three items of information are lacking, the planar equilibrium of a rigid body is indeterminate. The three equations (6.2) allow at most the determination of three force magnitudes, two magnitudes and one direction, or one magnitude and two directions.

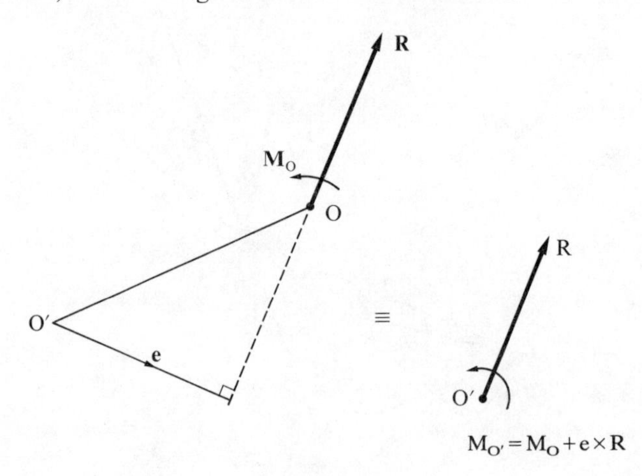

FIG. 2.12

As this result has been deduced from the parallelogram rule for vector addition it cannot tell us any more about a particular problem than could be deduced by direct application of the parallelogram or triangle rules. It can, however, often enable answers to be obtained more quickly and more readily. Let us look again at the example in Fig. 2.8, which is reproduced in Fig. 2.13. The problem posed was:

Example 2.2: A board ABCD, Fig. 2.13(a) hangs in a vertical plane supported by two strings AH and BJ. To maintain AB and DC horizontal, a horizontal force is applied at C. Determine the magnitude of **P**, *and the tensions in the strings in terms of the weight* **W**, *which acts through the centre of gravity of the board.*

Now suppose that the requirement to determine the tensions in the strings is dropped. The previous solution calculated these forces whether or not they were required. We are now free to determine **P** without at the same time evaluating $\mathbf{F}_A$ and $\mathbf{F}_B$. If the board is in equilibrium the sum of the moments of all the forces must be zero about any point. Let us therefore take moments about the point O where $\mathbf{F}_A$ and $\mathbf{F}_B$ intersect. The moments of these two forces about O is zero, and we are left with

$$w \times W = p \times P$$

or

$$P = Ww/p.$$

By drawing the diagram to scale and measuring w and p, P is determined very quickly. The price paid for this extra speed is that we no longer find $\mathbf{F}_A$ and $\mathbf{F}_B$. They are not required. But if HA and BJ are strings and so incapable of acting in compression our answer for P will be nonsense if it requires either $\mathbf{F}_A$ or $\mathbf{F}_B$ to be compressive. It is therefore

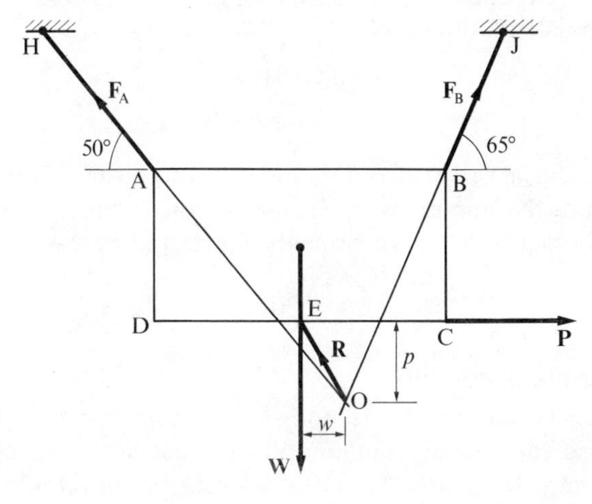

FIG. 2.13

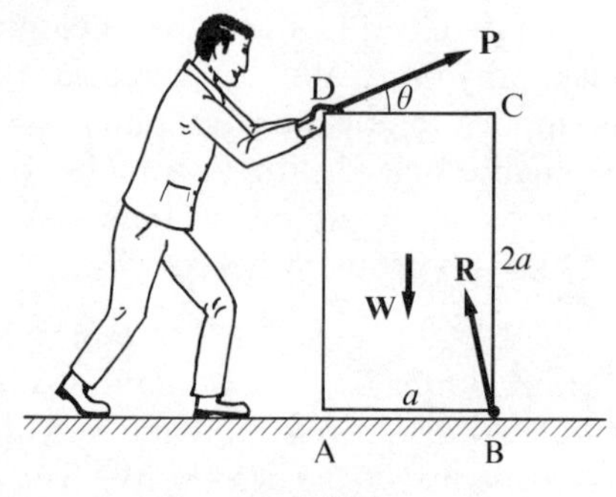

FIG. 2.14

prudent to note that the resultant of **W** and **P** lies along EO, and so $\mathbf{F}_A$ and $\mathbf{F}_B$ are tensile provided E lies between these two forces as in the figure.

Example 2.3: A heavy box, Fig. 2.14, standing on one end is to be turned on its side. The height is 2a, the width a, and the weight acts through the geometric centre. What is the minimum force required and where and in what direction should it be applied? It is to be assumed that the box does not slip on the ground.

Assume that the box is to be turned to the right pivoting about edge B. The instant that the box starts to tilt the reaction between the ground and the box must be confined to the point of contact at B. For equilibrium the moments of **W** and **P** about B must be zero, since the moment of the ground reaction about B is zero. It is evident that to minimize **P** it must be applied as far away from B as possible, i.e. at D, and in a direction perpendicular to DB. Any component of force along DB is wasted effort. For **P** to be perpendicular to DB, $\theta = \tan^{-1}(\frac{1}{2}) = 26{\cdot}6°$. Equating the moments about B of **P** and **W**.

$$P \times a\sqrt{5} = W \times a/2$$
$$P = W/2\sqrt{5}.$$

Experience shows that it is easier, and safer, to push at D rather than to pull at C where the minimum force is $W/4$ horizontally. Experience also shows that in practice our efforts may be frustrated by the box slipping on the ground.

2.3. Friction

2.3.1. The angle of friction

The resistance to sliding of one body over another is called friction. It occurs because surfaces are not smooth, so that asperities catch on one another and impede motion. It may also occur because the surfaces weld together momentarily at the points of contact and require force to tear

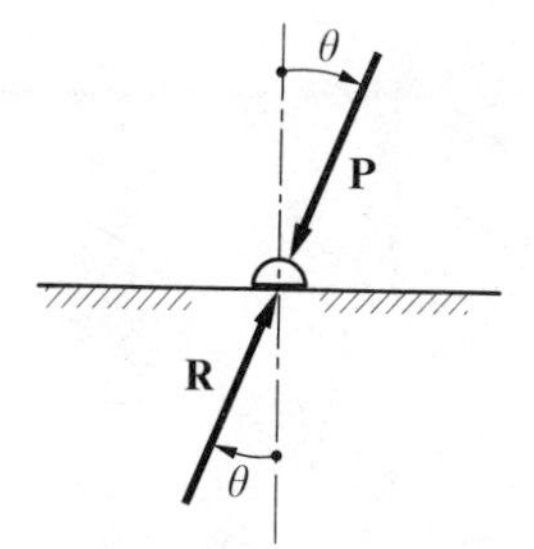

FIG. 2.15

them apart. The resistance to motion depends, therefore, on the degree of smoothness, the properties of the materials of the bodies, and whether there are contaminants such as oxide films, dirt, oil, etc. present that affect the properties. The purpose of lubrication is to separate the surfaces by a film of fluid. The only force then required is the force to shear the layers of fluid. We shall be concerned only with the friction between solid bodies known as Coulomb friction.

The amount of friction between two dry surfaces is difficult to predict with any degree of accuracy. Even if the degree of roughness is closely controlled, the presence of contaminants is usually uncontrollable. Calculations which involve friction must not be regarded as giving more than a reasonable estimate of what may be expected to happen in practice.

A particle is held in contact with a plane surface, Fig. 2.15, by a force **P** normal to the plane. The reaction **R** between the plane and the particle is equal and opposite to **P** giving

$$\mathbf{R}+\mathbf{P}=0.$$

Now imagine **P** to pivot round the particle inclined at an angle θ to the normal. **R** must pivot in unison to maintain equilibrium. Eventually a limiting value of θ is reached when **R** is unable to pivot further. Equilibrium is lost and the particle slips on the plane. The limiting angle ϕ for θ is called the *angle of friction.* If a number of forces $\mathbf{P}_1$, $\mathbf{P}_2$, etc., including its weight, act on a particle, the condition to be satisfied for equilibrium to be possible is that the angle θ between **R** as determined by

$$\sum \mathbf{P}+\mathbf{R}=0$$

and the normal to the surface at the point of contact is not greater than ϕ.

The angle of friction is a helpful concept in considering the equilibrium

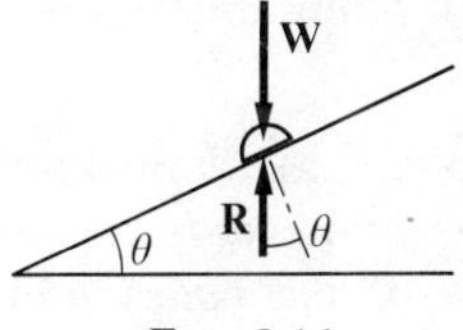

FIG. 2.16

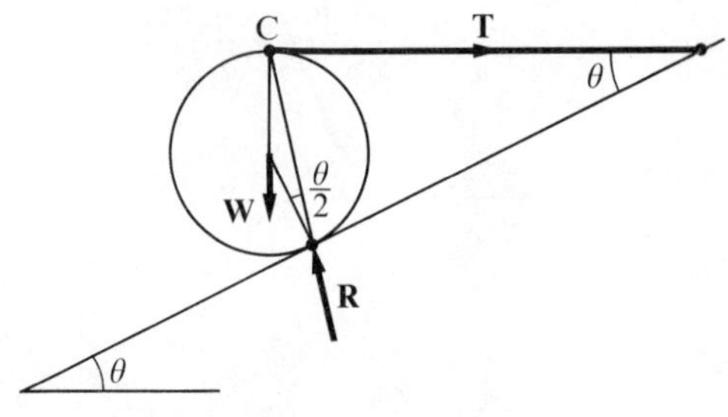

FIG. 2.17

of a particle resting on a slope, Fig. 2.16. If the only forces acting on the particle are its weight **W** and the reaction **R** from the plane, the angle between **R** and the normal to the plane equals the angle θ of the slope to the horizontal. The particle must slide down the slope if $\theta > \phi$.

Example 2.4: A uniform cylinder rests on an inclined plane with its axis perpendicular to the line of greatest slope. It is held in equilibrium by a horizontal string attached to the uppermost point halfway along its length. What is the greatest possible slope angle θ given that the angle of friction between the cylinder and the plane is ϕ?

Clearly, the argument developed in relation to contact between a particle and a plane applies along the line of contact between a cylinder and a plane.

The cylinder in Fig. 2.17 is acted on by three forces: the weight, in fact distributed along the central axis and having its resultant **W** acting halfway along the length; the reaction between the plane and the cylinder, likewise distributed and having a resultant **R** also halfway along the length; and the tension **T** is the string. The three forces **W**, **R**, and **T** are coplanar and for equilibrium must be concurrent.

As **W** and **T** intersect at C, the uppermost point of the cylinder, **R** must also pass through C.

The angle between **R** and the normal to the surface at the point of contact is $\theta/2$, as θ is the angle between the string and the plane. For equilibrium $\theta/2 \leqslant \phi$. The greatest possible value for θ is therefore 2ϕ.

In the conditions so far envisaged, the forces and the impending motion have been coplanar. If in these circumstances there is contact between two rigid bodies at two points, slip at one point of contact must be accompanied by slip at the other point unless, of course, the bodies simply separate at the second point.

Example 2.5: A light rigid plank rests horizontally between two slopes, both of which are at angles to the horizontal that are greater than the angle of friction ϕ between the ends of the board and the slope. Determine the possible range of motion for a man standing on the plank if it is not to slip.

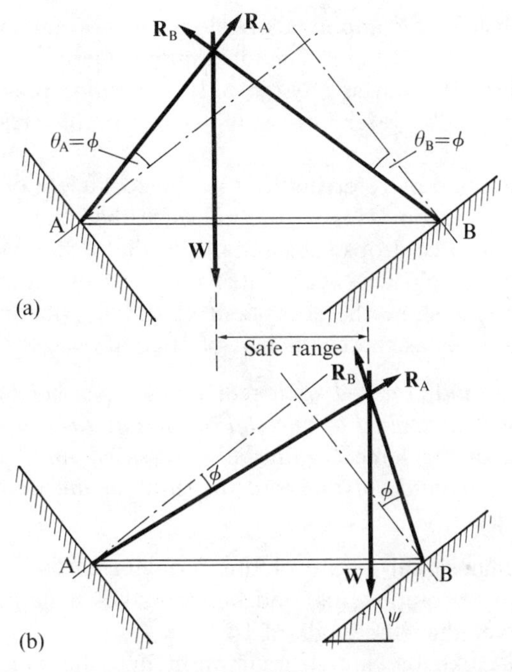

FIG. 2.18

If the plank slips it must do so at both ends simultaneously. Furthermore is one end A, say, slides down the slope on which rests, the other end B must slide up its slope. Assume that the plank is on the point of slipping in this way. The reactions at A and B must be inclined at angles ϕ to the normals to the slopes at these points, as shown in Fig. 2.18(a).

The weight **W** of the man must be so placed that it is concurrent with $\mathbf{R}_A$ and $\mathbf{R}_B$. If the weight moves to the left of this position then for concurrency of the three forces either θ_A or θ_B must be greater than ϕ so that equilibrium is no longer possible. A slight movement to the right, then either θ_A or θ_B must be less than ϕ and the possibility of slip recedes.

A substantial shift to the right brings about the set of forces depicted in Fig. 2.18(b) where end B is on the point of slipping down the slope and end A slipping up. The safe region for **W** is between the two limits just determined.

It can be seen that if the slope ψ at B (or at A) is less than ϕ then it will be safe for the man to stand right up to that end.

2.3.2. The coefficient of friction

In the above examples it was helpful to proceed on the basis that slip is impending when the angle θ between the normal to plane on which slipping is about to occur and the reaction **R** there equals the angle of friction ϕ. In other circumstances it is more convenient to recognize that

R may be resolved into components **N** along the normal and **T** tangential to the plane. Then $T = N \tan \theta$. The maximum value that T can have, and the value at which slip starts, is $N \tan \phi$. It is common practice to write μ in place of $\tan \phi$ and to refer to μ as the *coefficient of friction*. So, for slip to start, $T = \mu N$.

μ is often qualified by referring to it as the coefficient of static friction, or coefficient of stiction. This is to emphasize that once sliding actually starts the friction force drops substantially so that if the disturbing forces are maintained the particle accelerates rather than sliding at a steady speed. Experiments show the coefficient of sliding friction to be just as variable a parameter as the coefficient of stiction.

Example 2.6: A uniform rod of length l lies on a horizontal plane. A horizontal force **P** *is applied at one end of the rod and perpendicular to it. The magnitude of the force is gradually increased until it starts to slip. Determine the magnitude of this force in terms of the weight W and the coefficient of friction μ.*

The three-dimensional nature of this problem encourages the resolution of forces into components, and hence makes it natural to think in terms of μ rather than the angle of friction.

We have to make the initial assumption that the weight of the rod results in a vertical reaction between the rod and the plane that is uniformly distributed along the length of the rod. The extent to which this assumption is realized will depend on the relative rigidity of the rod and the plane. It implies that the limiting friction force is constant along the length of the rod and equals $\mu W/l$. To permit equilibrium between the resultant friction force and **P**, the direction of the distributed friction force must reverse part way along the rod, the distribution being as shown in Fig. 2.19. The initial motion will be a rotation about the point O where the direction of the friction force reverses. Assume that O is at a distance αl from the end remote from **P**.

For equilibrium the moment of the forces about any point must be zero. Taking moments about the point of application of P

$$(\alpha l \times \mu W/l) \times (l - \alpha l/2) - [(1-\alpha) l \times \mu W/l] \times (l - \alpha l)/2 = 0$$

$$\Rightarrow \qquad \alpha(1-\alpha/2) - (1-\alpha)^2/2 = 0$$

$$\Rightarrow \qquad \alpha^2 - 2\alpha + \tfrac{1}{2} = 0$$

$$\Rightarrow \qquad \alpha = 1 \pm 1/\sqrt{2}.$$

As $\alpha < 1$ the negative root holds and $\alpha = 0{\cdot}293$.

P is determined by equating the resultant force to zero.

$$\begin{aligned} P &= (1-\alpha) l \times \mu W/l - \alpha l \times \mu W/l \\ &= (1-2\alpha)\mu W \\ &= (\sqrt{2}-1)\mu W = 0{\cdot}414\,\mu W. \end{aligned}$$

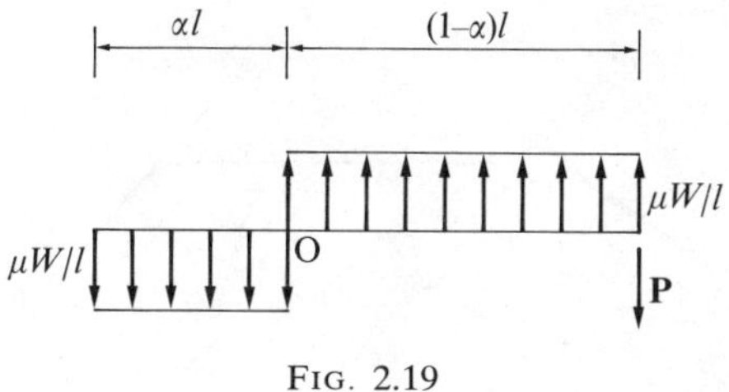

FIG. 2.19

The last example merely touches on equilibrium in three dimensions. More complex examples of three-dimensional statics, with or without friction, will be considered in Chapter 3.

2.4. Forces in mechanisms

We understand a structure to be a set of members connected together to form a rigid assembly. If the assembly is able to distort, other than by distortion of the individual members, in a constrained manner, we have a mechanism. If the assembly, whether a structure or a mechanism, is in equilibrium under a set of externally applied forces, including the weights of the members, each constituent member must also be in equilibrium. The forces acting on any one member may include external forces and certainly include forces applied by the other members which make direct contact with the member in question. The study of the equilibrium of a structure or a mechanism thus involves two considerations: the over-all equilibrium and the equilibrium of the individual members.

We will not study in detail the internal distribution of stresses imposed on the members, but it is necessary to understand the different types of 'force' to which a component may be subjected.

2.4.1. Bending moment, shear force, axial force, and torque

A bar, Fig. 2.20, is in equilibrium with three concurrent forces $\mathbf{P}_1$, $\mathbf{P}_2$, $\mathbf{P}_3$ applied as shown. If the bar were cut through at O between two of the loads, the two halves of the bar would separate and equilibrium would be lost. Evidently forces are transmitted within the bar across the section at which the cut is contemplated.

The force $\mathbf{P}_1$ applied at the left-hand end of the bar is statically equivalent to an equal force $\mathbf{P}_1$ to the bar at O and a couple $\mathbf{M}_\mathrm{O}$ equal to the moment of the original force $\mathbf{P}_1$ about O; the length of the bar to the right of O is in precisely the same state whether $\mathbf{P}_1$ is applied as originally or together with $\mathbf{M}_\mathrm{O}$ at O. The couple $\mathbf{M}_\mathrm{O}$ is the *bending moment* in the bar at O, its effect being to bend the bar. The force $\mathbf{P}_1$ can be resolved into two components, one $\mathbf{P}_\mathrm{n}$ along the bar and the other $\mathbf{P}_\mathrm{s}$ transversely, that is in the plane of the cut. The longitudinal component is tensile if its tendency is to stretch the bar, otherwise it is compressive. The transverse component is the *shear force*; it tends to cause sliding between the two

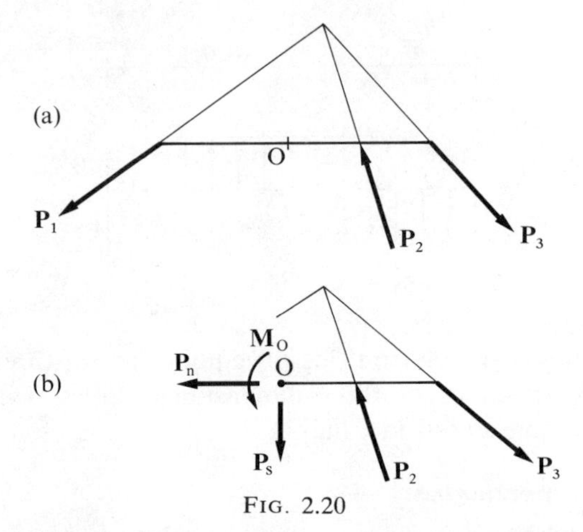

FIG. 2.20

faces of the imaginary cut. The bending moment $\mathbf{M}_O$, shear force $\mathbf{P}_s$, and longitudinal force $\mathbf{P}_n$ are in equilibrium with $\mathbf{P}_2$ and $\mathbf{P}_3$, and except for reversal of signs could have been deduced by considering equilibrium of the whole length of the bar to the right of O.

The shear force, longitudinal force, and bending moment vary along the length of the bar. If, as in this example, the bar is straight the shear force and longitudinal force are constant between the point of application of adjacent loads, and change only at these points. The bending moment varies continuously along the beam. Graphs showing the distribution of shear force and bending moment along a member are of particular importance when strength aspect are being considered in the design of engineering structures and mechanisms. These graphs are known as shear force and bending moment diagrams. For the bar in Fig. 2.20 the shear force and bending moment diagrams are as in Fig. 2.21. It will be noticed

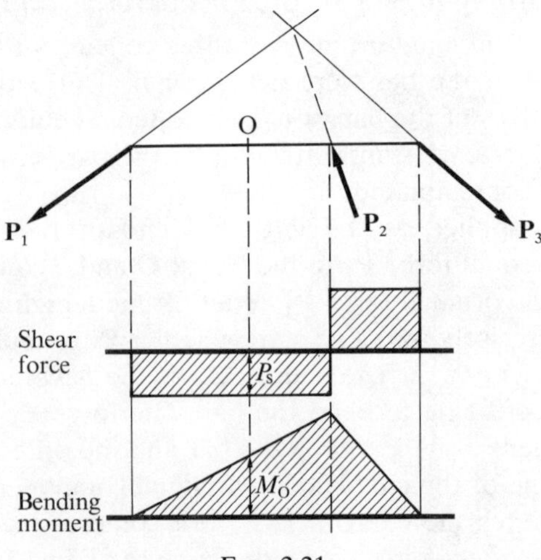

FIG. 2.21

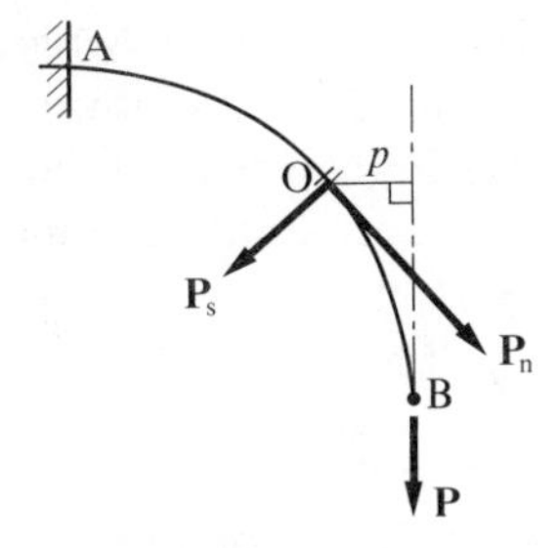

FIG. 2.22

that whereas the bending moment in this example has the same sign along the length of the beam, the shear force changes sign at the point of application of $\mathbf{P}_2$. The choice of signs is a matter of convention which need not be pursued here.

In a straight bar the bending moment and shear force depend only on the transverse components of the leads. The longitudinal components must be in equilibrium, but can otherwise be altered at will without affecting either the bending moment or the shear force. The force $\mathbf{P}_1$ in Fig. 2.21 can be split into the two components $\mathbf{P}_n$ longitudinally and $\mathbf{P}_s$ transversely. The transverse component $\mathbf{P}_s$ is the shear force at O and M_O can be seen to be equal to $AO \times \mathbf{P}_s$, whilst $\mathbf{P}_n$ has no moment about O. The shear force and bending moment are therefore interdependent.

The longitudinal component of the load has no effect on the bending moments only if the bar is straight. The curved bar AB in Fig. 2.22 is held firmly at A and a load $\mathbf{P}$ is applied axially at B. In this case there is no shear force at B. Nevertheless there is a bending moment at O of magnitude $p \times P$.

In Fig. 2.23 a bar CB is rigidly attached at B to another bar AB which is rigidly held at A. CB is at right angles to AB. A load $\mathbf{P}$ is applied at C perpendicular to the plane of ABC. Equilibrium of BC requires a reaction at B equal and opposite to $\mathbf{P}$ and a couple whose moment is equal and opposite to the moment of $\mathbf{P}$ about B. This moment is the bending moment in BC at B. Just round the corner bar AB experiences

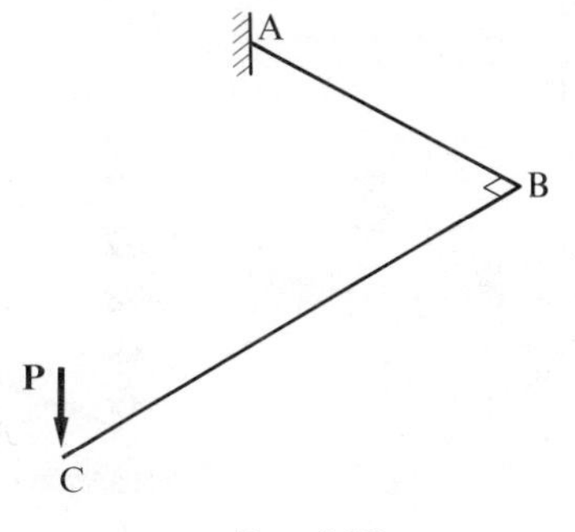

FIG. 2.23

the same moment but about an axis that is along the bar. The loading transmitted to AB at B comprises, therefore, a shear force **P** and a torsional moment $\mathbf{M}_B = \overrightarrow{CB} \times \mathbf{P}$. The reaction at A consists of a shearing force equal and opposite to **P** and a torsional moment or torque equal to $\overrightarrow{CA} \times \mathbf{P}$. The moment at A is more conveniently thought of in terms of two components, a torque $\overrightarrow{CB} \times \mathbf{P}$ and a bending moment $\overrightarrow{BA} \times \mathbf{P}$.

2.4.2. Statics of planar mechanisms

It has been observed earlier that equilibrium of a mechanism requires balance of the external forces applied to it, and equilibrium of the forces applied to the individual members. The forces on any one member will certainly include the forces from the other members with which it is in direct contact, and may include external forces. A number of examples will illustrate the method of analysis.

Example 2.7: A member ABC, Fig. 2.24, is hinged at A and B to blocks which run in fixed guides. A second member CD is hinged at D to a similar block. What force is required at A to overcome a resisting force **F** *at D? Friction is negligible at all points of contact.*

We have to consider equilibrium of the blocks at A, B, and D, and of members ABC and CD.

The block at A is subjected to the driving force **P** and a reaction from member ABC. As it is improbable that these two forces will be equal and opposite, and likely that there will be a sideways reaction $\mathbf{R}_A$ from the guide. In the absence of friction $\mathbf{R}_A$ is perpendicular to the axis of the guide.

Similar considerations lead to the assumption that there are reactions $\mathbf{R}_B$ and $\mathbf{R}_D$. Equilibrium of members ABC and CD taken together requires $\mathbf{R}_A$, $\mathbf{R}_B$, $\mathbf{R}_D$, **P** and **F** to be in equilibrium. The directions of all these forces are known, but only **F** is known in magnitude, and there is as yet insufficient information to determine the magnitudes of the other forces. The basic requirements for equilibrium, first, that the vector sum

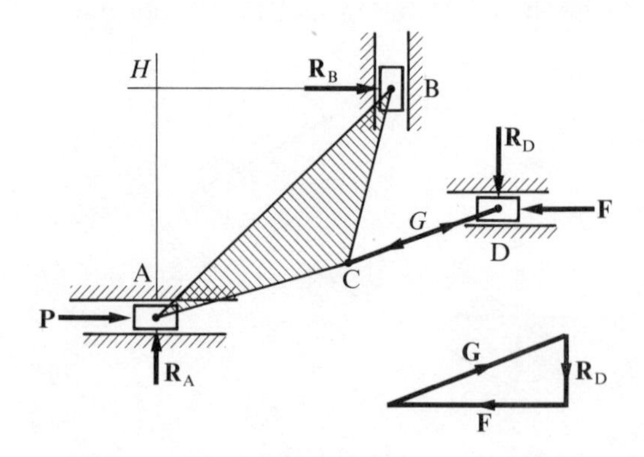

FIG. 2.24

of all the forces must be zero; second, that the sum of the moments about an arbitrary point must be zero, enable only three unknown quantities to be determined. The additional piece of information which in this case enables four quantities to be determined comes from consideration of member CD. As the forces acting on it are applied only at its two ends, the resultant force at either end must act along the line of the bar. Let the forces exerted by the bar be $\pm\mathbf{G}$.

G can be found immediately by equating the sum of the moments about H, Fig. 2.24, of the forces acting on member ABC. This gives G in terms of P.

F is now determined in terms of G, and hence in terms of P, by drawing a triangle of forces at D.

Example 2.8: In the mechanism of the previous example the coefficient of friction between each of the blocks A, B, and D, and their guides is μ. Determine the force $\mathbf{P}$ required at A to overcome a resisting force $\mathbf{F}$ at D. Friction in the pivots is negligible.

The direction of sliding is known at A and can be deduced for B and D by inspection or by drawing a velocity diagram.

The reaction between guide and block at A, B, and D is in each case inclined at an angle $\phi = \tan^{-1}\mu$ to the normal. Taking into account the directions of sliding, the reactions are as shown in Fig. 2.25.

The calculation now proceeds exactly as before first by taking moments about H′, the point of intersection of $\mathbf{R}'_A$ and $\mathbf{R}'_B$, to obtain G', then by drawing a triangle of forces at D to obtain F' in terms of P.

The simplicity of this particular mechanism enables all the forces to be depicted without ambiguity on a sketch of the complete mechanism. Very often this is difficult to achieve, and it is then advisable to take the mechanism apart to show how the forces act on each of the members.

Example 2.9: A pair of forces $\pm\mathbf{P}$ are applied to a 'lazy tongs' to balance a force $\mathbf{R}$, Fig. 2.26. Determine R in terms of P, and the forces in the members. Friction is negligible.

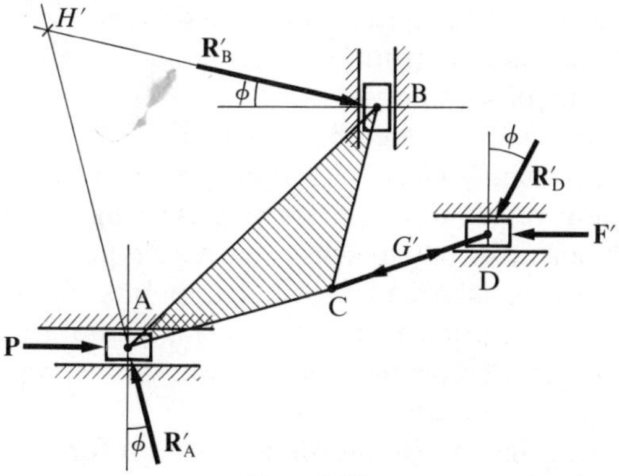

FIG. 2.25

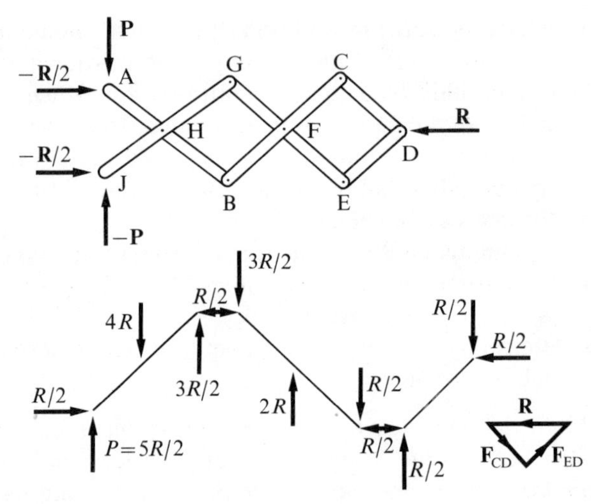

FIG. 2.26

Take the mechanism apart and consider the equilibrium of members DE, EG, and GJ in turn.

DE is subjected to forces at D and E only, so that the resultant forces at either end must be in the direction DE. By drawing a triangle of forces at D the forces in members DE and DC are found to have magnitude $R/\sqrt{2}$. However it is convenient to represent the forces at either end of DE in component form. At D the force transmitted by the pivot is one half the external load and a force of equal magnitude at right angles to the external load. There are similar components at E, with equal and opposite components acting on GE at E.

At G the component parallel to the load **R** must be of magnitude $R/2$, because together with a similar component at B there must be over-all equilibrium with **R**.

On taking moments about F of the forces acting on GE the magnitude of the transverse component of the force at G is found to be $3R/2$. To give over-all equilibrium of GE the force applied to it by the pin at F is of magnitude $2R$ in the direction shown. The direction of this force, since there must be equal and opposite force on member BC, is to be expected from considerations of symmetry.

Finally, a similar study of member GJ yields $P = 5R/2$.

It becomes increasingly clear with these examples that whilst there is advantage in using vector notation for representing external forces on a body or a mechanism, if only because the bold type clearly distinguishes letters for forces from labels for geometric points, there are difficulties when the forces are internal to a mechanism. There are also difficulties when it becomes either convenient or necessary to express the magnitude of one force in terms of another.

The first problem lies in distinguishing between forces that are applied to a body, such as a bar in a mechanism, and the forces that one body applies to others. In Fig. 2.24 the bar CD pushes on the joint at C, and in the opposite direction on the block at D. Thus the arrow-heads drawn on the bar indicate the directions of the forces applied by the bar. This is a standard convention for bars in tension or compression. It would be logical now to label one arrow-head $\mathbf{G}$ and the other $-\mathbf{G}$, but it would also be cumbersome. Simply to write $\pm\mathbf{G}$ would be unhelpful, because there would be no indication then of which sign was associated with which arrow-head. In fact, G denoting the magnitude of the two forces is sufficient.

In the mechanism of Fig. 2.26 the bars are not all subjected to simple compression or tension, and it is easier to think in terms of the forces applied to the bars through the joints. By drawing the arrows externally to a bar we clearly indicate that the forces are applied to the bar, rather than by the bar. We conclude very quickly that the horizontal and vertical components of the force transmitted through joint E, for example, both have magnitude $R/2$. However, only the horizontal force applied through E to EG could properly be labelled $\mathbf{R}/2$. Such a label would be quite wrong if applied to any of the other three force components shown at that joint, for in every case the direction would be wrong.

2.4.3. Statics of plane frames: Bow's notation

The statical analysis of framework structures in which all the members act either in tension or compression, that is as ties or struts, and are not subjected to bending moments, is much simplified by the adoption of a special technique to systematize the construction of triangles of forces and polygons of forces for the joints. The technique involves a particular notation, Bow's notation, for identifying force vectors.

If a set of forces $\mathbf{P}_1$, $\mathbf{P}_2$, etc., Fig. 2.27(a), are in equilibrium the force polygon closes and the vectors of that polygon may be labelled as in Fig. 2.27(b). This is the method we have so far adopted, except for drawing velocity diagrams for which another notation is particularly appropriate. Bow's notation involves the assignment of letters to the spaces between adjacent force vectors, A, B, C etc. in Fig. 2.27(c). The force polygon is now drawn by imagining a circuit round the point of application of the forces. Starting at A, say, the force vector $\mathbf{P}_1$ is crossed going to B and is represented by the vector $\overrightarrow{ab}$ in Fig. 2.27(d). On proceeding to C, bc is drawn, then cd. Finally, on completing the circuit, da is drawn. No arrowheads are needed on the force polygon provided it is understood that the circuit round the joint is always to be taken in the clockwise direction. The direction of the force that separates areas A and B is in the direction a to b. If this force is applied by a member of a framework to a joint, then the same member must apply an equal and opposite force to the joint at its other end. Nevertheless, $\overrightarrow{ab}$ still represents the force in the member.

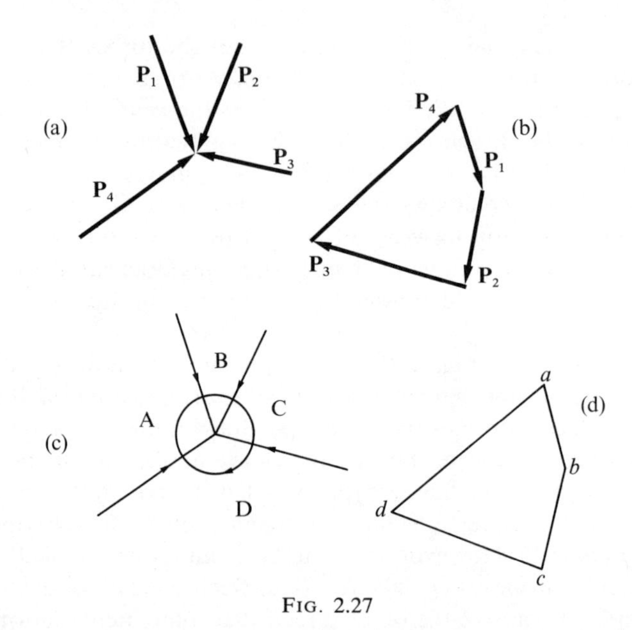

FIG. 2.27

Example 2.10: An equilateral triangular pin-jointed framework is attached to a fixed abutment at one corner X and is supported on a frictionless roller at another corner Z. A force **P** *is applied as shown in Fig. 2.28 to the third corner Y. Determine the forces in the members of the structure and the reactions at the supports.*

First, the areas between the external forces and the members and between the members themselves are labelled A, B, C, D as shown. The reaction $\mathbf{R}_2$ at Z must be perpendicular to the movement of the roller support if there is no friction there. The direction of the reaction $\mathbf{R}_1$ at X is unknown.

Let us start the construction of the force diagram at joint Y where the external force **P** is specified.

Using Bow's notation, the triangle *bcd* is drawn; side *bc* is of known length *P* and its direction is parallel to *P*. Sides *cd* and *db* are then drawn in the directions of members YZ and YX respectively. A circuit round the

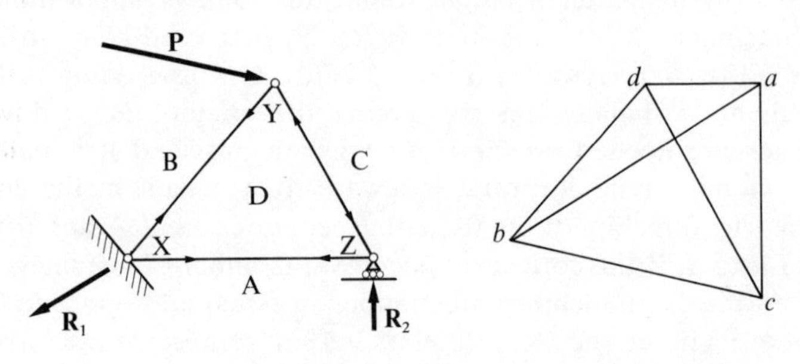

FIG. 2.28

triangle of forces corresponding to a clockwise circuit round X gives force directions as indicated by the arrowheads drawn close to joint Y in the diagram of the frame.

Now consider joint Z. Side *dc* of the relevant triangle of forces has already been drawn. *ca* is drawn parallel to $\mathbf{R}_2$ and *da* is parallel to ZX. A clockwise circuit round Z shows that the directions of the forces at Z are as indicated by the arrowheads. It is to be noted that the forces at either end of YZ are consistent with each other.

It is sufficient now to join points *a* and *b* in the force diagram to determine the magnitude and direction of $\mathbf{R}_1$. A clockwise circuit of the joint yields the directions of the forces at X as shown.

Triangle *abc* is the force triangle for the external forces on the structure. As a check on the work it can now be confirmed that the direction of $\mathbf{R}_1$ is such that $\mathbf{R}_1$, $\mathbf{R}_2$, and $\mathbf{P}$ are concurrent.

Example 2.11: Determine the forces in the members of the plane framework in Fig. 2.29 under the given loads.

The procedure must be slightly different from that used in the preceding example because on this occasion it is not possible to start drawing the force diagram at one of the joints at which specified loads act. Too many members meet at both *x* and *y*. A start is made by taking moments about *w* to determine R_2. The areas of the frame diagram are labelled as before.

Starting with *da*, force triangle *dae* is drawn. As $\mathbf{P}_2$ is known polygon *defc* can be drawn. Then considering joint *u*, *g* is located.

An attempt to draw triangle *ahg* shows that *g* and *h* are concurrent, and that consequently there is no force in member *vx*. Finally, *b* is added by considering a circuit round *x*. As a check, *bh* in the force diagram should be parallel to member *wx*. In the force diagram *a*, *b*, *c*, *d* are all on the same vertical line so that over-all equilibrium of the frame with

$$\mathbf{P}_1+\mathbf{P}_2=\mathbf{R}_1+\mathbf{R}_2$$

is achieved automatically.

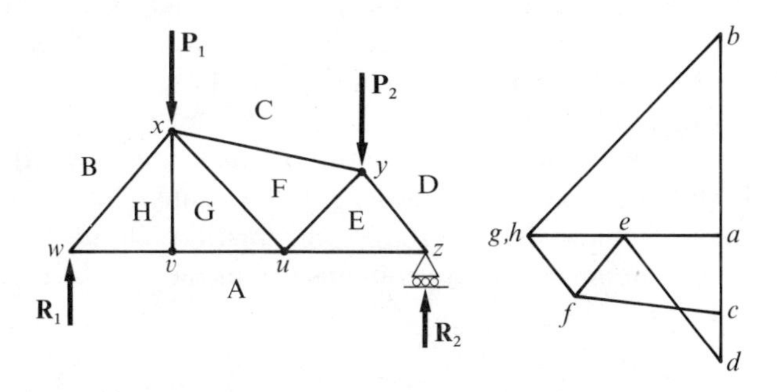

FIG. 2.29

2.5. Statical determinacy

It will be noted that with all the problems so far considered the conditions have been just right to obtain a solution. When a particular force has been calculated by two different routes one value has been a check on the other. In no case has it appeared that there might be more than one solution. There have always been sufficient forces available to ensure equilibrium of bodies, and there has never been more forces to be calculated than the available information permitted. The problems were, of course, chosen to achieve this. They were all *statically determinate.*

When the equilibrium of only one or two bodies is being studied it is usually obvious whether it is possible to calculate all the unknown forces from the information available. When a large number of bodies is involved, as with the members of a structural framework, it may be less obvious whether equilibrium is possible and whether all the forces are calculable.

The degree of statical determinacy of a system of rigid bodies is allied to its mobility as determined by Grübler's formula, p. 26. If conditions are such that there are just sufficient constraints to prevent motion, then the structure is statically determinate. Grübler's formula was derived by considering the loss of freedom of the members of a mechanism resulting from the interconnections between members. Equivalent conclusions can be reached by considering the equilibrium of the members.

Suppose that n members of arbitrary length are connected together in a plane by g joints, and that arbitrary coplanar loads may be applied at any one of the joints.

For equilibrium at a joint, the vector polygon must close. This requirement is equivalent to two scalar equations obtained by balance of the components of the forces in two arbitrary directions. This gives a total of $2g$ scalar equations. These are, however, not entirely independent because the forces must balance over all as well as at every joint. Over-all equilibrium imposes three conditions, moment balance and force balance in two directions. The number of unknown quantities for which the resulting set of equations can be solved is therefore, in general, $(2g-3)$. If this equals the number of unknown forces, that is the number of members n, the system is statically determinate with $n=2g-3$.

If $n<2g-3$ the system is a mechanism and there will be equilibrium only if the loads are chosen so as to achieve balance.

If $n>2g-3$, there are more members than are needed and the structure is statically indeterminate, or redundant. Calculation of the forces in the members then requires information about the material properties of the members. Such problems are in the province of 'Theory of Structures'.

Example 2.12: Verify that the framework in Fig. 2.30(a) is a statically determinate structure. Determine also whether the system in Fig. 2.30(b) is statically determinate.

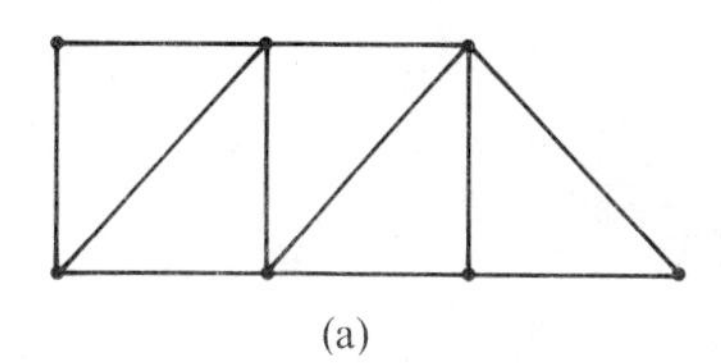

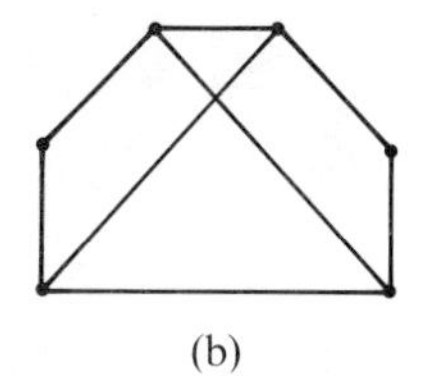

FIG. 2.30

The framework in Fig. 2.30(a) has $n=11$ and $g=7$.
But,

$$2g-3=11.$$

Hence the framework is statically determinate.

The answer is possibly less obvious for the framework in Fig. 2.30(b). Here $g=6$, so $(2g-3)=9$, but $n=8$ so that the framework has one bar too few for statical determinacy, and is a mechanism.

Unfortunately, the results of this analysis can be misleading. The test for statical determinacy starts with the assumption that all the members of the structure are of arbitrary length and that the loading is also arbitrary, except to the extent needed to achieve over-all equilibrium. With special geometries or special loadings, which may arise inadvertently, the formula fails. Fig. 2.31 gives two such examples.

In Fig. 2.31(a), $n=13$ and $g=8$, and so $n=2g-3$ is satisfied. But because of the equality of lengths of the members of the central panel, the framework can distort freely as shown. This means that it is incapable of supporting a general load system.

On the other hand in Fig. 2.31(b), where $n=4$ and $g=4$, the structure has one member too few for statical determinacy but it is quite capable of satisfactorily sustaining the loading shown.

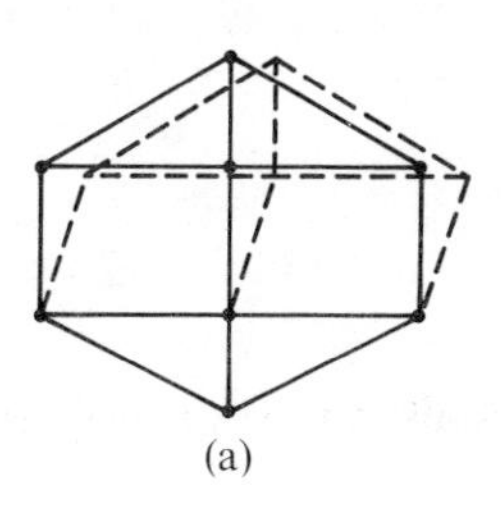

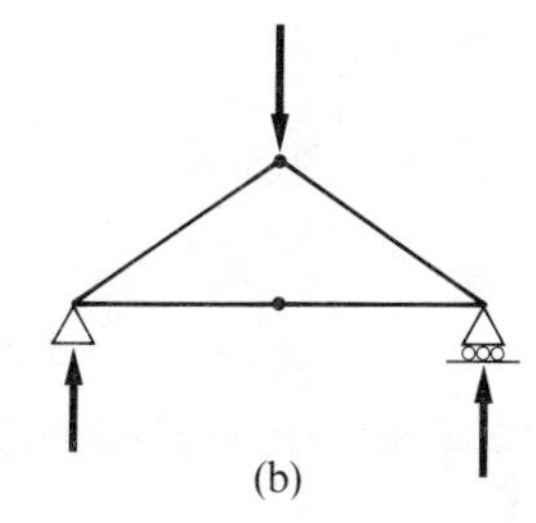

FIG. 2.31

2.6. Problems

1. Determine the resultant of the three coplanar forces shown in Fig. 2.32 by drawing a force polygon, and confirm that the answer is the same whatever the order in which they are added.

2. Determine the resultant of the forces in Fig. 2.32,
   (a) by resolving the forces in the direction Ox and Oy,
   (b) by resolving the forces parallel and perpendicular to the 2 N force, and
   (c) by resolving the forces in the directions of the 2 N and 1 N forces.

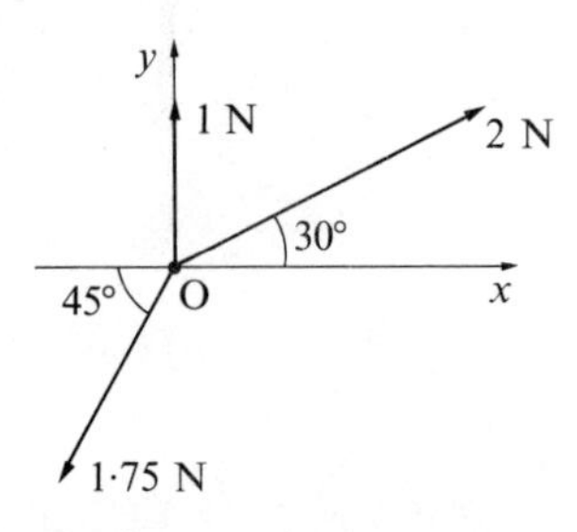

FIG. 2.32

3. The angle between a force **P** and the plane Oxz is seen to be ϕ_x when looking along Ox and ϕ_z when looking along Oz. Show that the direction cosines of **P** are

$$\cos\gamma_x = \frac{\sin\phi_x \cos\phi_z}{(1-\cos^2\phi_x \cos^2\phi_z)^{\frac{1}{2}}}; \qquad \cos\gamma_z = \frac{\sin\phi_z \cos\phi_x}{(1-\cos^2\phi_x \cos^2\phi_z)^{\frac{1}{2}}}.$$

4. A smooth ring C of weight 10 N is threaded on a light string, the ends of which are attached to the ceiling at A and B. A horizontal force acts at C such that the plane ACB makes an angle of 45° with the vertical, and the bisector of ACB makes an angle of 60° with AB.

Find the magnitude of the force.

5. The arrangement of the members of a swinging derrick is shown in Fig. 2.33. If the derrick is carrying a vertical load of 200 kN, find the loads in all the members.

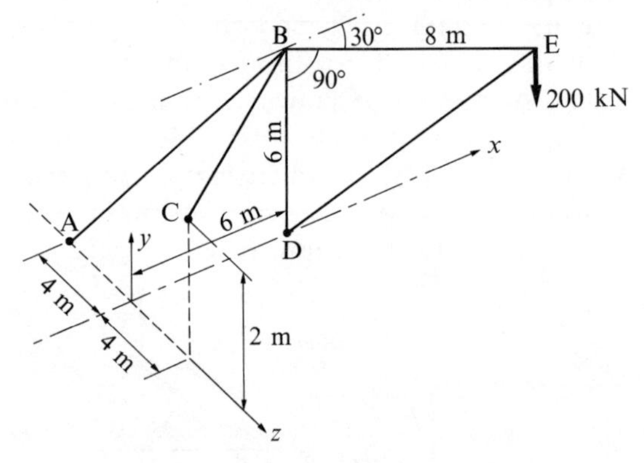

FIG. 2.33

6. Forces of 1, 2, 3, and 4 N act along the sides AB, BC, CD, DA respectively of a square of side length a.

Determine the resultant force and its line of action,

(i) by combining the forces one by one, first in the order 1, 2, 3, 4; then combining force 1 with 3, and then 2 with 4, and finally adding the two resultants. Verify that the answer is the same for the two orders of combination, and

(ii) by taking moments about one corner of the square, e.g. A.

Finally, verify that the total moment of the forces about an arbitrary point on the line of action of the resultant is zero.

7. A uniform rectangular lamina ABCD has sides AB = $2a$ and BC = $2b$. A string is connected to A and B, and it is found that the lamina hangs in equilibrium with the string passing over a small frictionless pulley, and with the edge AB at 60° to the vertical. Show

that the two parts of the string make equal angles with the vertical, and find the length of the string.
Note. The lamina can be in equilibrium in the manner described only if $b < a\sqrt{3}$

8. A uniform concrete block weighing 1.7 kN stands on a rough horizontal surface. It has a long, light cable attached to the central point of its top face, as shown in Fig. 2.34. Find the tension in the cable at which the block starts to lift. Show that if the tension in the cable remains constant, the block will topple over.
Note. The results of question 3 may be useful.

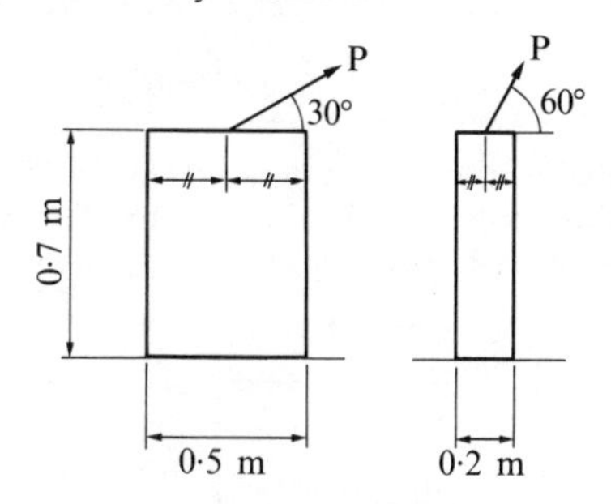

FIG. 2.34

9. Fig. 2.35 is a plan of a hammock AB slung from two perpendicular bulkheads. BD and AC are cords and AE is a horizontal strut, the inclined length CE being 1·5 m. The heights above the deck of the points C, A, and D are 2·1, 1·2, and 1·5 m respectively. The loaded hammock weighs 90 kg and its centre of gravity is at G. Find the forces in the two cords and the strut.

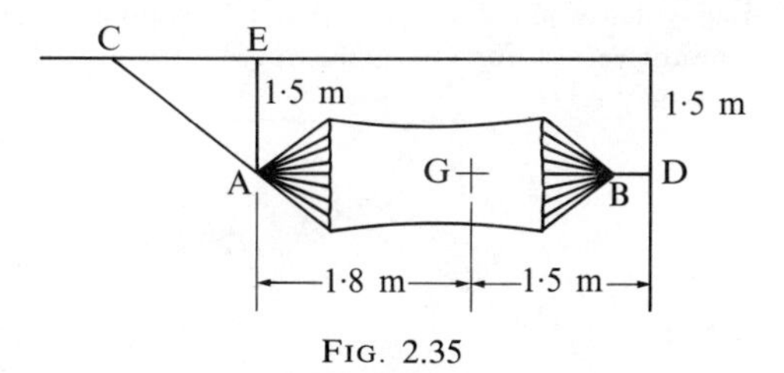

FIG. 2.35

10. Two light bars AB and BC of equal length are connected together at B by a pin joint. BC is connected at C to a rigid abutment and AB is connected to a slider at A to which a horizontal force of 5 kN is applied as shown in Fig. 2.36. A vertical force of 5 kN is applied gradually at B as shown. Determine the forces in AB and BC. The coefficient of friction is 0·25 for the slider. Friction in the pin joints is negligible.

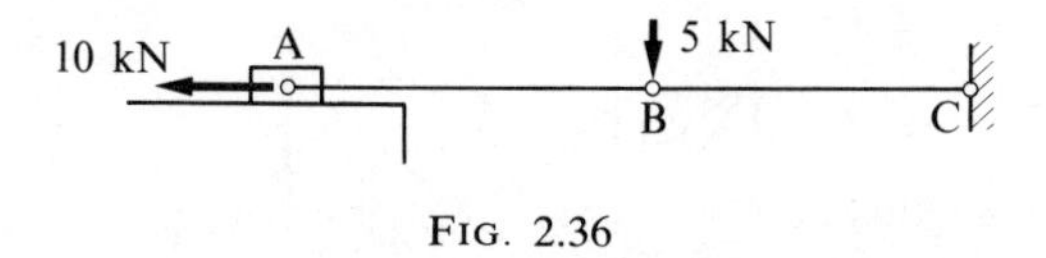

FIG. 2.36

11. A uniform cylinder of weight W lies with its axis horizontal on an inclined plane, resting against a light wedge. The angle between the working faces of the wedge is β. The plane makes an angle α with the horizontal and the angle of friction between the wedge and the plane is λ. There is no friction elsewhere.

If $\alpha < \lambda < \beta$, what is the least force applied to the wedge that will prevent slipping?

12. A uniform beam weighing 220 kN is being dragged into position by a horizontal rope as shown in Fig. 2.37. If the coefficient of friction at the ground is 0·2 and at the wall is 0·3, find the magnitude P of the rope force.

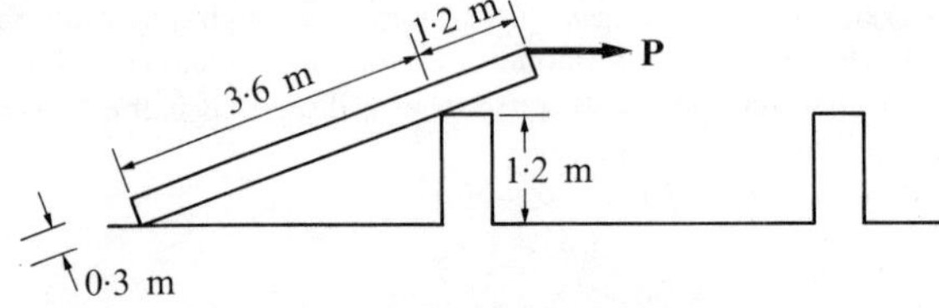

FIG. 2.37

13. A load of 2 kN is raised by means of wedges as shown in Fig. 2.38. The coefficient of friction at all rubbing surfaces is 0·25. Determine the force **F** that is just sufficient to cause movement. The weight of the wedges is negligible.

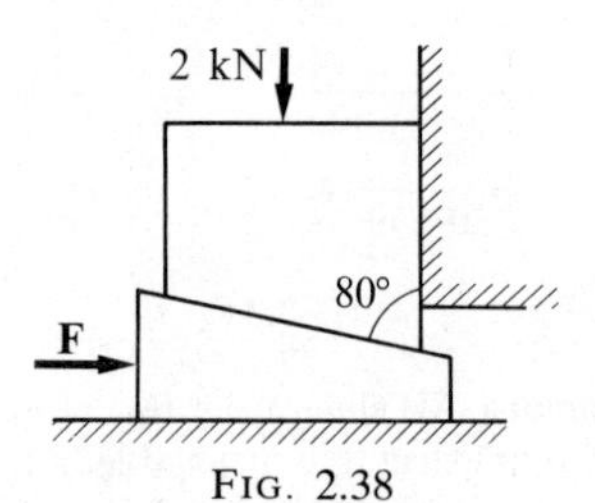

FIG. 2.38

14. A cylinder of weight **W** is attached by a horizontal string to a rough plane as shown in Fig. 2.39. If the cylinder is on the point of slipping, find the magnitude of the least upward vertical force applied at A that would cause the cylinder to slip up the plane. If, instead of the force at A, a downward vertical force is applied at B, show that slipping cannot occur however large that force may be

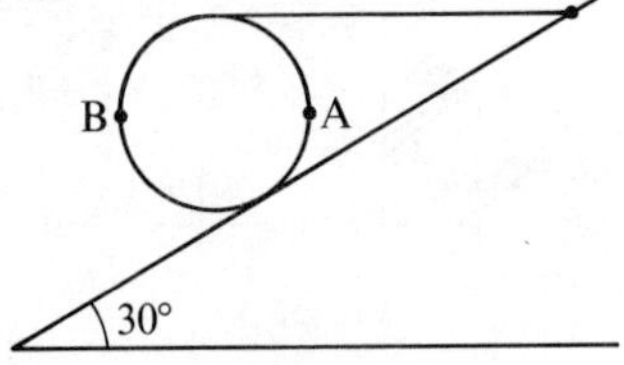

FIG. 2.39

15. Referring to Problem 11, chapter 1, what torque must be applied at D to overcome P = 500 N?

16. Verify that the Bow's notation diagram for a plane structure including two intersecting but non-connected bars can be correctly drawn by inserting a pin-joint at the intersection,

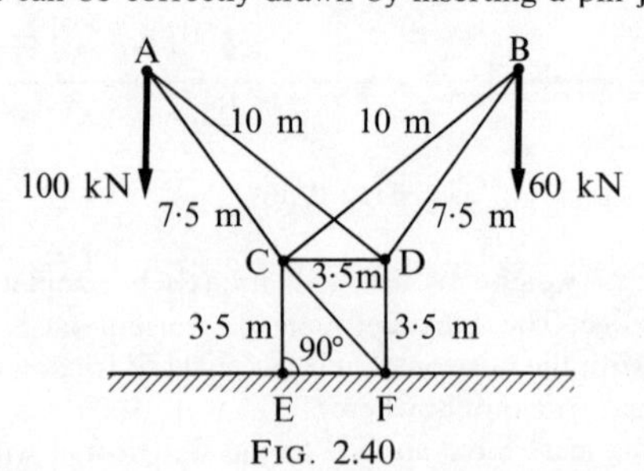

FIG. 2.40

leaving all the remaining conditions unchanged. Hence find the forces in the members of the framework shown in Fig. 2.40.

17. Determine the forces in the bars of the pin-jointed frame shown in Fig. 2.41 when the load applied to the rope is 5 kN. Friction at the pulley is negligible.

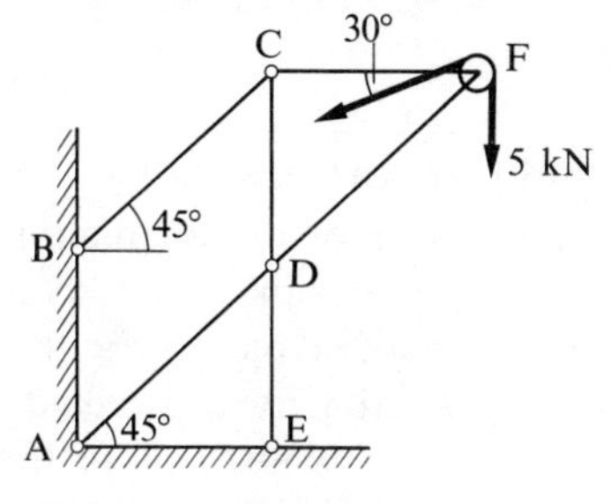

FIG. 2.41

18. State whether each of the three pin-jointed frameworks shown in Fig. 2.42 is statically determinate or not. The frameworks are supported as shown.

Structures (a) and (b) may be loaded by any combination of vertical downward loads acting at their lower chord joints. On sketches of these two frameworks indicate whether the force in each member is: statically determinate (S),

always tensile (+),
always compressive (−),
tensile or compressive (±),
or always zero (0).

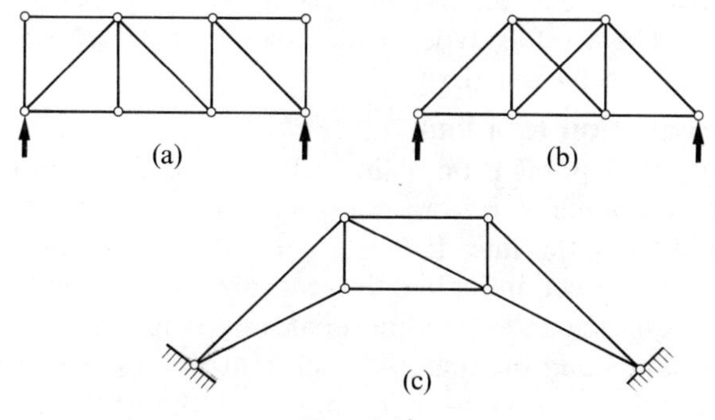

FIG. 2.42

3. Spatial kinematics and statics

3.1. Bound, sliding, and free vectors

WE have observed that the nature of vector quantities can vary according to circumstances. The velocity at a point in a body is specific to that point. In a similar way, the position of a particle with respect to a point of reference, or origin, is represented by a directed line, a vector, drawn from the origin to the particle. It is intrinsic to the notion of a position vector that it emanates from a particular origin. Vectors such as these are referred to as *bound vectors.*

A force can be thought of as operating anywhere along its line of action, irrespective of the point at which it may be physically applied to a body, and is a *sliding vector.*

The weather forecast for a particular area may include the statement that the wind velocity will be S.W., force 4. Here a vector is associated neither with a particular point nor with a particular line of action, and is an example of a *free vector.*

Distinctions between bound, free, and sliding vectors do not give rise to any difficulties in practice, and indeed it is rarely necessary to give particular thought to the type of a given vector. Nevertheless, the distinction ought to be noted.

3.2. Vector equation to a line

The location $\mathbf{r}$ of a point P on a line AP, Fig. 3.1, can be expressed in terms of two components, $\mathbf{a}$ from the origin to a convenient datum A on the line and $\mathbf{p}$ along the line. If P is a general point on the line, so that whilst $\mathbf{p}$ has fixed direction its length is variable, it is often convenient to express the magnitude and direction separately. This is done by defining a *unit vector* $\mathbf{e}$, say, along the line and such that $e = |\mathbf{e}| = 1$ and then letting $\mathbf{p} = p\mathbf{e}$ where p is the variable that determines the distance AP:

$$\mathbf{r} = \mathbf{a} + p\mathbf{e}$$

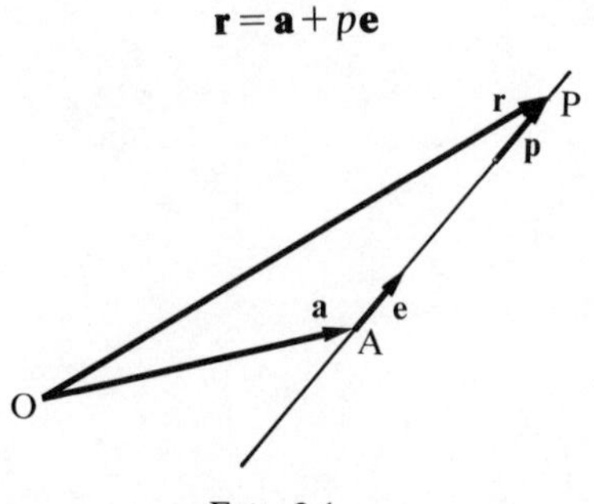

FIG. 3.1

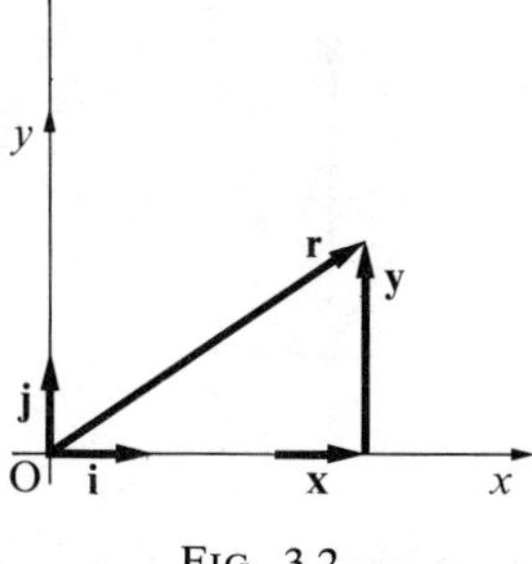

FIG. 3.2

is then the vector equation of the line.

It is self-evident that the velocity and acceleration of a particle along the line can be written as $\dot{p}\mathbf{e}$ and $\ddot{p}\mathbf{e}$ respectively ($\dot{p} \equiv \mathrm{d}p/\mathrm{d}t$, $\ddot{p} \equiv \mathrm{d}^2p/\mathrm{d}t^2$).

3.3. Frames of reference

Just as a single vector can be used to locate a point on a line, two vectors are used to locate a point in a given plane. In Fig. 3.2

$$\mathbf{r} = \mathbf{x} + \mathbf{y}$$
$$= x\mathbf{i} + y\mathbf{j}$$

where $\mathbf{i}$ and $\mathbf{j}$ are unit vectors. Such a pair of unit vectors need not be perpendicular to each other, but it is invariably convenient to have them so.

It is often useful to set up the vector equations for a system in terms of one set of unit vectors and then to refer to a different set of unit vectors to facilitate visualization of the outcome. Different sets of unit vectors are related by the parallelogram rule.

In Fig. 3.3 the unit vector $\mathbf{e}$ resolves into lengths $1 \,.\, \cos\theta$ and $1 \,.\, \sin\theta$ in the directions of the unit vectors $\mathbf{i}$ and $\mathbf{j}$ respectively. It follows that

$$\mathbf{e} = \mathbf{i}\cos\theta + \mathbf{j}\sin\theta.$$

Fig. 3.4 shows two sets of orthogonal unit vectors ($\mathbf{i}, \mathbf{j}$) and ($\mathbf{n}, \mathbf{t}$). The unit vectors $\mathbf{i}$ and $\mathbf{j}$ may, for example, be used to locate a particle relative to a fixed origin and $\mathbf{n}$ and $\mathbf{t}$ may give the directions of the normal and

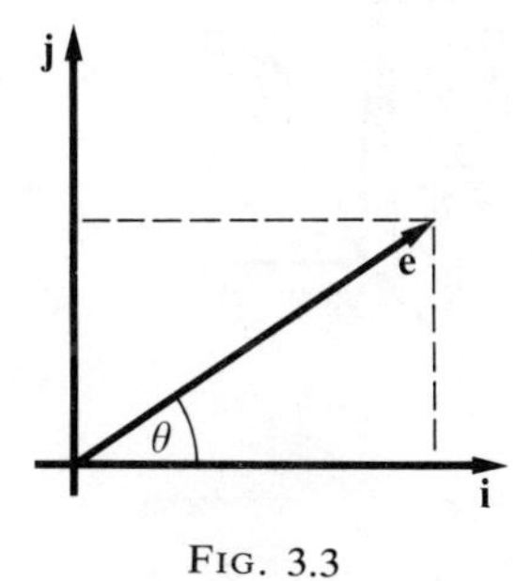

FIG. 3.3

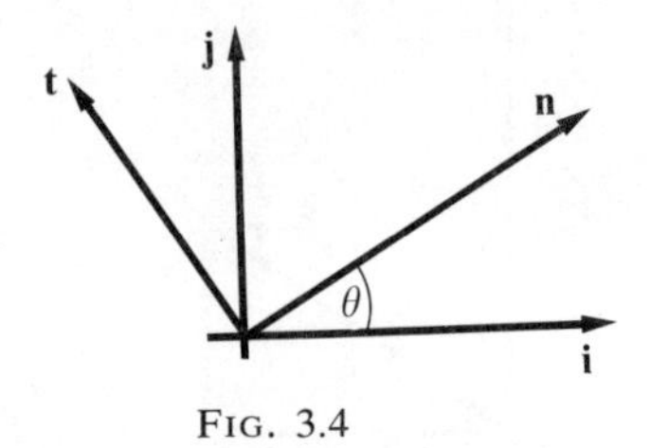

FIG. 3.4

tangent to the path of the particle. Hence

$$\mathbf{n} = \mathbf{i}\cos\theta + \mathbf{j}\sin\theta$$

$$\mathbf{t} = -\mathbf{i}\sin\theta + \mathbf{j}\cos\theta.$$

In matrix form

$$\begin{bmatrix}\mathbf{n}\\ \mathbf{t}\end{bmatrix} = \begin{bmatrix}\cos\theta & \sin\theta\\ -\sin\theta & \cos\theta\end{bmatrix}\begin{bmatrix}\mathbf{i}\\ \mathbf{j}\end{bmatrix}.$$

The transformation between these two sets of unit vectors can be effected in either direction so that

$$\begin{bmatrix}\mathbf{i}\\ \mathbf{j}\end{bmatrix} = \begin{bmatrix}\cos\theta & -\sin\theta\\ \sin\theta & \cos\theta\end{bmatrix}\begin{bmatrix}\mathbf{n}\\ \mathbf{t}\end{bmatrix}.$$

The set of unit vectors associated with a three-dimensional set of Cartesian co-ordinate axes Oxyz is by convention **i**, **j**, **k**, Fig. 3.5. It is also conventional to use a right-handed set, that is to say, when looking along Oz from O, a rotation from Ox to Oy through 90° is in the clockwise direction. Though it does not matter here, this convention is important later on. The single unit vector **e** is given in terms of **i**, **j**, and **k** by

$$\mathbf{e} = \mathbf{i}\cos\theta_1 + \mathbf{j}\cos\theta_2 + \mathbf{k}\cos\theta_3.$$

$\cos\theta_1$, $\cos\theta_2$, and $\cos\theta_3$ are the *direction cosines* of **e** with respect to the Oxyz axes.

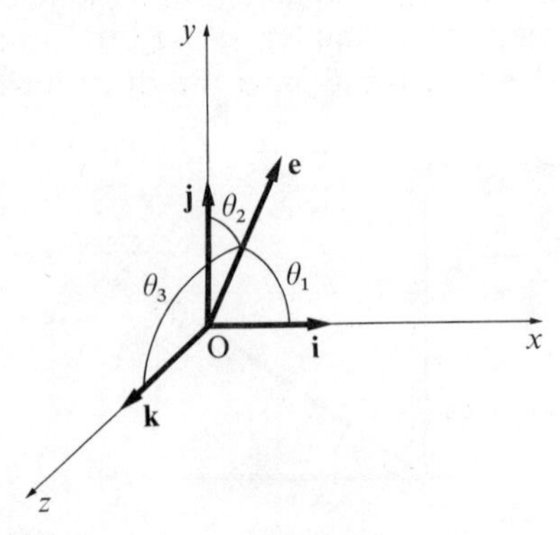

FIG. 3.5

3.4. The scalar product

So far our study of kinematics has involved only the addition of vectors. There has been no evident call for their multiplication. Let us now see how this need can arise.

We observed earlier that angular velocity has some of the attributes of a vector and multiplied it by a length to obtain a velocity. Because our study was then restricted to two-dimensional motion it was not necessary to worry about the implication of multiplying two vectors; common sense showed what the result must be. Likewise, taking moments requires the multiplication of a force by a distance, but again we have hitherto been concerned only with two-dimensional problems. This restriction is in part justified by pedagogic considerations: the development of our subject in easy stages. It is also justified by the fact that practical engineering problems can, more often than not, properly be regarded as two-dimensional even though they relate to solid objects. Nevertheless the real world is three-dimensional and we must take this into account. The generalization from two to three dimensions makes it necessary to reconsider some of the ideas developed in the first two chapters. First we must give thought to the matter of multiplying vectors. It turns out that there are two ways in which this can be done. The first type of product to be considered is the scalar product.

3.4.1. Definition of scalar product

The scalar product of two vectors $\mathbf{a}$ and $\mathbf{b}$ is defined as $ab \cos \theta$ where θ is the angle between the directions of $\mathbf{a}$ and $\mathbf{b}$. This product is often referred to as the *dot product* and we write

$$\mathbf{a} \cdot \mathbf{b} = ab \cos \theta.$$

If one of the vectors is a unit vector $\mathbf{e}$, say, then

$$\mathbf{a} \cdot \mathbf{e} = a \cos \theta$$

and can be recognized as the magnitude of the component of $\mathbf{a}$ in the direction of $\mathbf{e}$.

$\mathbf{a} \cdot \mathbf{b}$ can be regarded as either

$$(\text{the component of } \mathbf{a} \text{ in the direction of } \mathbf{b}) \times b$$

or as

$$(\text{the component of } \mathbf{b} \text{ in the direction of } \mathbf{b}) \times a.$$

It is to be noted that neither the formal definition of the scalar product nor this interpretation implies that $\mathbf{a}$ and $\mathbf{b}$ intersect. In Fig. 3.6, $\mathbf{a} \cdot \mathbf{b} = ab \cos \theta$ irrespective of the perpendicular distance h between $\mathbf{a}$ and $\mathbf{b}$. This product is called the *scalar product* because the result is a scalar quantity.

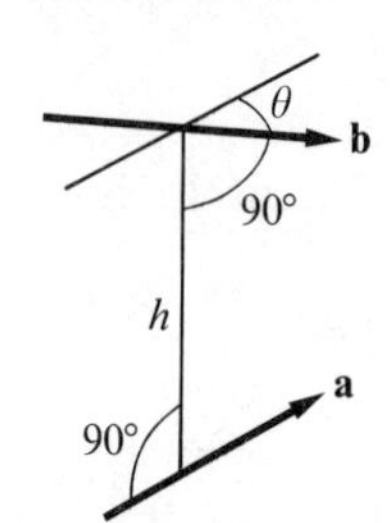

FIG. 3.6

3.4.2. Various corollaries of the scalar product

If two vectors **a** and **b** are parallel to each other $\mathbf{a}\,.\,\mathbf{b}=ab$. In particular $\mathbf{a}\,.\,\mathbf{a}=a^2$.

Conversely, if **a** and **b** are perpendicular to each other then $\mathbf{a}\,.\,\mathbf{b}=0$.

The following relationships hold between the cartesian unit vectors **i**, **j**, **k**:

$$\mathbf{i}\,.\,\mathbf{i}=\mathbf{j}\,.\,\mathbf{j}=\mathbf{k}\,.\,\mathbf{k}=1$$

$$\mathbf{i}\,.\,\mathbf{j}=\mathbf{j}\,.\,\mathbf{k}=\mathbf{k}\,.\,\mathbf{i}=0.$$

It follows from the definition of the scalar product that the order of the terms has no significance and that $\mathbf{a}\,.\,\mathbf{b}=\mathbf{b}\,.\,\mathbf{a}$. The multiplication is said to be *commutative.*

Doubling the magnitude of one vector in a scalar product obviously doubles the product and, in general,

$$(\lambda\mathbf{a})\,.\,(\mu\mathbf{b})=\lambda\mu(\mathbf{a}\,.\,\mathbf{b}).$$

Following the interpretation of $\mathbf{a}\,.\,\mathbf{c}$ as c times the component of **a** in the direction of **c**, it can be seen from Fig. 3.7 that

$$(\mathbf{a}+\mathbf{b})\,.\,\mathbf{c}=\mathbf{a}\,.\,\mathbf{c}+\mathbf{b}\,.\,\mathbf{c}.$$

This is the *distributive law.*

When numerical evaluation of a scalar product $\mathbf{a}\,.\,\mathbf{b}$ is required, it will usually be convenient to express **a** and **b** in terms of the unit vectors **i**, **j**, **k**.

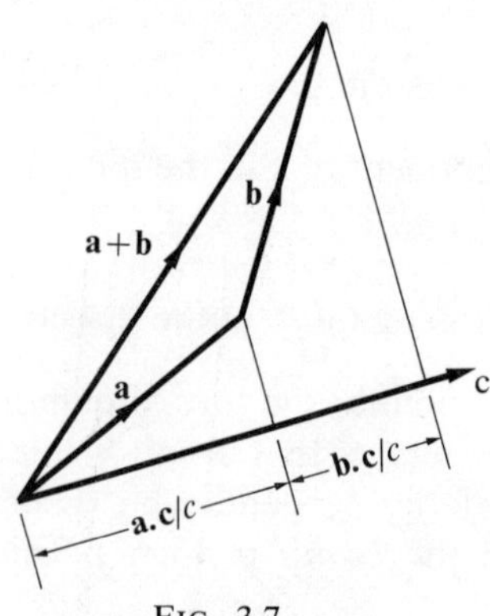

FIG. 3.7

Then

$$\mathbf{a} \cdot \mathbf{b} = (a_x\mathbf{i} + a_y\mathbf{j} + a_z\mathbf{k}) \cdot (b_x\mathbf{i} + b_y\mathbf{j} + b_z\mathbf{k})$$
$$= a_x b_x + a_y b_y + a_z b_z.$$

In particular

$$\mathbf{a} \cdot \mathbf{a} = a_x^2 + a_y^2 + a_z^2 = a^2$$

as has already been pointed out above.

A unit vector in the direction of $\mathbf{a}$ is $\mathbf{a}/(\mathbf{a} \cdot \mathbf{a})^{\frac{1}{2}}$.

A continued scalar product $\mathbf{a} \cdot \mathbf{b} \cdot \mathbf{c}$ is meaningless and can result only from an error in analysis. Probably either $\mathbf{a}(\mathbf{b} \cdot \mathbf{c})$ or $(\mathbf{a} \cdot \mathbf{b})\mathbf{c}$ is intended. It must be noted, however, that these are two different interpretations.

Example 3.1: Determine the angle between the two lines OA and OB where the co-ordinates of A and B are A = (1, 0, 2) and B = (2, 2, 0).

Let $\mathbf{a} = \overrightarrow{\mathrm{OA}}$ and $\mathbf{b} = \overrightarrow{\mathrm{OB}}$ and let θ be the angle between $\mathbf{a}$ and $\mathbf{b}$ so that

$$\mathbf{a} \cdot \mathbf{b} = ab \cos\theta \quad \text{and } \cos\theta = \mathbf{a} \cdot \mathbf{b}/ab.$$
$$\mathbf{a} = (\mathbf{i} + 0\mathbf{j} + 2\mathbf{k}), \qquad a^2 = \mathbf{a} \cdot \mathbf{a} = 1^2 + 0^2 + 2^2 = 5$$
$$\mathbf{b} = (2\mathbf{i} + 2\mathbf{j} + 0\mathbf{k}), \qquad b^2 = \mathbf{b} \cdot \mathbf{b} = 2^2 + 2^2 + 0^2 = 8$$
$$\mathbf{a} \cdot \mathbf{b} = (\mathbf{i} + 0\mathbf{j} + 2\mathbf{k}) \cdot (2\mathbf{i} + 2\mathbf{j} + 0\mathbf{k}) = 2 + 0 + 0 = 2.$$

Hence $\cos\theta = 2/(\sqrt{5} \times \sqrt{8}) = 1/\sqrt{10}$.

Example 3.2: What is the shortest distance from the origin to any point on the line joining A and B in the previous example?

The vector joining A and B is

$$\overrightarrow{\mathrm{AB}} = (\mathbf{b} - \mathbf{a}) = \mathbf{i} + 2\mathbf{j} - 2\mathbf{k}.$$

The position vector of a general point P on the line is

$$\mathbf{r} = \mathbf{a} + \lambda(\mathbf{b} - \mathbf{a})$$

where λ is a parameter such that P moves from A to B as λ varies from 0 to 1. For a particular value of λ, $\mathbf{r}$ has minimum length.

$$r^2 = \mathbf{r} \cdot \mathbf{r} = [\mathbf{a} + \lambda(\mathbf{b} - \mathbf{a})] \cdot [\mathbf{a} + \lambda(\mathbf{b} - \mathbf{a})]$$
$$= \mathbf{a} \cdot \mathbf{a} + 2\lambda\mathbf{a} \cdot (\mathbf{b} - \mathbf{a}) + \lambda^2(\mathbf{b} - \mathbf{a}) \cdot (\mathbf{b} - \mathbf{a}).$$

On differentiating both sides of this equation with respect to λ and equating the result to zero, we obtain the value of λ for which r^2 has its minimum value, and hence r has its minimum value:

$$\mathrm{d}(r^2)/\mathrm{d}\lambda = 2\mathbf{a} \cdot (\mathbf{b} - \mathbf{a}) + 2\lambda(\mathbf{b} - \mathbf{a}) \cdot (\mathbf{b} - \mathbf{a}) = 0$$

$$\Rightarrow \qquad \lambda = \frac{-\mathbf{a} \cdot (\mathbf{b} - \mathbf{a})}{(\mathbf{b} - \mathbf{a}) \cdot (\mathbf{b} - \mathbf{a})} = \frac{\mathbf{a}^2 - \mathbf{a} \cdot \mathbf{b}}{\mathbf{b}^2 - 2\mathbf{a} \cdot \mathbf{b} + \mathbf{a}^2}.$$

On substituting the values of the scalar products from the previous example we find

$$\lambda = \frac{5-2}{8-2\times 2+5} = \frac{1}{3}$$

$$\Rightarrow \qquad \mathbf{r} = (\mathbf{i}+0\mathbf{j}+2\mathbf{k}) + \tfrac{1}{3}(\mathbf{i}+2\mathbf{j}-2\mathbf{k})$$

$$= \tfrac{2}{3}(2\mathbf{i}+\mathbf{j}+2\mathbf{k})$$

$$\Rightarrow \qquad \mathbf{r}^2 = \tfrac{4}{9}(4+1+4) = 4.$$

The minimum distance from the origin to any point on AB is therefore 2.

The preliminary analysis could have been shortened somewhat by noting that when P is closest to O, OP is perpendicular to AB, that is $\mathbf{r}\,.\,(\mathbf{b}-\mathbf{a})=0$

$$\Rightarrow \qquad [\mathbf{a}+\lambda(\mathbf{b}-\mathbf{a})]\,.\,(\mathbf{b}-\mathbf{a})=0$$

$$\Rightarrow \qquad \mathbf{a}\,.\,(\mathbf{b}-\mathbf{a})+\lambda(\mathbf{b}-\mathbf{a})\,.\,(\mathbf{b}-\mathbf{a})=0.$$

This is precisely the same result as that obtained by equating $\mathrm{d}(r^2)/\mathrm{d}\lambda$ to zero.

Example 3.3: Fig. 3.8 shows a simple mechanism in which two sliders are connected by ball and socket joints to a rigid push-rod AB. The motion of A is along a line parallel to Ox and B moves along Oy. If the velocity of A is $-v\mathbf{i}$*, what is the velocity of B?*

We will make use of the fact that if AB is rigid, the components along the rod of the velocities at the two ends must be equal to each other.

The component of the velocity $\mathbf{v}_A$ at A along the rod is $\mathbf{v}_A\,.\,\mathbf{e}$, where $\mathbf{e}$ is a unit vector along the rod, $\mathbf{e}=\overrightarrow{AB}/AB$.

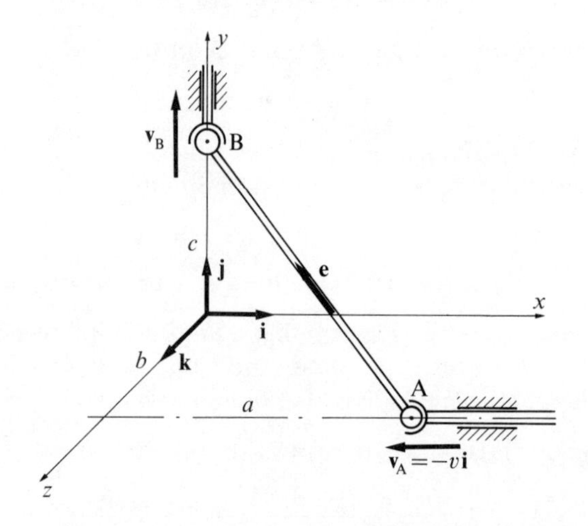

FIG. 3.8

Similarly the component velocity along the rod at B is $\mathbf{v}_B \cdot \mathbf{e}$. Hence

$$\mathbf{v}_B \cdot \mathbf{e} = \mathbf{v}_A \cdot \mathbf{e}$$

$$\Rightarrow \quad \mathbf{v}_B \cdot \overrightarrow{AB} = \mathbf{v}_A \cdot \overrightarrow{AB}$$

$$\Rightarrow \quad v_B\mathbf{j} \cdot (-a\mathbf{i} - b\mathbf{k} + c\mathbf{j}) = -v\mathbf{i} \cdot (-a\mathbf{i} - b\mathbf{k} + c\mathbf{j})$$

$$\Rightarrow \quad v_B = va/c.$$

3.5. The vector product

The vector product of two vectors **a** and **b** is defined as the vector $ab \sin \theta \mathbf{e}$ where **e** is a unit vector that is perpendicular to **a** and **b** such that the rotation θ from **a** to **b** is clockwise when seen in the positive direction of **e**, Fig. 3.9. The vector product is written

$$\mathbf{a} \times \mathbf{b} = ab \sin \theta \mathbf{e}$$

and is spoken of as '**a**' cross '**b**', and referred to alternatively as the cross product of **a** and **b**. It is written alternatively as $\mathbf{a} \wedge \mathbf{b}$.

The vector product may be relevant in those physical situations where two vectors interact to cause an effect that may be represented by a vector that is perpendicular to the original pair of vectors. For example, the velocity **v** at a point P in a rigid body spinning with angular velocity $\boldsymbol{\omega}$ about a fixed axis will be shown to be $\boldsymbol{\omega} \times \mathbf{r}$, where **r** is the vector drawn from any point O on the axis of spin. The moment of a force P about a point O will be shown to be $\mathbf{r} \times \mathbf{P}$, where **r** is a vector drawn from O to any point on the line of action of P. And so on.

3.5.1. Various corollaries of the vector product

If $\theta = 0$, $\mathbf{a} \times \mathbf{b} = 0$, hence $\mathbf{a} \times \mathbf{a} = 0$. This result might be expected on physical grounds, because if **a** is parallel to **b** there is no unique direction perpendicular to **a** and **b**.

Referring to Fig. 3.10, the following relationships hold between the unit vectors **i**, **j**, **k**:

$$\mathbf{i} \times \mathbf{j} = \mathbf{j} \times \mathbf{j} = \mathbf{k} \times \mathbf{k} = 0$$

$$\mathbf{i} \times \mathbf{j} = \mathbf{k}, \qquad \mathbf{j} \times \mathbf{k} = \mathbf{i}, \qquad \mathbf{k} \times \mathbf{i} = \mathbf{j}.$$

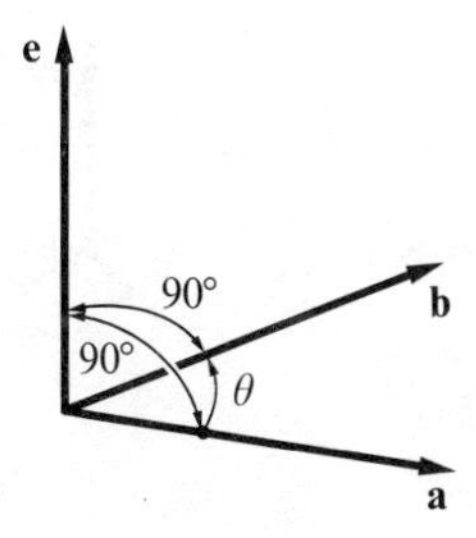

FIG. 3.9

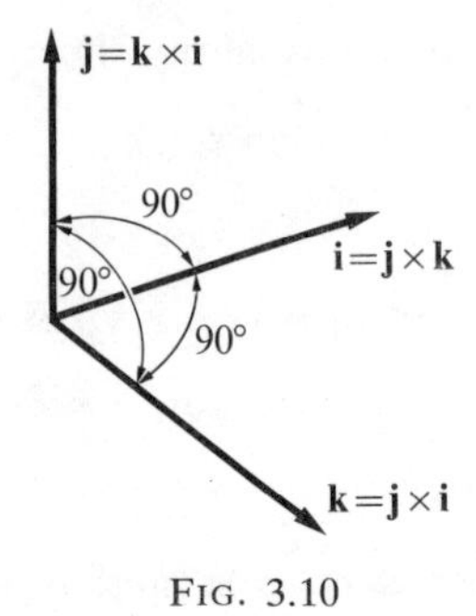

FIG. 3.10

It follows from the definition of the vector product that if $\mathbf{a}\times\mathbf{b}=\mathbf{d}$, then $\mathbf{b}\times\mathbf{a}=-\mathbf{d}$, and hence that $\mathbf{a}\times\mathbf{b}=-\mathbf{b}\times\mathbf{a}$. The commutative law does not apply to vector products, and neither does the associative law. A particular example, Fig. 3.11 shows that

$$(\mathbf{a}\times\mathbf{b})\times\mathbf{c}\neq\mathbf{a}\times(\mathbf{b}\times\mathbf{c}).$$

However, the distributive law does apply and

$$\mathbf{a}\times(\mathbf{b}+\mathbf{c})=\mathbf{a}\times\mathbf{b}+\mathbf{a}\times\mathbf{c},$$

see Section 1 of Appendix 1 (p. 247).

When two vectors are expressed in terms of **i**, **j**, **k** their vector product is most compactly expressed as a determinant

$$\mathbf{a}\times\mathbf{b}=\begin{vmatrix}\mathbf{i} & \mathbf{j} & \mathbf{k}\\ a_x & a_y & a_z\\ b_x & b_y & b_z\end{vmatrix}$$

see Section 2 of Appendix 1 (p. 247).

3.5.2. Geometrical interpretation of the vector product

In Fig. 3.12 the area of the shaded parallelogram is $ab\sin\theta$, so that $\mathbf{a}\times\mathbf{b}$ can be regarded as a vector representing that area. The magnitude of the cross product is equal to the area, and its direction is in the direction of the normal to the area.

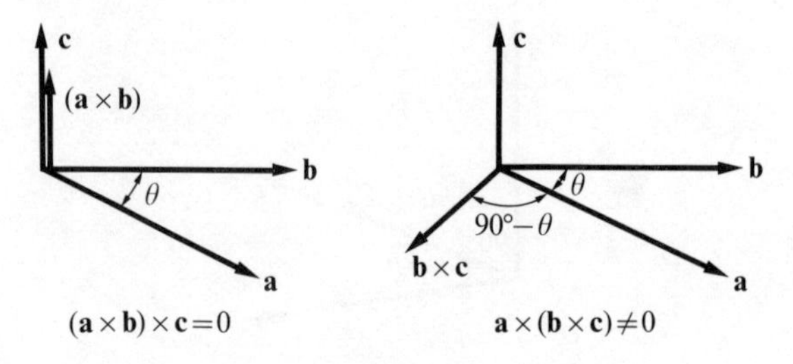

FIG. 3.11

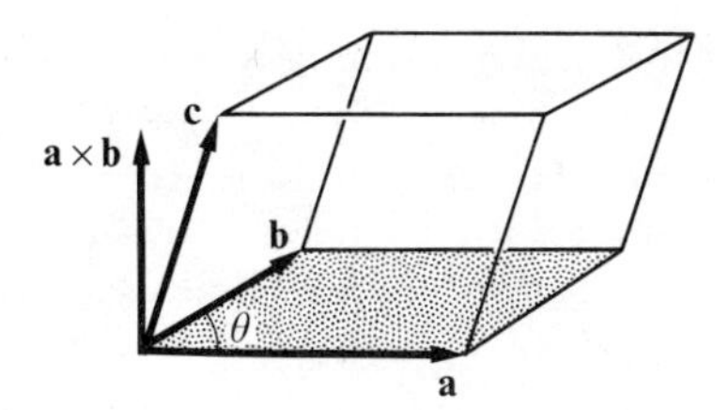

FIG. 3.12

3.5.3. The scalar triple-product

The product $(\mathbf{a} \times \mathbf{b}) \cdot \mathbf{c}$ equals the product of $|\mathbf{a} \times \mathbf{b}|$ and the component of $\mathbf{c}$ in the direction of $\mathbf{a} \times \mathbf{b}$. As $\mathbf{a} \times \mathbf{b}$ is the area of the shaded parallelogram in Fig. 3.12, $(\mathbf{a} \times \mathbf{b}) \cdot \mathbf{c}$ is the volume of the parallelepiped that has $\mathbf{a}$, $\mathbf{b}$, and $\mathbf{c}$ for adjacent edges. It follows that

$$(\mathbf{a} \times \mathbf{b}) \cdot \mathbf{c} = (\mathbf{c} \times \mathbf{a}) \cdot \mathbf{b} = (\mathbf{b} \times \mathbf{c}) \cdot \mathbf{a}.$$

This particular form of compound product is known as the scalar triple-product. The ability to cycle the factors in this identity proves in due course to be very useful.

It is to be noted that if $(\mathbf{a} \times \mathbf{b}) \cdot \mathbf{c} = 0$, the volume of the parallelepiped is zero and that $\mathbf{a} \times \mathbf{b}$ is perpendicular to $\mathbf{c}$. This can often be interpreted to mean, as in this example, that $\mathbf{a}$, $\mathbf{b}$, and $\mathbf{c}$ are coplanar.

It follows that

$$(\mathbf{a} \times \mathbf{b}) \cdot \mathbf{a} = (\mathbf{a} \times \mathbf{b}) \cdot \mathbf{b} = 0.$$

3.5.4. The triple-vector product

The compound product $\mathbf{a} \times (\mathbf{b} \times \mathbf{c})$ is known as the triple-vector product. It will be found to arise from time to time but less commonly than the scalar triple-product. It is shown in Section 3 of Appendix 1 (p. 248) that

$$\mathbf{a} \times (\mathbf{b} \times \mathbf{c}) = \mathbf{b}(\mathbf{a} \cdot \mathbf{c}) - \mathbf{c}(\mathbf{a} \cdot \mathbf{b}).$$

3.5.5. The common perpendicular between two lines

There is nothing in the definition of $\mathbf{a} \times \mathbf{b}$ that requires $\mathbf{a}$ and $\mathbf{b}$ to be concurrent, the vector product being, like the scalar product, independent of the perpendicular distance h between the vectors $\mathbf{a}$ and $\mathbf{b}$, Fig. 3.13.

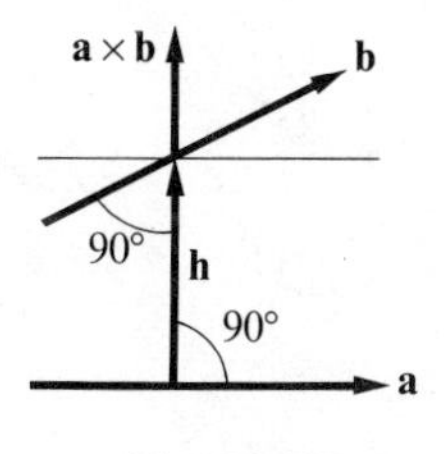

FIG. 3.13

There are various ways of determining $\mathbf{h}$, the common normal between two lines $\mathbf{r}_a = \mathbf{a}_O + \alpha\mathbf{a}$ and $\mathbf{r}_b = \mathbf{b}_O + \beta\mathbf{b}$.

(i) A vector between the two lines is $(\mathbf{r}_a - \mathbf{r}_b)$. To determine the conditions that this vector shall be $\mathbf{h}$, we use the fact that if $\mathbf{h}$ is perpendicular to both lines then $\mathbf{a} \cdot \mathbf{h} = 0$ and $\mathbf{b} \cdot \mathbf{h} = 0$. These yield the pair of simultaneous equations

$$\mathbf{a} \cdot (\mathbf{r}_a - \mathbf{r}_b) = 0$$

and

$$\mathbf{b} \cdot (\mathbf{r}_a - \mathbf{r}_b) = 0$$

for the determination of α and β corresponding to the two ends of $\mathbf{h}$. Hence $\mathbf{h}$ is determined.

(ii) All vectors joining the two lines have the same component $\mathbf{h}$ in the direction perpendicular to $\mathbf{a}$ and $\mathbf{b}$. Hence if $\mathbf{e}$ is a unit vector of $\mathbf{a} \times \mathbf{b}$

$$\begin{aligned} h &= (\mathbf{r}_a - \mathbf{r}_b) \cdot \mathbf{e} \\ &= (\mathbf{a}_O + \alpha\mathbf{a} - \mathbf{b}_O - \beta\mathbf{b}) \cdot \mathbf{e} \\ &= (\mathbf{a}_O - \mathbf{b}_O) \cdot \mathbf{e}. \end{aligned}$$

The last step follows because $\mathbf{a}$ and $\mathbf{b}$ are both perpendicular to $\mathbf{e}$. This equation yields the length h of the common perpendicular but does not fix its location.

3.6. Angular velocity

A body turns with speed ω about a fixed axis whose direction is defined by a unit vector $\mathbf{e}$, Fig. 3.14. Let O be an arbitrary point on the axis of rotation and P be any point in the body other than one on the axis. The normal PN to the axis of rotation has length a. The velocity of P has magnitude $a\omega$ and is in the direction normal to ONP. So $v_P = \omega r \sin\theta$, $\mathbf{v}_P$ is perpendicular to $\mathbf{r}$ and $\mathbf{e}$, and θ is the angle between $\mathbf{r}$ and $\mathbf{e}$. By inspection, the direction of $\mathbf{v}_P$ is such that $\mathbf{v}_P = \omega\mathbf{e} \times \mathbf{r}$.

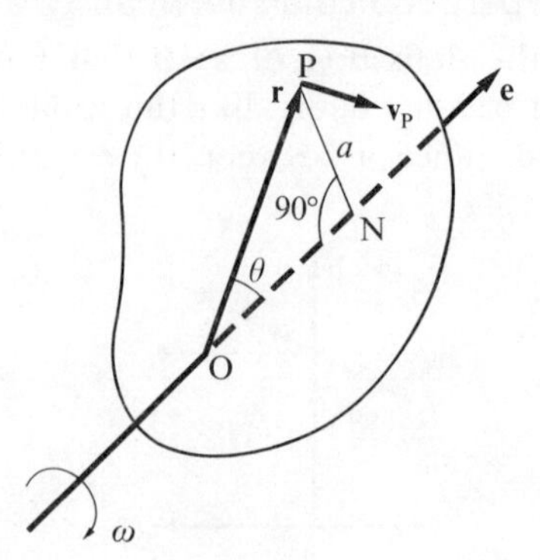

FIG. 3.14

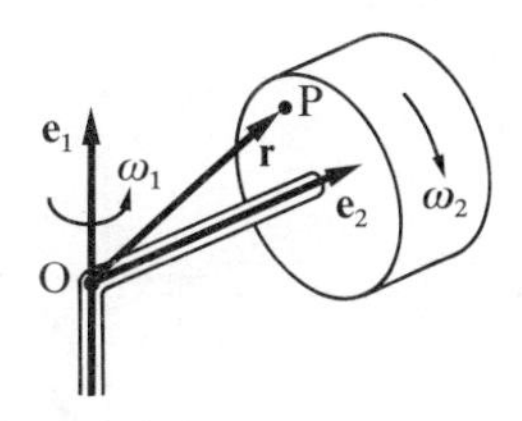

FIG. 3.15

Now consider the body to be turning simultaneously about two axes which intersect at a fixed point O, as in Fig. 3.15. The velocities at P add to give $\mathbf{v}_P$ so

$$\mathbf{v}_P = \mathbf{v}_1 + \mathbf{v}_2 = \omega_1\mathbf{e}_1 \times \mathbf{r} + \omega_2\mathbf{e}_2 \times \mathbf{r}$$
$$= (\omega_1\mathbf{e}_1 + \omega_2\mathbf{e}_2) \times \mathbf{r}.$$

For points on the instantaneous axis of rotation $\mathbf{v}_P = 0$. This happens when $\mathbf{r}$ is parallel to $(\omega_1\mathbf{e}_1 + \omega_2\mathbf{e}_2)$. Let the unit vector in this direction be $\mathbf{e}$ and let the resulting angular speed be ω.

For points other than those on the instantaneous axis

$$(\omega_1\mathbf{e}_1 + \omega_2\mathbf{e}_2) \times \mathbf{r} = \omega\mathbf{e} \times \mathbf{r}$$

or

$$(\omega_1\mathbf{e}_1 + \omega_2\mathbf{e}_2 - \omega\mathbf{e}) \times \mathbf{r} = 0.$$

As $\mathbf{r}$ is arbitrary this equation holds in general only if

$$\omega_1\mathbf{e}_1 + \omega_2\mathbf{e}_2 = \omega\mathbf{e}.$$

It follows that the angular velocities $\boldsymbol{\omega}_1 = \omega_1\mathbf{e}_1$ and $\boldsymbol{\omega}_2 = \omega_2\mathbf{e}_2$ add in accordance with the parallelogram rule, and that angular velocity is a vector quantity. Henceforth we may write

$$\boldsymbol{\omega} = \boldsymbol{\omega}_1 + \boldsymbol{\omega}_2 \quad \text{and} \quad \mathbf{v} = \boldsymbol{\omega} \times \mathbf{r}. \tag{3.1}$$

The value of eqn (3.1) is limited in that it refers to the motion of a body which has one point fixed in position. This means that the body always rotates about an axis through the fixed point. We will see below that the existence of such a fixed point is a necessary condition, in general, for there to be an instantaneous axis of rotation.

Although it is a restricted type of motion, rotation of a body about a fixed point can arise in many ways and ideas associated with this motion are of considerable practical importance. Consider, for example, the mechanism shown in Fig. 3.16.

A wheel turns freely on a horizontal axle which is fixed at one end to a vertical rotating shaft. As a consequence the wheel rolls along a circular path on the ground. As all points on the wheel are at fixed distances from O the wheel rotates about O. If there is no slip between the wheel and the ground at C, the velocity of the wheel is zero at C and OC must be

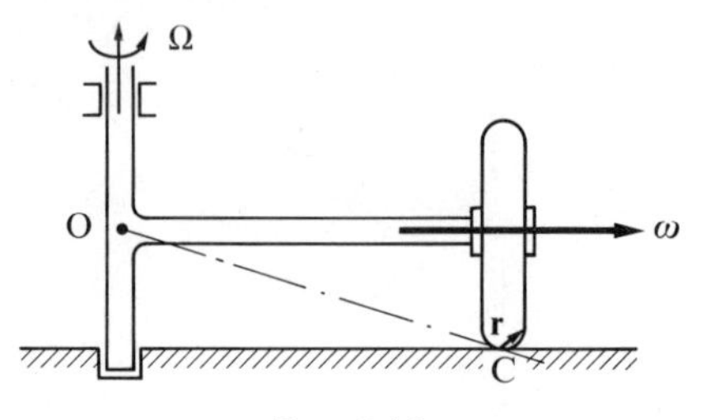

FIG. 3.16

the instantaneous axis of rotation for the wheel. The angular velocity of the wheel is $\mathbf{\Omega}+\boldsymbol{\omega}$, where $\mathbf{\Omega}$ is the angular velocity of the vertical shaft and $\boldsymbol{\omega}$ is the angular velocity of the wheel relative to its axle. The wheel is considered to have a rolling velocity $\boldsymbol{\omega}$ and a spin velocity $\mathbf{\Omega}$. The spin velocity is an important factor in wear: if the wheel were rigid and had a perfectly smooth surface and the ground were perfectly flat, contact would be at a single point and there would be no wear due to rubbing between the two surfaces. This ideal situation really presupposes that there is no load on the wheel. The velocity at a point on the wheel with position $\mathbf{r}$ relative to C is $(\mathbf{\Omega}+\boldsymbol{\omega})\times\mathbf{r}$. Close to C the wheel surface glides over the ground with a velocity $\mathbf{\Omega}\times\mathbf{r}$. If the surfaces were not perfectly smooth, and in practice they never are, asperities on the two surfaces would collide and deform with resulting wear. This mechanism is accentuated by loading, which spreads contact between the wheel and the ground over a finite area.

Now imagine a second identical wheel mounted on the same axle at a different radius from the first. If there is no slip between it and the ground its instantaneous axis of rotation must be different from that of the first wheel. The wheels will have the same spin velocity but different rolling velocities. If the wheels were connected together to form a single body, they would be unable to move along a circular path on the ground without gross slip at one or other point of contact. It is for this reason that one of the rear wheels of a pedal tricycle must be allowed to turn freely on its axle. Such a one-sided drive would be unsatisfactory on a motor car, but the problem remains. It is solved in a car with a rear-wheel drive by introducing a differential gear between the drive shaft from the engine/gear-box and the two halves of the rear axle. In this way power is transmitted through both rear wheels which nevertheless are able to have different rolling velocities.

Fig. 3.17 shows a mechanism in which all the parts rotate about a common point O. It is known as a universal or Hooke's joint and is a standard connection between shafts whose axes intersect. A rigid cruciform member is joined by swivel joints to a pair of yokes, one on the end of each shaft. It is found that if one shaft rotates with uniform angular velocity $\boldsymbol{\omega}_1$ the other shaft has a varying velocity $\boldsymbol{\omega}_2$. The fluctuations in $\boldsymbol{\omega}_2$ are quite small provided that the angle ϕ between the

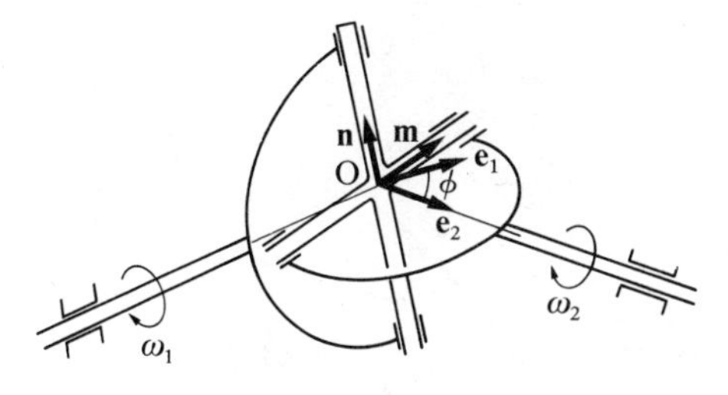

FIG. 3.17

axes of the two shafts is not too large. It is of interest to derive a relationship for the speed ratio ω_2/ω_1.

Let $\mathbf{e}_1$ and $\mathbf{e}_2$ be unit vectors along the axes of the two shafts, and let **m** and **n** be unit vectors along the arms of the cross, as shown. Let $\boldsymbol{\Omega}$ be the angular velocity of the cross.

$\boldsymbol{\Omega}$ can be regarded as having two components: ω_1 due to the rotation of shaft 1 and $\alpha\mathbf{n}$ due to the rotation of the cross in the bearings whereby it is attached to shaft 1. Thus

$$\boldsymbol{\Omega} = \boldsymbol{\omega}_1 + \alpha\mathbf{n}.$$

Similarly

$$\boldsymbol{\Omega} = \boldsymbol{\omega}_2 + \beta\mathbf{m}.$$

Hence

$$\boldsymbol{\omega}_1 + \alpha\mathbf{n} = \boldsymbol{\omega}_2 + \beta\mathbf{m}$$

where α and β are as yet unknown. In fact they are of no particular interest and can be eliminated simultaneously from the equation by taking the scalar product of both side with the vector product $\mathbf{m}\times\mathbf{n}$. This is because $(\mathbf{m}\times\mathbf{n})\cdot\mathbf{m} = (\mathbf{m}\times\mathbf{n})\cdot\mathbf{n} = 0$. Hence $\boldsymbol{\omega}_1\cdot(\mathbf{m}\times\mathbf{n}) = \boldsymbol{\omega}_2\cdot(\mathbf{m}\times\mathbf{n})$. Putting $\boldsymbol{\omega}_1 = \omega_1\mathbf{e}_1$ and $\boldsymbol{\omega}_2 = \omega_2\mathbf{e}_2$ we have

$$\frac{\omega_1}{\omega_2} = \frac{\mathbf{e}_1\cdot(\mathbf{m}\times\mathbf{n})}{\mathbf{e}_2\cdot(\mathbf{m}\times\mathbf{n})}.$$

The evaluation of this expression in terms of the rotation θ of the shaft from the position when **n** is perpendicular to the plane of the two shafts and the angle β between the shafts is a moderately complicated exercise in vector algebra which need not be pursued here, where our interest is primarily in the kinematics.

The fluctuation in the speed of the output shaft can give rise to troublesome vibrations. Consequently, universal joints are very often used in pairs so that the second joint cancels out the speed fluctuation that the first one has introduced.

3.6.1. Angular velocity as a free vector

It is natural to think of the body in Fig. 3.15 as having an angular velocity $\boldsymbol{\omega}$ about the point O, thus implying that $\boldsymbol{\omega}$ is a vector tied to O or a

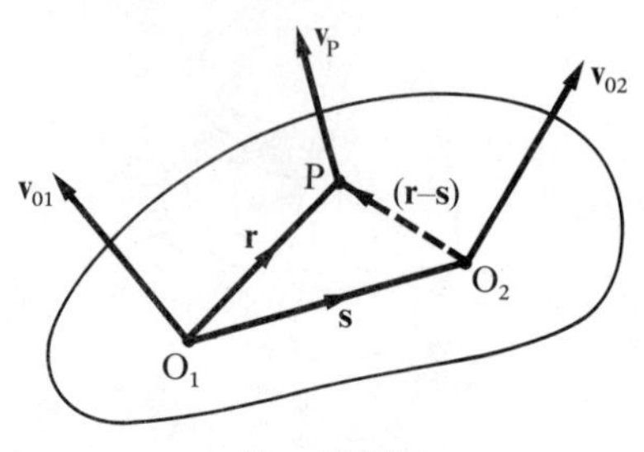

FIG. 3.18

sliding vector on a line through O. Consider now the solid body in Fig. 3.18 which contains two particular points O_1 and O_2 separated by the vector **s**, and a general point P. The body has spatial motion such that O_1, O_2, and P have velocities $\mathbf{v}_{O1}$, $\mathbf{v}_{O2}$, and $\mathbf{v}_P$ respectively. As the body is rigid the three points are fixed distances apart. This means that the velocity difference between P and O_1, say, in the direction **r** is zero. Hence

$$(\mathbf{v}_P - \mathbf{v}_{O1}) \cdot \mathbf{r} = 0.$$

In other words $(\mathbf{v}_P - \mathbf{v}_{O1})$ must be perpendicular to **r**. An alternative way of expressing this is to introduce a vector $\boldsymbol{\omega}_1$, which is associated in some unspecified way with O_1 such that $(\mathbf{v}_P - \mathbf{v}_{O1}) = \boldsymbol{\omega}_1 \times \mathbf{r}$.

We will leave on one side for the moment the problem of finding such a vector.

We can equally argue that as the distance O_2P is constant $(\mathbf{v}_P - \mathbf{v}_{O2}) = \boldsymbol{\omega}_2 \times (\mathbf{r} - \mathbf{s})$, where $\boldsymbol{\omega}_2$ is associated with O_2. Eliminating $\mathbf{v}_P$ from these two equations we find

$$\begin{aligned}\mathbf{v}_{O2} - \mathbf{v}_{O1} &= \boldsymbol{\omega}_2 \times (\mathbf{r} - \mathbf{s}) - \boldsymbol{\omega}_1 \times \mathbf{r} \\ &= (\boldsymbol{\omega}_2 - \boldsymbol{\omega}_2) \times \mathbf{r} - \boldsymbol{\omega}_2 \times \mathbf{s}.\end{aligned}$$

But as the distance O_1O_2 is constant

$$(\mathbf{v}_{O2} - \mathbf{v}_{O1}) \cdot \mathbf{s} = 0.$$

Hence

$$[(\boldsymbol{\omega}_2 - \boldsymbol{\omega}_1) \times \mathbf{r}] \cdot \mathbf{s} - (\boldsymbol{\omega}_2 \times \mathbf{s}) \cdot \mathbf{s} = 0.$$

But

$$(\boldsymbol{\omega}_2 \times \mathbf{s}) \cdot \mathbf{s} = 0$$

so

$$[(\boldsymbol{\omega}_2 - \boldsymbol{\omega}_1) \times \mathbf{r}] \cdot \mathbf{s} = 0.$$

Remembering that the terms of a scalar triple product can be cycled we see that $(\mathbf{r} \times \mathbf{s}) \cdot (\boldsymbol{\omega}_1 - \boldsymbol{\omega}_2) = 0$.

Now as **r** is an arbitrary vector $\mathbf{r} \times \mathbf{s}$ is not, in general, zero. It follows that $(\boldsymbol{\omega}_1 - \boldsymbol{\omega}_2)$ is zero and that $\boldsymbol{\omega}_1 = \boldsymbol{\omega}_2 = \boldsymbol{\omega}$, say, and

$$\mathbf{v}_\mathrm{P} - \mathbf{v}_\mathrm{O} = \boldsymbol{\omega} \times \mathbf{r}$$

or

$$\mathbf{v}_\mathrm{P} = \mathbf{v}_\mathrm{O} + \boldsymbol{\omega} \times \mathbf{r}. \tag{3.2}$$

where O and P are any two points in the body.

We can define $\boldsymbol{\omega}$ as the *angular velocity of the body*. As it is not associated with a particular point in the body, it is a free vector.

If O is stationary, $\mathbf{v}_\mathrm{P} = \boldsymbol{\omega} \times \mathbf{r}$, as in the previous section, but we now see that $\boldsymbol{\omega}$ does not have to be thought of as being along the axis about which the motion is physically constrained to take place.

Eqn (3.2) is identical to eqn (1.2) which was deduced in Chapter 1 in relation to planar motion. In dealing with planar motion $\boldsymbol{\omega} \times \mathbf{r}$ was introduced as a convention whereby a particular vector was specified. The idea of a vector product was not then introduced because it would have been an unnecessary complication. It can now be seen that the planar kinematics of Chapter 1 is a special case of the spatial kinematics we are now considering. Care was taken in presenting Chapter 1 to ensure that the equations were always written so as to be consistent with our present study. The next example gives a straightforward application of eqn (3.2) and leads on to show that it would be dangerous, simply on the bases of the identification of eqns (1.2) and (3.2), to assume that the conclusions of Chapter 1 on planar motion can readily be translated into three-dimensional terms.

Example 3.4: A lamina in the form of an equilateral triangle ABC of side length 'a' rests with one edge BC on the floor and the other two edges in contact with adjacent walls at the corner of a room, the floor and the walls being mutually perpendicular. Corner A starts to slide downwards with speed V whilst maintaining contact with the two walls. Corner B slides and maintains contact with the floor and one of the walls. Edge BC maintains contact with the floor. What is the velocity of C, and what is the angular velocity of the triangle?

In Fig. 3.19 unit vectors $\mathbf{i}, \mathbf{j}, \mathbf{k}$ have been introduced at the corner of the room, and the sides of the lamina have been designated by vectors $\mathbf{a}, \mathbf{b}, \mathbf{c}$.

The two unknown velocities are given by

$$\mathbf{v}_\mathrm{B} = \mathbf{v}_\mathrm{A} + \boldsymbol{\omega} \times \mathbf{a}$$

and

$$\mathbf{v}_\mathrm{C} = \mathbf{v}_\mathrm{A} + \boldsymbol{\omega} \times \mathbf{c}.$$

As the direction of $\mathbf{v}_\mathrm{B}$ is known its determination from the first of these two equations is a simple matter. Let $\mathbf{v}_\mathrm{B} = v_\mathrm{B}\mathbf{k}$, then

$$v_\mathrm{B}\mathbf{k} = -V\mathbf{j} + \boldsymbol{\omega} \times \mathbf{a}.$$

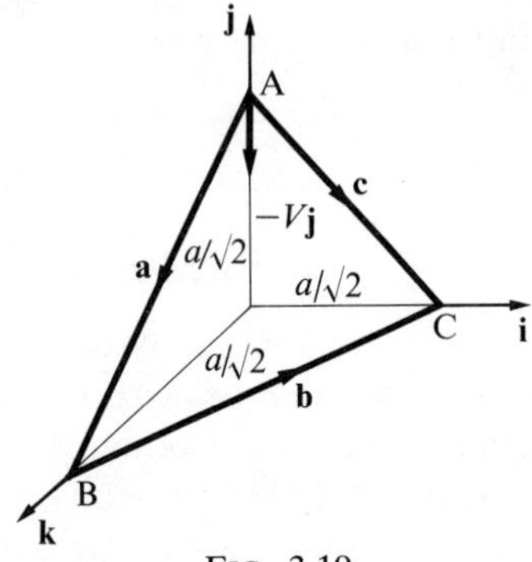

FIG. 3.19

The unknown $\boldsymbol{\omega}$ is now eliminated by taking the scalar product of all the terms with $\mathbf{a}$, $\mathbf{a}\,.\,(\boldsymbol{\omega}\times\mathbf{a})$ being zero,

$$v_{\mathrm{B}}\mathbf{k}\,.\,\mathbf{a}=-V\mathbf{j}\,.\,\mathbf{a}$$

$$\Rightarrow \qquad v_{\mathrm{B}}=V.$$

This result could have been derived more directly by noting that the component of the velocity difference between A and B is zero in the direction of $\mathbf{a}$. But appeal to our first equation is necessary to determine $\boldsymbol{\omega}$.

Substitution of the now known v_{B} in the equation for $\mathbf{v}_{\mathrm{B}}$ gives

$$V(\mathbf{k}+\mathbf{j})=\boldsymbol{\omega}\times\mathbf{a}.$$

This equation is insufficient to determine $\boldsymbol{\omega}$ fully, but it does yield some information. Taking the vector product of both sides with $\mathbf{a}$ we find

$$V\mathbf{a}\times(\mathbf{k}+\mathbf{j})=\mathbf{a}\times(\boldsymbol{\omega}\times\mathbf{a})$$

$$\Rightarrow \qquad \frac{aV}{\sqrt{2}}(-\mathbf{j}+\mathbf{k})\times(\mathbf{k}+\mathbf{j})=\boldsymbol{\omega}(\mathbf{a}\,.\,\mathbf{a})-\mathbf{a}(\boldsymbol{\omega}\,.\,\mathbf{a})$$

$$\Rightarrow \qquad -(\sqrt{2})aV\mathbf{i}=a^2\boldsymbol{\omega}-\mathbf{a}(\boldsymbol{\omega}\,.\,\mathbf{a}).$$

So

$$\boldsymbol{\omega}=\sqrt{2}(-V\mathbf{i}+\alpha\mathbf{a})/a$$

where α is scalar constant that has still to be found. The information to be determined from the equation for $\mathbf{v}_{\mathrm{B}}$ now being exhausted, we turn to

$$\mathbf{v}_{\mathrm{C}}=\mathbf{v}_{\mathrm{A}}+\boldsymbol{\omega}\times\mathbf{c}$$

and substitute for $\boldsymbol{\omega}$, giving

$$\mathbf{v}_{\mathrm{C}}=\mathbf{v}_{\mathrm{A}}+\sqrt{2}(-V\mathbf{i}+\alpha\mathbf{a})\times\mathbf{c}/a.$$

By inspection, $\mathbf{i}\times\mathbf{c}=-a\mathbf{k}/\sqrt{2}$, and $\mathbf{a}\times\mathbf{c}$ is found to be $\frac{1}{2}a^2(\mathbf{i}+\mathbf{j}+\mathbf{k})$ so that

$$\mathbf{v}_{\mathrm{C}}=-V\mathbf{j}+V\mathbf{k}+\frac{\alpha a}{\sqrt{2}}(\mathbf{i}+\mathbf{j}+\mathbf{k}).$$

But $\mathbf{v}_C$ has no component in the $\mathbf{j}$-direction so that $\alpha a/\sqrt{2} = V$ giving $\alpha = V\sqrt{2}/a$ with,

$$\mathbf{v}_C = V(\mathbf{i} + 2\mathbf{k}),$$

and

$$\boldsymbol{\omega} = V\sqrt{2}(-\mathbf{i} - \mathbf{j} + \mathbf{k})/a.$$

3.7. The instantaneous screw axis

The velocity distribution in a rigid body is fully known if the velocity at any point O and the angular velocity $\boldsymbol{\omega}$ are known. At any other point P located relative to O by $\mathbf{r}$ the velcoity is given by eqn (3.2) as

$$\mathbf{v}_P = \mathbf{v}_O + \boldsymbol{\omega} \times \mathbf{r} \tag{3.2}$$

It is natural to enquire whether there is a point in the body where the velocity is zero: does a body in spatial motion have an instantaneous centre?

Taking the scalar product of all the terms in eqn (3.2) with $\boldsymbol{\omega}$ we find

$$\mathbf{v}_P \cdot \boldsymbol{\omega} = \mathbf{v}_O \cdot \boldsymbol{\omega}$$

from which it follows that all points in the body have the same component of velocity in the direction of $\boldsymbol{\omega}$. Unless this component happens to be zero there can be no point in the body where the resultant velocity is zero. Hence, in general, a body in spatial motion does not have an instantaneous centre.

We might enquire next whether it is possible for two points of a body in spatial motion to have equal velocities. Suppose that

$$\mathbf{v}_{P'} = \mathbf{v}_O + \boldsymbol{\omega} \times \mathbf{r}' = \mathbf{v}_P.$$

Then

$$\boldsymbol{\omega} \times \mathbf{r} = \boldsymbol{\omega} \times \mathbf{r}'$$

or

$$\boldsymbol{\omega} \times (\mathbf{r}' - \mathbf{r}) = 0.$$

Hence $\mathbf{v}_{P'} = \mathbf{v}_P$ if PP′ is parallel to $\boldsymbol{\omega}$. It follows that on any line parallel to $\boldsymbol{\omega}$ all points have the same velocity.

Our final question must be whether there exists a line where the velocity is parallel to $\boldsymbol{\omega}$. If so then the instantaneous motion can be seen to be a screwing motion about the instantaneous screw axis (ISA).

Let $\mathbf{r}$ be the normal from the reference point O to the ISA (assuming that the ISA exists), and let $\lambda\boldsymbol{\omega}$ be the velocity of points on the ISA. Then at point P on the ISA

$$\mathbf{v}_P = \lambda\boldsymbol{\omega} = \mathbf{v}_O + \boldsymbol{\omega} \times \mathbf{n},$$

where λ and $\mathbf{n}$ are both unknown.

λ can readily be eliminated by taking the vector product of all terms with $\boldsymbol{\omega}$,

$$\lambda\boldsymbol{\omega}\times\boldsymbol{\omega}=0=\boldsymbol{\omega}\times\mathbf{v}_O+\boldsymbol{\omega}\times(\boldsymbol{\omega}\times\mathbf{n})$$
$$=\boldsymbol{\omega}\times\mathbf{v}_O+\boldsymbol{\omega}(\boldsymbol{\omega}\,.\,\mathbf{n})-\mathbf{n}(\boldsymbol{\omega}\,.\,\boldsymbol{\omega}).$$

By definition $\mathbf{n}$ is perpendicular to $\boldsymbol{\omega}$, hence $\boldsymbol{\omega}\,.\,\mathbf{n}=0$ and

$$\mathbf{n}=\boldsymbol{\omega}\times\mathbf{v}_O/\boldsymbol{\omega}\,.\,\boldsymbol{\omega}. \tag{3.3}$$

So the instantaneous screw axis exists unless $\boldsymbol{\omega}=0$.
To find λ, and hence the velocity of points on the ISA, we go back to the original equation and take the scalar product of all terms with $\boldsymbol{\omega}$. This eliminates the term in $\mathbf{n}$ and gives

$$\lambda=\boldsymbol{\omega}\,.\,\mathbf{v}_O/\boldsymbol{\omega}\,.\,\boldsymbol{\omega}.$$

λ is known as the *pitch* of the motion. It is the displacement along the ISA per radian of rotation. If $\boldsymbol{\omega}=0$ the body is in a state of pure translational motion, the instantaneous screw axis being at infinity. If $\lambda=0$ the motion is a pure rotation about the instantaneous axis of rotation, which seen end-on becomes the instantaneous centre.

It may be thought that screwing motion arises only when a body is constrained to move in a relatively complicated way. This is certainly not so (see Example 3.5).

Example 3.5: A disc rotates with angular velocity $\boldsymbol{\omega}$ relative to an arm on which it is mounted. The arm rotates about a fixed axis with angular velocity $\boldsymbol{\Omega}$, as shown in Fig. 3.20. The common perpendicular between $\boldsymbol{\Omega}$ and $\boldsymbol{\omega}$ is $\mathbf{s}$. Determine the location of the instantaneous screw axis for the disc and the pitch of its motion.

The angular velocity of the disc is $(\boldsymbol{\Omega}+\boldsymbol{\omega})$.
The velocity at the centre O of the disc is $\boldsymbol{\Omega}\times\mathbf{s}$.
Hence

$$\lambda=\frac{(\boldsymbol{\Omega}+\boldsymbol{\omega})\,.\,(\boldsymbol{\Omega}\times\mathbf{s})}{(\boldsymbol{\Omega}+\boldsymbol{\omega})\,.\,(\boldsymbol{\Omega}+\boldsymbol{\omega})}$$
$$=\frac{\boldsymbol{\omega}\,.\,(\boldsymbol{\Omega}\times\mathbf{s})}{\boldsymbol{\Omega}^2+\boldsymbol{\omega}^2}=\frac{\omega\Omega\mathbf{s}}{\boldsymbol{\Omega}^2+\boldsymbol{\omega}^2}.$$

$$\mathbf{n}=\frac{(\boldsymbol{\Omega}+\boldsymbol{\omega})\times(\boldsymbol{\Omega}\times\mathbf{s})}{(\boldsymbol{\Omega}+\boldsymbol{\omega})\,.\,(\boldsymbol{\Omega}+\boldsymbol{\omega})}=\frac{-\mathbf{s}\boldsymbol{\Omega}^2}{\boldsymbol{\Omega}^2+\boldsymbol{\omega}^2}.$$

Example 3.6: Two bodies rotate about fixed axes with angular velocities $\boldsymbol{\omega}_1$ and $\boldsymbol{\omega}_2$ respectively. The common perpendicular between the axes is $\mathbf{s}$. The bodies make contact at a point P. Show that, in general, there must be sliding between the two surfaces at P.

Let O be at the foot of the perpendicular $\mathbf{s}$ from one axis to the other, Fig. 3.21, and let $\mathbf{r}$ be the vector from O to P. The velocity of sliding at P

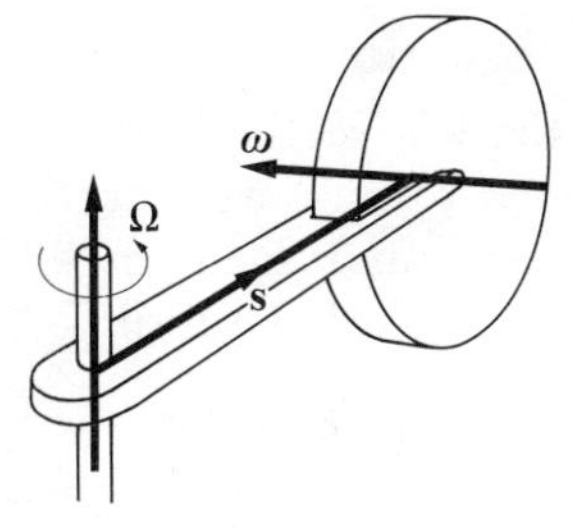

FIG. 3.20

is then

$$\mathbf{v} = \boldsymbol{\omega}_1 \times \mathbf{r} - \boldsymbol{\omega}_2 \times (\mathbf{r} - \mathbf{s})$$
$$= (\boldsymbol{\omega}_1 - \boldsymbol{\omega}_2) \times \mathbf{r} + \boldsymbol{\omega}_2 \times \mathbf{s}.$$

Now take the scalar product with $(\boldsymbol{\omega}_1 - \boldsymbol{\omega}_2)$. This eliminates $\mathbf{r}$ from the equation and leaves a result which must hold for all possible points of contact.

$$\mathbf{v} \cdot (\boldsymbol{\omega}_1 - \boldsymbol{\omega}_2) = \boldsymbol{\omega}_1 \cdot (\boldsymbol{\omega}_2 \times \mathbf{s})$$
$$= \mathbf{s} \cdot (\boldsymbol{\omega}_1 \times \boldsymbol{\omega}_2)$$
$$= \omega_1 \omega_2 \mathbf{s} \sin \beta,$$

where β is the angle between the axes.

Hence wherever the point of contact is located there must be a component of sliding velocity in the direction of $(\boldsymbol{\omega}_1 - \boldsymbol{\omega}_2)$ unless (a) $s = 0$, so that the axes intersect, or (b) $\beta = 0$, so that the axes are parallel. Neither of these two conditions by itself ensures slip-free contact, it is necessary also for the two bodies to turn at appropriate relative speeds.

The practical consequence of this result is that it is impracticable to have a satisfactory direct-friction drive between a pair of skew shafts.

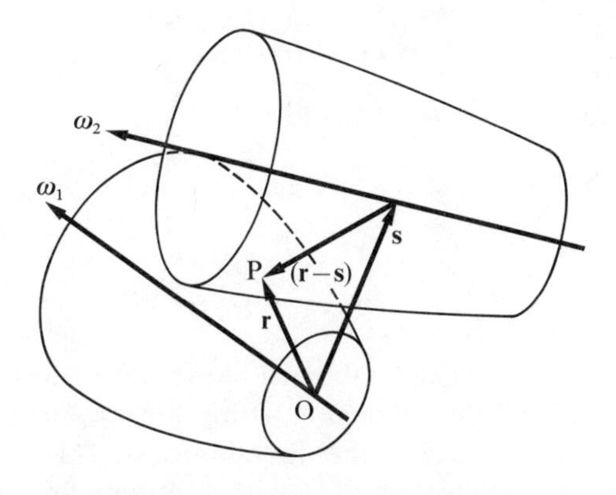

FIG. 3.21

3.8. Solution of vector equations

The techniques used in the last section for solving vector equations are not simply tricks developed to deal with particular problems; the same types of equation arise time and again in different guises. The basic equation to be solved is usually of the form

$$\mathbf{a}+\mathbf{b}\times\mathbf{c}=\mathbf{d}$$

and it is common for two of the vectors to be unknown to some extent. It has to be remembered that a vector equation is equivalent to three scalar equations, so that to be fully solvable a vector equation must not involve more than three unknown scalar quantities. It may happen that **a**, for example, is of unknown magnitude but is known to be perpendicular to a vector **e**, say, (two unknown scalar quantities) and that **c** is of unknown magnitude (the third unknown scalar quantity), but known direction.

By taking the scalar product with **e** we eliminate the unknown **a**, thus

$$\mathbf{a}\,.\,\mathbf{e}+(\mathbf{b}\times\mathbf{c})\,.\,\mathbf{e}=\mathbf{d}\,.\,\mathbf{e}$$

$$\Rightarrow \qquad (\mathbf{e}\times\mathbf{b})\,.\,\mathbf{c}=\mathbf{d}\,.\,\mathbf{e}.$$

Writing $\mathbf{c}=c\mathbf{f}$, where **f** is a unit vector in the known direction of **c** we find $c=\mathbf{d}\,.\,\mathbf{e}/(\mathbf{e}\times\mathbf{b})\,.\,\mathbf{f}$.

a can now be determined by straight substitution for **c** in the original equation.

It would be inappropriate to consider in detail all the possible combinations of unknown quantities. Another standard form is, however, of particular interest. Suppose that

$$\mathbf{a}\times\mathbf{b}=\mathbf{c}$$

where **b** and **c** are known but **a** is unknown. Take the vector product of both sides of the equation with **b**

$$\mathbf{b}\times(\mathbf{a}\times\mathbf{b})=\mathbf{b}\times\mathbf{c}$$

$$\Rightarrow \qquad \mathbf{a}(\mathbf{b}\,.\,\mathbf{b})-\mathbf{b}(\mathbf{a}\,.\,\mathbf{b})=\mathbf{b}\times\mathbf{c}$$

$$\Rightarrow \qquad \mathbf{a}=\frac{\mathbf{b}\times\mathbf{c}}{\mathbf{b}\,.\,\mathbf{b}}-\mathbf{b}\frac{\mathbf{a}\,.\,\mathbf{b}}{\mathbf{b}\,.\,\mathbf{b}}$$

$$=\frac{\mathbf{b}\times\mathbf{c}}{\mathbf{b}\,.\,\mathbf{b}}-\alpha\mathbf{b}, \qquad (3.5)$$

where α is an unknown scalar; unknown because it involves the unknown **a**. Thus we have solved the original equation for **a** up to a point, there is an unknown component of **a** in the direction of **b**. This is because in the original equation the vector product with **b** automatically excludes any component that **a** may have in the direction of **b**.

3.9. The moment of a vector

The moment of a vector P about a point O is defined as $\mathbf{r}\times\mathbf{P}$, where $\mathbf{r}$ is a position vector drawn from O to any point on the line of action of P. It follows from this definition that the direction of the moment is perpendicular to both $\mathbf{r}$ and $\mathbf{P}$. This definition is consistent with the definition of the moment of a force that is given in Chapter 2, though the idea of a moment being a vector is not brought out there. The reason for this is that Chapter 2 is concerned primarily with systems of coplanar forces. The associated moments are all normal to the plane of the forces so that the addition of moments involves no more than the simple addition of their magnitudes.

3.9.1. Equilibrium of a general set of forces

A set of forces $\mathbf{P}$ are in equilibrium if

$$\sum \mathbf{P}=0$$

and

$$\sum \mathbf{M}_{\text{O}}=\sum \mathbf{r}\times\mathbf{P}=0$$

where $\mathbf{M}_{\text{O}}$ is the sum of the moments of the forces about an arbitrary point O.

For the second condition it is sufficient to show that there is one point about which the sum of the moments is zero. For if $\sum \mathbf{P}=0$ it is assured that the sum of the moments will be zero about any other point.

Let $\sum \mathbf{r}\times\mathbf{P}=0$ where $\mathbf{r}$ is measured from a particular point O to the line of action of associated force $\mathbf{P}$. Let the vector from O′ to O be $\boldsymbol{\rho}$. Then

$$\begin{aligned}\mathbf{M}_{\text{O}'}&=\sum (\boldsymbol{\rho}+\mathbf{r})\times\mathbf{P}\\&=\sum \boldsymbol{\rho}\times\mathbf{P}+\sum \mathbf{r}\times\mathbf{P}\\&=\boldsymbol{\rho}\times\sum \mathbf{P}+0=0.\end{aligned}$$

It follows that in considering equilibrium, one is free to take moments about the point that is thought to be most convenient.

In considering particular cases it is usually most convenient to express the forces in terms of components in the directions of a set of suitably chosen cartesian axes. Moments are naturally expressed in terms of a similar set of components spoken of as the moments about the axes. The moment of a force $\mathbf{P}$ about an axis defined by a unit vector $\mathbf{e}$ is $\mathbf{e}\cdot(\mathbf{r}\times\mathbf{P})$, where $\mathbf{r}$ is any vector between $\mathbf{e}$ and $\mathbf{P}$. Physically it can be thought of as the component of $\mathbf{P}$ perpendicular to $\mathbf{e}$ multiplied by the perpendicular distance between $\mathbf{e}$ and $\mathbf{P}$.

Example 3.7: Fig. 3.22 shows a simple space-frame consisting of three members such that the enclosed space forms a regular tetrahedron of edge length a. A force $\mathbf{P}$ *applied to the apex where the three members meet has*

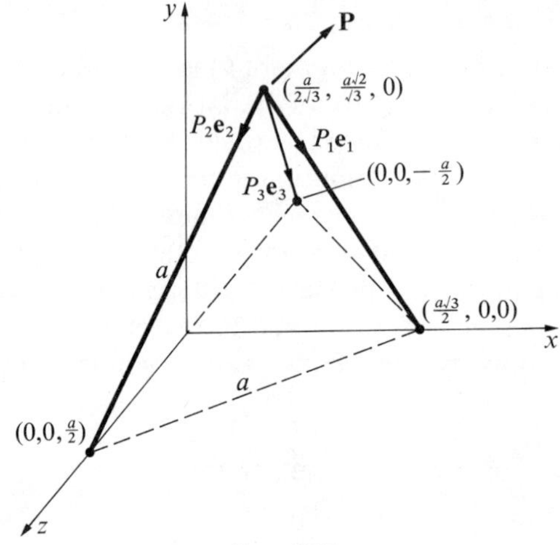

FIG. 3.22

components $\mathbf{P}_x$, $\mathbf{P}_y$, *and* $\mathbf{P}_z$ *in the directions of the set of co-ordinate axes shown. What are the forces in the three members?*

The co-ordinates of the vertices of the tetrahedron are as shown in the figure. The three unit vectors for the directions of the three members are therefore,

$$\mathbf{e}_1 = \mathbf{i}/\sqrt{3} - \mathbf{j}\sqrt{2}/\sqrt{3}; \qquad \mathbf{e}_2 = -\mathbf{i}/2\sqrt{3} - \mathbf{j}\sqrt{2}/\sqrt{3} + \mathbf{k}/2$$

$$\mathbf{e}_3 = -\mathbf{i}/2\sqrt{3} - \mathbf{j}\sqrt{2}/\sqrt{3} - \mathbf{k}2.$$

If the forces in the members are $\mathbf{P}_1$, $\mathbf{P}_2$, and $\mathbf{P}_3$, then

$$\mathbf{P}_1 + \mathbf{P}_2 + \mathbf{P}_3 + \mathbf{P} = 0$$

$$P_1\mathbf{e}_1 + P_2\mathbf{e}_2 + P_3\mathbf{e}_3 + P_x\mathbf{i} + P_y\mathbf{j} + P_z\mathbf{k} = 0.$$

On substituting for $\mathbf{e}_1$, $\mathbf{e}_2$ and $\mathbf{e}_3$ and equating the coefficients of $\mathbf{i}$, $\mathbf{j}$, and $\mathbf{k}$ we find

$$-P_1/\sqrt{3} + (P_2 + P_3)/2\sqrt{3} = P_x$$

$$(P_1 + P_2 + P_3)\sqrt{2}/\sqrt{3} = P_y$$

$$(-P_2 + P_3)/2 = P_z.$$

This set of equations yields

$$P_1 = -\frac{2P_x}{\sqrt{3}} + \frac{P_y}{\sqrt{6}}; \qquad P_2 = \frac{P_x}{\sqrt{3}} + \frac{P_y}{\sqrt{6}} - P_z; \qquad P_3 = \frac{P_x}{\sqrt{3}} + \frac{P_y}{\sqrt{6}} + P_z$$

Example 3.8: A circular table of radius a stands on three legs of length h which are evenly spaced round its rim. A horizontal force $\mathbf{P}$ *is applied tangentially to the rim of the table at the opposite end of the diameter drawn from the top of one of the legs, Fig. 3.23(a). Determine the condition for the table to tilt before it starts to slide on the floor, assuming the same coefficient of friction at the foot of each leg.*

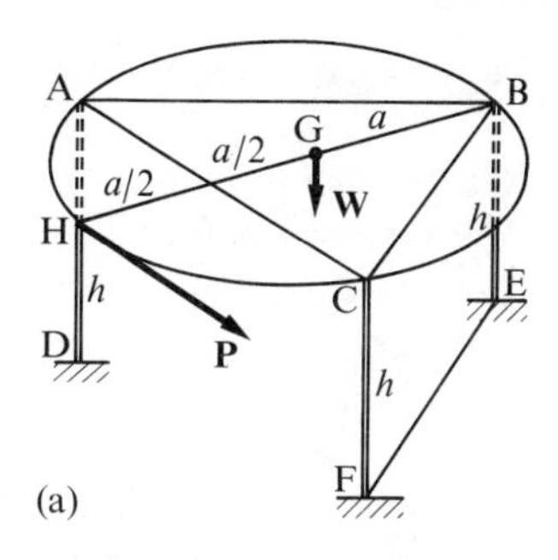

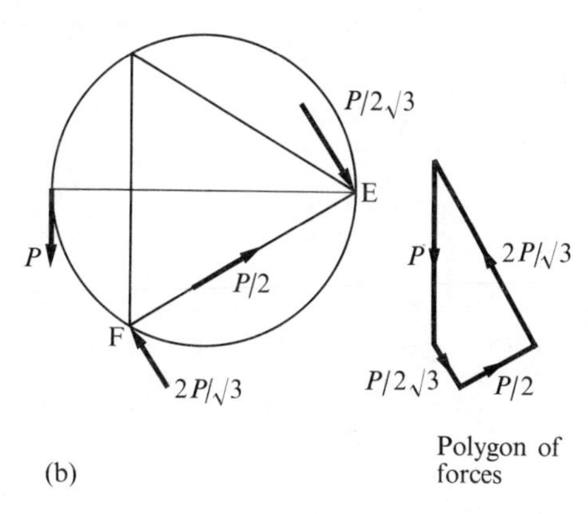

FIG. 3.23

By inspection, the table will start to tilt about EF, thus making the reaction $\mathbf{R}_\mathrm{D}$ at D zero. We will assume that simultaneously the table is just on the point of slipping. It cannot be seen by inspection whether the impending motion will be a pirouette about F, or about E, or whether there will be simultaneous sliding at both points. Whatever happens, the relationship between **P** and **W** is unaffected, for the reactions at E and F have no moment about the line EF, and the moments of **P** and **W** must balance.

As a first step **P** is split into components parallel and perpendicular to EF. These components are of magnitudes $P/2$ and $P\sqrt{3}/2$ respectively. The moment of **P** about EF is now readily seen to be $h \times P\sqrt{3}/2$. Equating this to the moment of **W** about EF we find

$$P = Wa/h\sqrt{3}.$$

The vertical component $\mathbf{R}_\mathrm{E}$ of the reaction at E is best found by balance of the moments about the line FD.

$$(a/2) \times W = (3a/2) \times R_\mathrm{E} \Rightarrow R_\mathrm{E} = W/3.$$

P, being parallel to FD, does not enter into this equation.

Forces must balance in the vertical direction so that $\mathbf{R}_E + \mathbf{R}_F = \mathbf{W}$. Hence $R_F = 2W/3$.

By taking moments about CF the component perpendicular to EF of the horizontal component of the reaction is found to be $P/2\sqrt{3}$. Balance of forces perpendicular to EF gives a similar component of magnitude $2P/\sqrt{3}$ at F. Over-all equilibrium requires a force of $P/2$ along EF. The distribution of this force between E and F cannot be determined by statics alone, Fig. 3.23(b).

If the table pivots about E the direction of sliding at F is perpendicular to EF and that must be the direction of the friction force there. This means that the whole of the $P/2$ component of force along EF must be taken at E. This is possible provided that the resultant horizontal force at E is less than the vertical component of the reaction multiplied by the coefficient of friction. A similar set of considerations apply to the possibility of pivoting about F. The two cases give rise to forces and minimum coefficients of friction as follows:

		At E	At F
Vertical reactions		$W/3$	$2W/3$
Pivoting at E	Horizontal reaction	$P/\sqrt{3} = Wa/3h$	$2P/\sqrt{3} = 2Wa/3h$
	Min. μ required	a/h	a/h
Pivoting at F	Horizontal reaction	$P/2\sqrt{3} = Wa/6h$	$P\sqrt{(19/12)} = a\sqrt{19}/6h$
	Min. μ required	$a/2h$	$a\sqrt{19}/4h$

It can be seen that if tilting and sliding take place simultaneously, the sliding will be at F with pivoting at E, and that the condition for this to happen is $\mu = a/h$. By coincidence, slip is also impending at E. The possibility of pivoting about F must be discounted for that would require a greater coefficient of friction at F than would permit slip to take place at E.

The argument used in solving this problem differs from that used in the preceding one. It is not convenient here to express all the forces in terms of a single set of orthogonal unit vectors. It would have been possible to do so but the solution would have been clumsier. No general rules can be given for selecting the most elegant method in any one case, and it may well be that different people would have differing views on what is the 'best' method.

3.10. Forces transmitted by a bar

Fig. 3.24(a) shows a curved member, perhaps a ship's davit, subjected to a force $\mathbf{F}$ at one end, and fixed at the other. At O the member has to

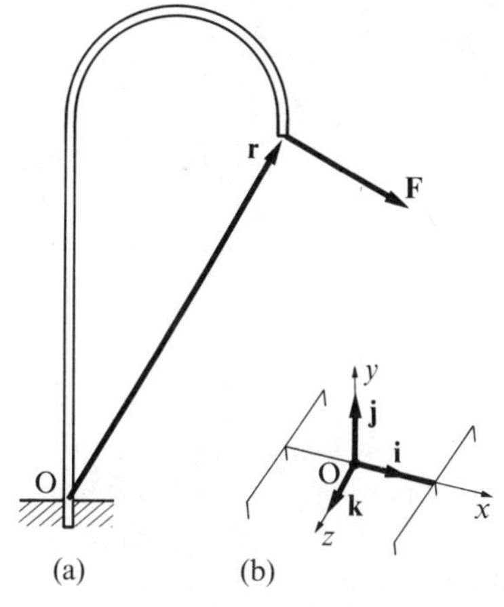

FIG. 3.24

transmit to the foundation the force **F** and the moment $\mathbf{M}_O = \mathbf{r} \times \mathbf{F}$, where **r** is drawn from the centroid of the cross-section of the member to **F**. The resulting stresses in the member at O are best determined by splitting **F** and $\mathbf{M}_O$ up into components each of which gives rise to a readily calculable set of stresses.

O is located at the centroid of the section so that the axial component of **F** causes a uniform tensile or compressive stress. In Fig. 3.24(b) **j** is a unit vector in the axial direction, and **i** and **k** are unit vectors along the principal axes of the cross-section. The components of **F** and $\mathbf{M}_O$ are then:

$\mathbf{F} \cdot \mathbf{j}$ = Resultant direct force, compressive or tensile on the cross section.
$\mathbf{F} \cdot \mathbf{i}$ = Shear force in the x-direction.
$\mathbf{F} \cdot \mathbf{k}$ = Shear force in the y-direction.
$(\mathbf{r} \times \mathbf{F}) \cdot \mathbf{j}$ = Torsional moment on the member.
$(\mathbf{r} \times \mathbf{F}) \cdot \mathbf{i}$ = Bending moment about Ox.
$(\mathbf{r} \times \mathbf{F}) \cdot \mathbf{k}$ = Bending moment about Oz.

3.11. Equivalent force systems—the wrench

It was shown for planar systems of forces that a single force is equivalent to an equal force acting through an arbitrary point together with a couple equal to the moment of the original force about the arbitrary point, Fig. 3.25. Several forces reduce to a single resultant force through an arbitrary point together with a resultant couple. The same reasoning applied to a set of forces in space. The only difference is that for the three-dimensional set the resultant force and moment vectors will not usually be at right angles to each other, as they are for a planar set of forces.

If the resultant force is zero, the system of forces reduces to a couple, and this is true for both spatial and planar sets of forces. If the resultant

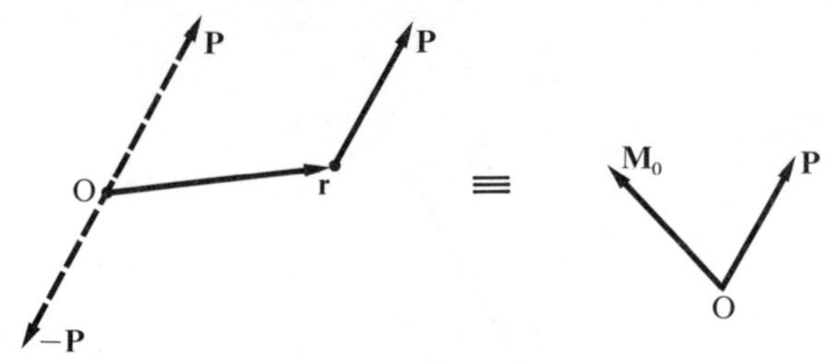

FIG. 3.25

force is not zero, then it is always possible in the planar case to choose a point in the plane for which the couple is zero. In general, this is not possible for a set of forces in three dimensions.

Let $\mathbf{R}=\sum\mathbf{P}$ be the resultant of a set of forces, and let $\mathbf{M}_O$ be the moment of the forces about O. If O′ is a point that is located relative to O by $\mathbf{r}$, then

$$\mathbf{M}_{O'}=\mathbf{M}_O-\mathbf{r}\times\mathbf{R}.$$

The two force systems $(\mathbf{R}, \mathbf{M}_O)$ and $(\mathbf{R}, \mathbf{M}_{O'})$ are equivalent in the sense that they would have the same external effect on any rigid body on which they acted.

It will not, in general, be possible to choose O′ such that $\mathbf{M}_{O'}=0$. The best that can be done is to make $|\mathbf{M}_{O'}|$ as small as possible, and this happens when $\mathbf{M}_{O'}$ is parallel to $\mathbf{R}$, i.e. $\mathbf{R}\times\mathbf{M}_{O'}=0$. The system of forces is then said to have been reduced to *a wrench.* Assuming this to be possible we can write $\mathbf{M}_{O'}=\lambda\mathbf{R}$ so that

$$\lambda\mathbf{R}=\mathbf{M}_O-\mathbf{r}\times\mathbf{R}.$$

This possibility is proved if we can solve this equation for the pitch λ and for $\mathbf{r}$. First take the scalar product with $\mathbf{R}$

$$\lambda\mathbf{R}\,.\,\mathbf{R}=\mathbf{R}\,.\,\mathbf{M}_O-\mathbf{R}\,.\,(\mathbf{r}\times\mathbf{R})$$

$$\Rightarrow \qquad \lambda=\frac{\mathbf{R}\,.\,\mathbf{M}_O}{\mathbf{R}\,.\,\mathbf{R}}.$$

Using eqn (3.5)

$$\mathbf{r}=\frac{\mathbf{R}\times(\mathbf{M}_O-\lambda\mathbf{R})}{\mathbf{R}\,.\,\mathbf{R}}+(\text{a component parallel to }\mathbf{R}).$$

The first term on the right-hand side of this equation must be the normal $\mathbf{n}$ from O to $\mathbf{R}$ for the wrench. As $\mathbf{R}\times\mathbf{R}=0$ we can write

$$\mathbf{n}=\frac{\mathbf{R}\times\mathbf{M}_O}{\mathbf{R}\,.\,\mathbf{R}}.$$

The reduction of a set of forces to a wrench is not an exercise of great practical application. It is, however, important to realize that in the

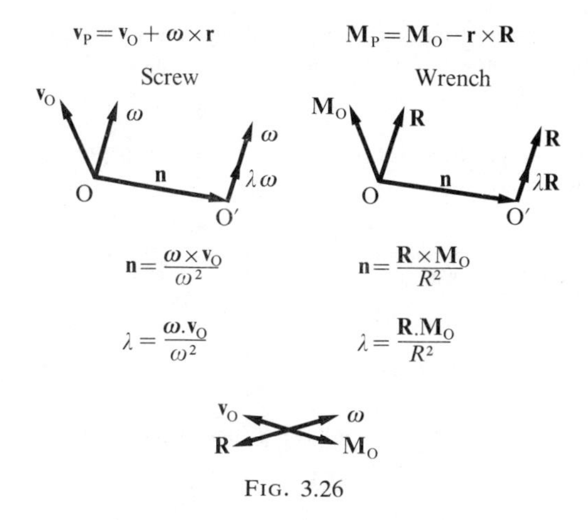

FIG. 3.26

spatial case reduction of a set of forces to a single force or couple is usually not possible.

The wrench is the force analogue of the instantaneous screw axis for spatial motion. It is worth setting the two sets of results side by side as in Fig. 3.26 because there is a certain skewness in the analogy which may be unexpected.

3.12. Problems

1. Vectors **a** and **b** are drawn from the origin O to A and B in the Oxy plane. Unit vectors **i** and **j** are in the directions Ox and Oy respectively. A second pair of unit vectors **n** and **t**, also in the Oxy plane, are at 60° anticlockwise to Ox and Oy respectively.

If A = (2, 1) and B = (1, 3), express **a** and **b** in terms of **n** and **t** by taking measurements off a drawing, or by using simple trigonometry. Thereafter verify that

$$a_n = a_x \cos\theta + a_y \sin\theta$$

and

$$a_t = -a_x \sin\theta + a_y \cos\theta,$$

and similarly for **b**.

Verify also that **a** . **b** for **a** and **b** expressed in terms of **n** and **t** has the same value as it does when they are given in terms of **i** and **j**.

2. The difference in position between two aircraft at a given instant is $\mathbf{r}_O$ measured from aircraft 1 to aircraft 2. Their velocities $\mathbf{v}_1$ and $\mathbf{v}_2$ are constant. Show that a collision will occur if $\mathbf{r}_O = \lambda(\mathbf{v}_1 - \mathbf{v}_2)$, where λ is a positive scalar quantity, and it will happen at time $t = \lambda$.

Show that the minimum distance between non-colliding aircraft occurs at

$$t = \frac{\mathbf{r}_O \cdot (\mathbf{v}_1 - \mathbf{v}_2)}{(\mathbf{v}_1 - \mathbf{v}_2) \cdot (\mathbf{v}_1 - \mathbf{v}_2)}$$

after the given instant.

In a particular case, aircraft 1 is 10 km due west of aircraft 2 and 1500 m higher. Both aircraft maintain the same altitude whilst aircraft 1 flies with constant speed and bearing 600 km h^{-1} NE, and aircraft 2 flies with constant speed and bearing 650 km h^{-1} NW. What will be the minimum distance separating the two aircraft?

3. $\mathbf{u}$ is the perpendicular from the origin to a plane. If $\mathbf{r}$ is the vector from the origin to any point in the plane, show that the equation to the plane is

$$\mathbf{u}\cdot\mathbf{r}=u^2.$$

If $\mathbf{u}$ and $\mathbf{v}$ are the perpendiculars from the origin to two planes, show that the line of intersection of the planes is

$$\mathbf{r}=a\mathbf{u}+b\mathbf{v}+\lambda(\mathbf{u}\times\mathbf{v})$$

where a and b are scalar constants and λ is a scalar variable. Find a and b and evaluate $(\mathbf{u}\times\mathbf{v})$ given

$$\mathbf{u}=\mathbf{i}+2\mathbf{j}-\mathbf{k}$$

$$\mathbf{v}=2\mathbf{i}+\mathbf{j}+2\mathbf{k}.$$

4. Four points A, B, C, D have the following components:

	i	**j**	**k**
A	2·5	0	0
B	0	0	1
C	0	0	0
D	2	3	2

Find the length of the common perpendicular between AB and CD, and where it cuts the plane Oxz.

5. A string with end A is drawn through a small eye situated at O, the length OA being kept taut, Fig. 3.27. If $\mathbf{v}$ is the velocity of A, and $\boldsymbol{\rho}$ is the vector from O to A at a particular instant,

(a) show that the speed S with which the string slides through the eye is $\mathbf{v}\cdot\boldsymbol{\rho}/\rho$,

(b) show also that, ignoring twisting of the string, the angular velocity $\boldsymbol{\omega}$ of OA is $\boldsymbol{\rho}\times\mathbf{v}/\rho^2$, and

(c) if A is attached to the end of a crank BA that turns with angular velocity Ω so that $\mathbf{v}=\boldsymbol{\Omega}\times\mathbf{r}$ what is the geometrical relationship between $\boldsymbol{\Omega}$, $\mathbf{r}$, and $\boldsymbol{\rho}$ if $S=0$?

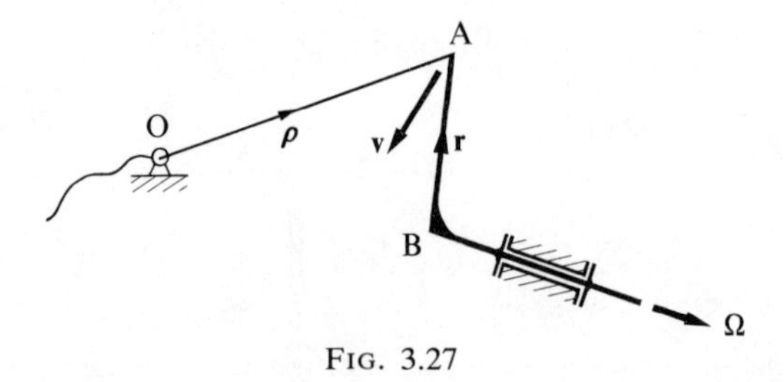

FIG. 3.27

6. In a particular system, O in question 5 is at the origin of a set of Cartesian co-ordinate axes. The shaft to which BA is attached lies in the Oxz plane at 45° to the positive directions of both Ox and Oz axes. The length of the crank BA is $\sqrt{5}$ and the centre of rotation B of the crank is located at (3, 0, 2).

At a given instant the shaft is turning at 10 rad s^{-1} and A is located at the point (3, 2, 1). What is the speed of the string through the eye at O, and what is the angular velocity of OA?

7. A flat, rigid lamina is rotating with an angular velocity ω about the x-axis which lies in its plane. At the instant when it is inclined at $\tan^{-1}(1/2)$ to the x–y plane, what is the velocity of the point (4, 2, 1) on the lamina?

A second body is hinged about a line having direction cosines $(1/\sqrt{3}, 1/\sqrt{3}, 1/\sqrt{3})$ and passing through the point (0, 0, 2). At this instant the two bodies are touching at the point (4, 2, 1) and at no other point. What is the instantaneous angular velocity of the second body and what is the magnitude of the velocity of sliding between the two bodies at their point of contact?

8. Fig. 3.28 shows a thrust bearing in which the lower race A is stationary and the upper B rotates about the XY. The balls run in symmetrical grooves in each race. Show that, for a given ball, rolling is not possible without slipping at one or more of the four points of contact.

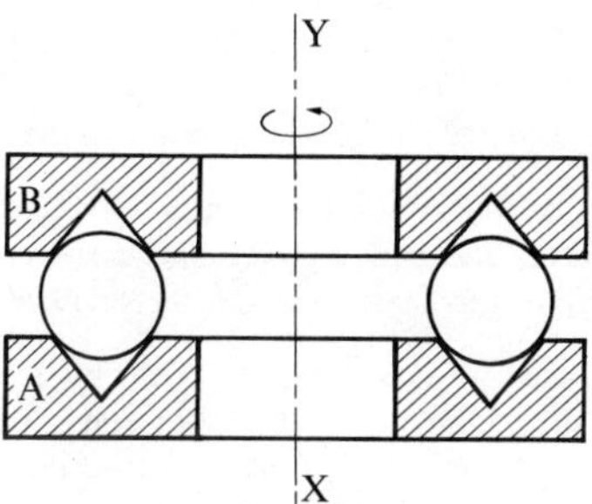

FIG. 3.28

9. Two shafts A and B enter a gearbox as shown in Fig. 3.29. To shaft A is applied a torque of 0·5 kN m and to shaft B a torque of 0·2 kN m, both anticlockwise as viewed towards the gearbox.

Find the magnitude and direction of the torque which must be applied to the gearbox to maintain equilibrium.

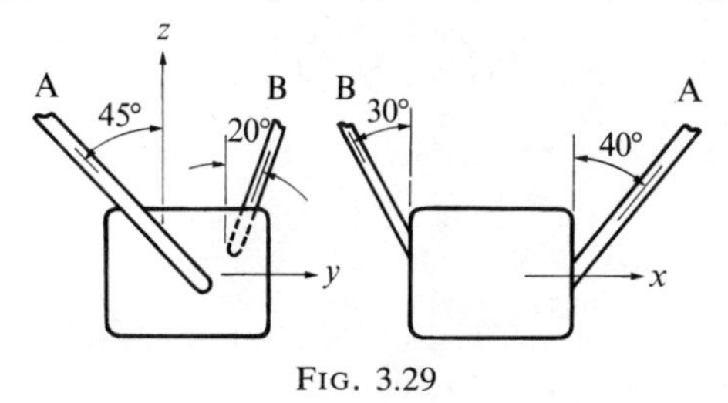

FIG. 3.29

10. Fig. 3.30 shows, in plan, a base which supports a tripod such as is used in chemical laboratories when heating a flask over a Bunsen burner. The tripod is light and has smooth

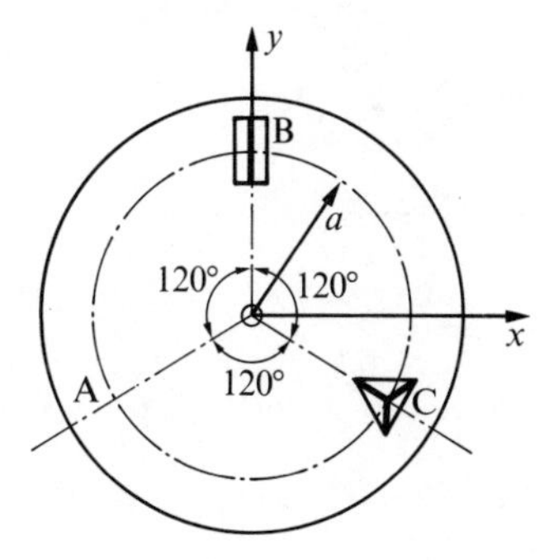

FIG. 3.30

ball-ended feet with centres equispaced on a horizontal circle of radius a. One foot rests on the horizontal upper surface of the base at A, another in a V-groove at B and the third in a tetrahedral indentation at C.

The tripod is subjected to a force $(3\mathbf{i}+4\mathbf{j}-12\mathbf{k})$ acting at a point with co-ordinates $(0, -\frac{1}{4}a, a)$ referred to the right-handed set of Cartesian axes shown in Fig. 3.30, with the origin O in the plane of the base. Find the x, y, and z components of the reactions on the feet of the tripod. The possibility that the tripod may tip up can be discounted.

11. A uniform rectangular plate OBCD is freely hinged along the side OB to a fixed line which passes through O and a point (2, 2, 1) referred to $Oxyz$ axes, with Oz vertically upwards. The corner C is at the point (6, 2, 3) and is supported by a light tie connected between C and a fixed point at (9, 4, 5).

Find the dimensions of the plate, and the tension in the tie if the plate weighs 44 N.

12. If at the given instant the tension in the string referred to in question 5 is 40 N, what are the axial force, shear force, and bending moment in the crank at B?

Determine also the torque in the shaft to which the crank is attached.

13. An umbrella-shaped canopy is supported solely by a single straight member OP which is rigidly attached to the canopy at P. The lower end O of the member is built in to a wall. Taking O as the origin of a rectangular Cartesian co-ordinate system (x, y, z), the point P lies in the horizontal plane $z=1$, where lengths are given in metres.

The weight of the canopy is 7 kN and acts vertically down through the point (1, 0, 0). The design wind force has components (4, 1, 1) kN and acts effectively through the point (−1, 1, 1). Find the co-ordinates of the point P such that there is no bending moment in the member at O. For the member in this position determine the axial force and shear force at O.

14. Twisting couples are distributed uniformly round a semicircle of radius R. They have a moment T per unit circumferential length. Find their resultant.

A ring of mean radius R has a uniformly distributed twisting moment T per unit length applied to it around its periphery.

What is (a) the torque and (b) the bending moment transmitted across any cross-section of the ring?

15. A smooth uniform rod AB of length l and weight w is pivoted at A to a fixed point. It is also supported at a point along its length by a horizontal smooth rail such that it makes an angle α with the vertical and β with the rail. It is kept in equilibrium by the application of a force F that is parallel to the rail and applied at point B.

Using unit vectors that are parallel to the direction of the rod and rail respectively, show that

$$F=\frac{w}{2}\frac{\cos\alpha\cos\beta}{\sin^2\beta}.$$

16.

	i	**j**	**k**	Acts through
P···	1	0	0	(0, 0, 0)
Q···	0	2	2	(0, 0, 1)

Forces **P** and **Q** act as given above (kN and m). Express their resultant as a force **R** combined with a couple **C** about an axis parallel to **R** (i.e. a 'wrench'). Give your answer in the following form:

Direction cosines of **R** and **C**.

Magnitudes of **R** and **C**.

Point where line of action of **R** cuts the Oxy plane.

4. Virtual work

4.1. Work and power

WHEN a force $\mathbf{P}$ moves its point of application through a distance $\delta\mathbf{x}$, the force is said to do mechanical work of an amount $\mathbf{P} \cdot \delta\mathbf{x}$. If the point of application moves with velocity $\mathbf{v}$, the rate of doing work, the power, is $\mathbf{P} \cdot \mathbf{v}$.

It will be noted that because of the scalar product in this definition, a force does work by virtue of movement in its own direction. So in Fig. 4.1 the force $\mathbf{P}$, which equals the weight $\mathbf{W}$, does work in lifting the weight. As the displacements at $\mathbf{P}$ and $\mathbf{W}$ are equal and opposite the net work is zero, and the work done by $\mathbf{P}$ equals the work done in overcoming $\mathbf{W}$. A small sideways movement at the point of application of $\mathbf{P}$ causes no movement at $\mathbf{W}$ and no work is done by $\mathbf{P}$.

The unit of work is the joule (J). It is the amount of work or energy expended by a force of one newton in a collinear displacement of one metre. The basic unit of power is the watt (W) and is equivalent to a rate of work of one joule per second. This definition is consistent with the definition of the unit in an electrical context as being the power provided by a current of one ampere flowing through a potential drop of one volt.

4.2. Virtual work

If two forces $\mathbf{P}$ and $-\mathbf{P}$ meet at a point they are in equilibrium. The total work done in a small displacement $\delta\mathbf{x}$ of the point is $\mathbf{P} \cdot \delta\mathbf{x} + (-\mathbf{P}) \cdot \delta\mathbf{x} = 0$. If several forces act at the point the total work done is $\sum \mathbf{P} \cdot \delta\mathbf{x}$, and this is self-evidently zero if $\sum \mathbf{P} = 0$, and the forces are in equilibrium. This must be true whether or not the displacement $\delta\mathbf{x}$ is physically possible in the given circumstances. The particle in Fig. 4.2 is in equilibrium under the forces $\mathbf{P}_1$, $\mathbf{P}_2$, and $\mathbf{R}$, the latter being the reaction between the particle and the plane. As $\mathbf{P}_1 + \mathbf{P}_2 + \mathbf{R} = 0$ it must be true that $(\mathbf{P}_1 + \mathbf{P}_2 + \mathbf{R}) \cdot \delta\mathbf{x} = 0$ whatever $\delta\mathbf{x}$ is visualized, even though the plane itself is immovable. Such an imaginary displacement is called a *virtual displacement* to distinguish it from displacements which actually occur. The work in a virtual displacement is virtual work. *The principle of virtual work states: the net (virtual) work done in a virtual displacement of a set of forces which are in equilibrium is zero.*

Though the principle of virtual work seems hardly worth stating in its application to a single particle, it is, in fact, extremely important and far-reaching in that it holds for all systems of forces whether applied to single bodies or systems of bodies.

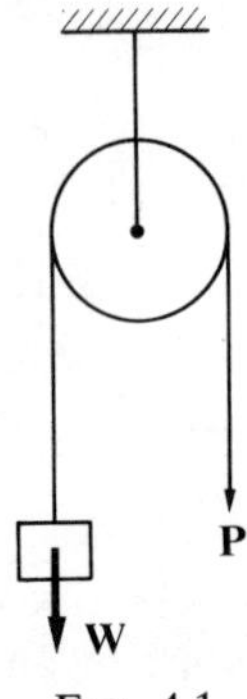

FIG. 4.1

4.2.1. Virtual work equation for a single rigid body

We start with the principle of virtual work as applied to a set of concurrent forces and proceed by way of the general equations of equilibrium and kinematics for a rigid body to show that the net work done in the displacement of a rigid body is zero.

A series of forces $\mathbf{P}_1, \mathbf{P}_2, \ldots$ is applied to points in the rigid body located at $\mathbf{r}_1, \mathbf{r}_2, \ldots$ in relation to an arbitrary origin O. Consider the body to have a virtual velocity field that is defined by $\mathbf{v}_O$ and $\boldsymbol{\omega}$. The virtual work done by a typical force in a small time interval δt is

$$\mathbf{P} \cdot (\mathbf{v}_O + \boldsymbol{\omega} \times \mathbf{r})\, \delta t = \{\mathbf{P} \cdot \mathbf{v}_O + \boldsymbol{\omega} \cdot (\mathbf{r} \times \mathbf{P})\}\, \delta t.$$

The total work done by all the forces is

$$\{\sum \mathbf{P} \cdot \mathbf{v}_O + \sum \boldsymbol{\omega} \cdot (\mathbf{r} \times \mathbf{P})\}\, \delta t.$$

As $\mathbf{v}_O$ and $\boldsymbol{\omega}$ are the same whatever particle is being looked at, we can write the virtual work as

$$\{(\sum \mathbf{P}) \cdot \mathbf{v}_O + \boldsymbol{\omega} \cdot (\sum \mathbf{r} \times \mathbf{P})\}\, \delta t = \{\mathbf{R} \cdot \mathbf{v}_O + \boldsymbol{\omega} \cdot \mathbf{M}_O\}\, \delta t.$$

If the body is in equilibrium $\mathbf{R} = \sum \mathbf{P} = 0$ and $\mathbf{M}_O = \sum \mathbf{r} \times \mathbf{P} = 0$ and the net virtual work is zero. The principle is thus extended from single particles to single bodies. Clearly it also applies to a set of rigid bodies which is in equilibrium, for each body must individually be in equilibrium so that the total virtual work is zero.

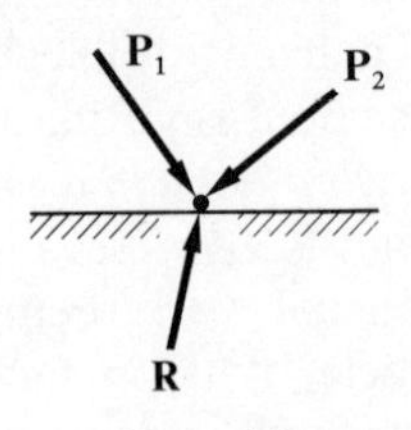

FIG. 4.2

Example 4.1: A light ladder of length l stands against a vertical wall at an angle θ to the vertical. The angle of friction between the ladder and the wall, and the ladder and the ground is ϕ. How far up the ladder can a man climb before the ladder slips if $\phi < \theta$?

The ladder is in equilibrium under the weight of the man and the reactions at either end of the ladder. In the limiting case the reactions $\mathbf{R}_1$ and $\mathbf{R}_2$ at either end of the ladder are inclined at angles ϕ to the normals to the ground and wall respectively. We have to remember that the angle between the reaction and the normal cannot exceed ϕ, and that if the ladder slips it must do so at both ends simultaneously. If the ladder is in equilibrium the net work done is a small displacement by $\mathbf{W}$, $\mathbf{R}_1$, and $\mathbf{R}_2$ is zero. As we are not restricted to displacements that are physically possible, let us allow the bottom end of the ladder to move with velocity $\mathbf{v}_1$ perpendicular to $\mathbf{R}_1$ and the upper end with velocity $\mathbf{v}_2$ perpendicular to $\mathbf{R}_2$. In such a movement the work done by the reactions is zero so that, if the ladder is in equilibrium, the work done by $\mathbf{W}$ must also be zero. Hence the movement at $\mathbf{W}$ must be horizontal.

The velocities at the ends of the ladder must be compatible with the ladder being a rigid body. This is obviously possible as the relationship between $\mathbf{v}_1$ and $\mathbf{v}_2$ is independent of any other considerations, and all that is necessary is to assume a virtual instantaneous centre at I where $\mathbf{R}_1$ and $\mathbf{R}_2$ intersect, Fig. 4.3. $\mathbf{W}$ moves horizontally and does no work if the man stands vertically under I.

The determination of x, the distance of the man from the foot of the ladder, is now a matter of simple trigonometry:

$$\frac{x}{\sin\phi} = \frac{\text{IM}}{\sin(\theta-\phi)} \quad \text{and} \quad \frac{l-x}{\cos\phi} = \frac{\text{IM}}{\cos(\theta-\phi)}$$

$$\Rightarrow \qquad \frac{x}{l-x} = \cot(\theta-\phi)\cos\phi$$

$$\Rightarrow \qquad \frac{x}{l} = \frac{1}{1+\tan\phi\tan(\theta-\phi)}.$$

We have studied the limiting case in which slip is impending. If the man is lower down the ladder $\mathbf{W}\,.\,\mathbf{v}\,\delta t$ is negative, which would only be possible if there were some source of power other than gravity. There is no such source, so there is no movement. If the man climbs above the limiting position, $\mathbf{W}\,.\,\mathbf{v}\,\delta t$ is positive, and as work is put into the system so the man and the ladder acquire *real* velocities.

It may be observed that if the need for the three forces acting on the ladder to be concurrent is recognized, the same answer to this problem is reached more directly. It is hoped, however, that this simple example will give the reader confidence to apply the principle of virtual work in situations where the alternative statical solution is less obvious.

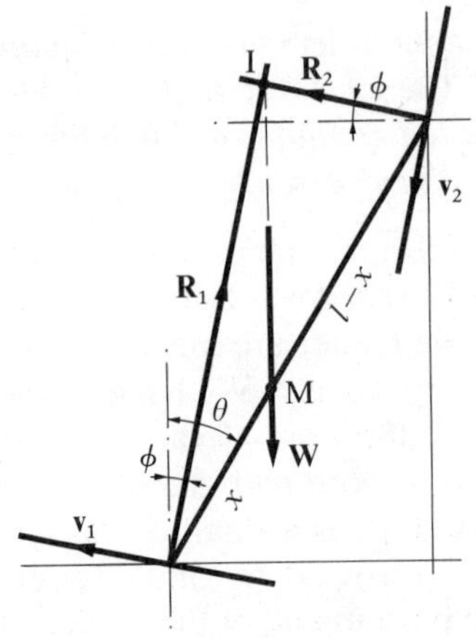

FIG. 4.3

This happens particularly where the equilibrium of several bodies is involved.

4.2.2. Virtual work applied to determine forces in mechanisms

Example 4.2: Fig. 4.4 shows schematically a six-bar mechanism, a basic type of mechanism that was used by both Watt and Stephenson in their engines. The mechanism is in equilibrium with torques $\mathbf{Q}_1$ *and* $\mathbf{Q}_2$ *applied as shown. Determine the relationship between* $\mathbf{Q}_1$ *and* $\mathbf{Q}_2$*, and the force in EF.*

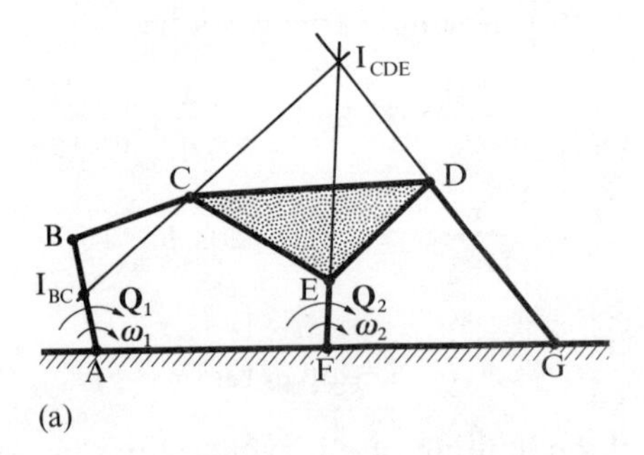

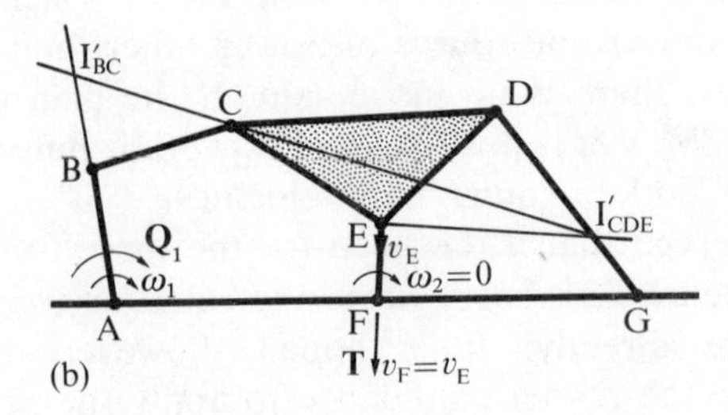

FIG. 4.4

In normal operation only the torques $\mathbf{Q}_1$ and $\mathbf{Q}_2$ do work on the mechanism. The relationship between them is therefore most readily found by allowing the mechanism to deform naturally and equating the net work done to zero,

$$(Q_1\omega_1 + Q_2\omega_2)\,\delta t = 0.$$

The instantaneous centre for CDE is at the intersection of DG and EF. Producing $\mathrm{I_{CDE}C}$ to intersect AB locates $\mathrm{I_{BC}}$.

$$v_\mathrm{C} = \omega_1 \mathrm{AB}(\mathrm{I_{BC}C/I_{BC}B}) = v_\mathrm{E}(\mathrm{I_{CDE}C/I_{CDE}E}).$$

But $v_\mathrm{E} = \omega_2\mathrm{EF}$, and so ω_2/ω_1 and hence Q_2/Q_1 are determined.

The external forces on the mechanism are the bearing reactions at A, F, and G, and the torques $\mathbf{Q}_1$ and $\mathbf{Q}_2$. The reactions at A, F, and G do not enter into the real-work equation because they are applied at fixed points. Now allow a virtual displacement in which bearing F is allowed to move a small amount Δ in the direction EF. The virtual work thereby done is $T_\mathrm{EF}\Delta$, where T_EF, assumed tensile, is the force in EF. Let us assume arbitrarily that this displacement involves no rotation of EF. There will naturally be rotations of the other members, which can be calculated in terms of $\Delta = v_\mathrm{F}\,\delta t$.

The instantaneous centres for CDE and AB are now located as shown in Fig. 4.4(b). v_E is now readily determined in terms of ω_1, and the virtual work equation

$$(Q_1\omega_1 + T_\mathrm{EF}v_\mathrm{F})\,\delta t = 0$$

yields T_EF in terms of Q_1.

It would have been just as easy to allow EF to rotate while AB was kept fixed. This would have yielded T_EF in terms of Q_2.

These two examples are concerned essentially with determining the forces for equilibrium in a given configuration. There is no reason in principle why the process should not be reversed and used to determine the geometrical configuration of a body or a set of bodies which are subjected to a given set of forces. In practice this converse problem is often awkward to solve.

4.3. Use of virtual work to find an equilibrium configuration

The principle of virtual work cannot reveal anything about the equilibrium of a set of bodies that could not be deduced by direct use of the equations of equilibrium. It can help to circumvent a lot of effort, particularly when several bodies are involved, and it can also help by bringing out results that might otherwise not be spotted.

Example 4.3: Two identical uniform rods AB and AC, each of length l, are freely hinged together at A. The rods rest in equilibrium on a horizontal cylinder of radius r as shown in Fig. 4.5. Determine the relationship

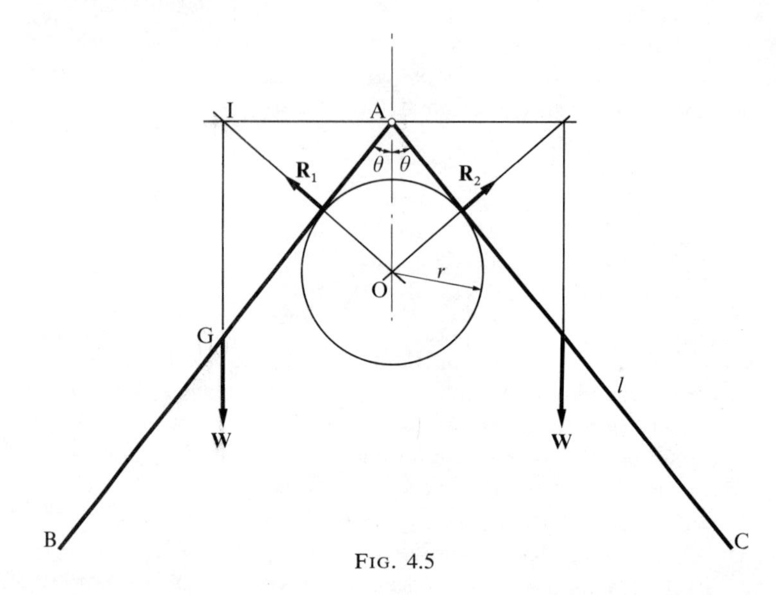

FIG. 4.5

between l, r, and θ, the angle that the rods make with the vertical. Friction is negligible.

The only forces that act on the rods are their weights **W** and the reactions $\mathbf{R}_1$ and $\mathbf{R}_2$ from the cylinder. Allow a virtual displacement of A vertically. The reactions $\mathbf{R}_1$ and $\mathbf{R}_2$ do no work so that for equilibrium the forces **W** must also do no work. The displacements of the centre of mass G must therefore be horizontal. The instantaneous centre of AB lies at I, the intersection of the horizontal line drawn from A with the normal to the cylinder at the point of contact with the rod. G moves horizontally if it is vertically below I. Hence

$$\mathrm{IA} = (l/2)\sin\theta = \mathrm{OA}\cot\theta$$
$$= r\operatorname{cosec}\theta\cot\theta$$
$$\Rightarrow \qquad r/l = \sin^2\theta\tan\theta.$$

This solution takes it for granted that the two rods will not slide bodily round the cylinder and fall off. This will, in fact, not happen for the system as it is drawn in Fig. 4.5. A formal study of stability is made in Chapter 6, but it may be that in this case the condition for the rods to be in stable equilibrium is obvious.

The system in Fig. 4.5 has two degrees of freedom, but because of symmetry only one degree of freedom has been considered. This very much simplifies the problem. When two variables have to be evaluated it will often be found impossible to obtain solutions in simple algebraic or trigonometric terms, even when the system is physically simple. This holds whether the equations of equilibrium are applied directly or by way of the principle of virtual work.

4.4. Problems

1. What driving torque must be applied to the crank DB in Fig. 1.33 (p. 28) to overcome a force of 100 N resisting motion at A together with a force of 120 N applied at C in the direction from A to C?

2. Fig. 4.6 shows a mechanism in which AD is fixed and CD rotates with an angular velocity of 10 rad s^{-1} clockwise. BCE is a rigid link.

Determine for the configuration shown the driving torque required to overcome the force of 500 N at F, friction and inertia being negligible.

Note. This problem is a continuation of question 11 in Chapter 1.

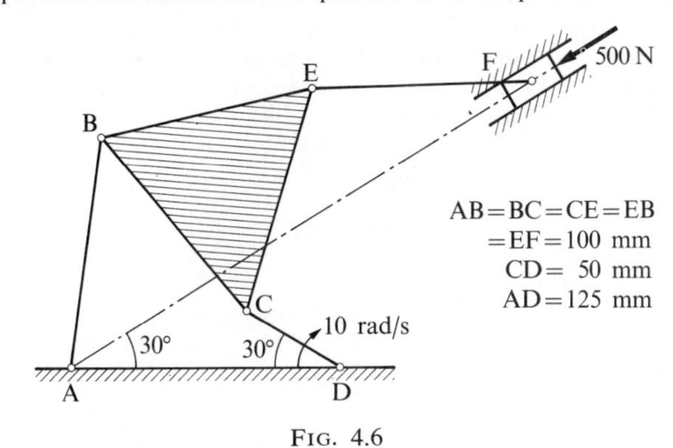

FIG. 4.6

3. The quick-return mechanism shown in Fig. 4.7 is used in a shaping machine. In the position shown the cutting speed is 0·5 m s^{-1} and the cutting force is 400 N.

Find the velocities of sliding at the sliders B and C. Find also the instantaneous torque which must be applied to the crank AC.

AC has a length 200 mm. Other dimensions may be scaled from the figure. Neglect friction and inertia.

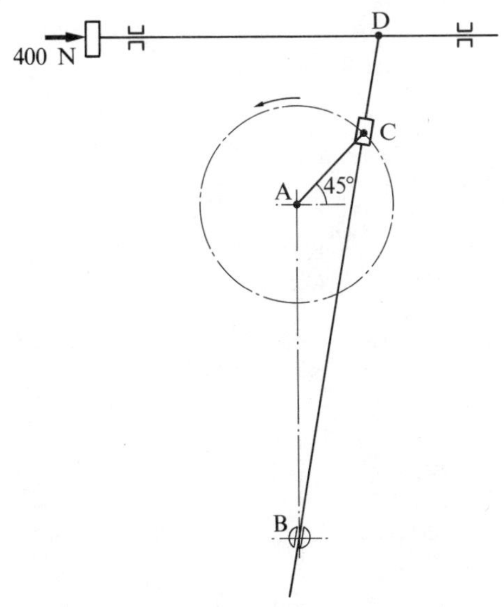

FIG. 4.7

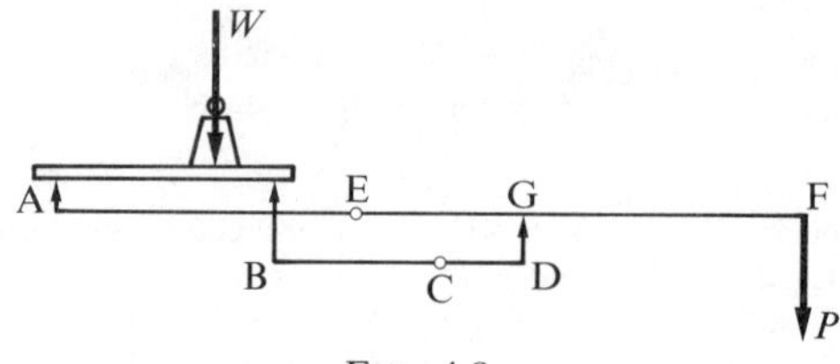

FIG. 4.8

4. Fig. 4.8 shows the mechanism of a platform weighing machine. AEF is a lever pivoted at E and BCD is a lever pivoted at C. The platform is supported on the ends A and B of the two levers by means of the short struts shown in the diagram and another short strut at D presses upwards on the lever AEF at G. It is required that a force P applied as shown at F shall balance the load W irrespective of its position on the platform. Show that this requirement is met by making

$$\frac{\text{AE}}{\text{EG}}=\frac{\text{BC}}{\text{CD}}.$$

If $P=\frac{1}{4}W$, show that $\text{AE}=\frac{1}{5}\text{AF}$.

5. Fig. 4.9 shows the mechanism of a 'Counterpoise' table lamp. There are four straight light links AOBC, DEF, EB, and FCG, pin-pointed as shown. OB = DE, EB = FC and EF = BC.

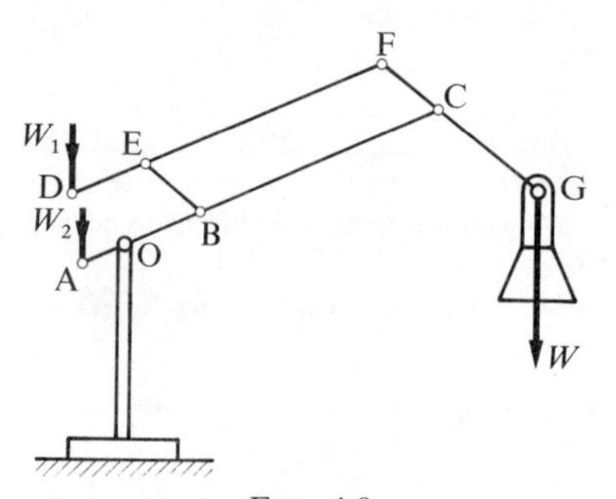

FIG. 4.9

Show that the weights which must be attached at D and A to counterbalance the weight W of the lamp *in any position* are

$$W_1=\frac{\text{CG}}{\text{FC}}W \quad \text{and} \quad W_2=\frac{\text{OC}}{\text{OA}}W.$$

Neglect friction at the pivots.

6. Determine the torque in the shaft to which the crank is attached in question 12, Chapter 3 by equating the instantaneous power input and output.

5. Further kinematics

5.1. Linear motion

SUPERFICIALLY it is a simple matter to describe the motion of a particle along a line; it would seem to be merely a matter of plotting distance moved along the line against time. However, information is not always available to enable this to be done without preliminary calculations, and there are occasions when it is desirable to present the information differently. A passenger in a car might note the times at which the vehicle passes milestones. In this case a direct plot of distance against time is possible. At night, or when crossing country without distinguishing features, he might have to be content with noting speed at convenient times. Then, a plot of speed against time would have to be integrated to give the distance travelled. Inertial guidance systems work by continually monitoring acceleration, which has to be integrated twice with respect to time to give distance travelled.

Lacking a watch the passenger in the motor car may note the speed as the car passes successive milestones. The information thus gained is speed as a function of distance:

$$v = \mathrm{d}x/\mathrm{d}t = f(x).$$

Distance as a function of time is now obtained by integration after separation of the variables:

$$t = \int_0^x \frac{\mathrm{d}x}{v} \qquad (5.1)$$

It is difficult to visualize on the basis of our present example acceleration being known as a function of displacement, but this quite often arises in dealing with the dynamics of particles. For the moment we will merely note the procedure for integration of

$$\frac{\mathrm{d}^2x}{\mathrm{d}t^2} = f(x).$$

We note first that

$$\frac{\mathrm{d}^2x}{\mathrm{d}t^2} = \frac{\mathrm{d}}{\mathrm{d}t}\left(\frac{\mathrm{d}x}{\mathrm{d}t}\right) = \frac{\mathrm{d}x}{\mathrm{d}t}\frac{\mathrm{d}}{\mathrm{d}x}\left(\frac{\mathrm{d}x}{\mathrm{d}t}\right)$$

$$= v\frac{\mathrm{d}v}{\mathrm{d}x}$$

$$= \frac{\mathrm{d}}{\mathrm{d}x}\left(\tfrac{1}{2}v^2\right),$$

so that

$$v^2 = 2\int_0^x f(x)\,\mathrm{d}x. \tag{5.2}$$

The integral can be evaluated algebraically if $f(x)$ is expressed in a suitable form; if not it must be integrated graphically. By taking the square root, $v = \mathrm{d}x/\mathrm{d}t$ is then known as a function of x and it can be integrated to give x as a function of t in the manner already indicated.

It should be noted that the path of motion in our example may twist and turn in any way, except in relation to the brief reference to inertial navigation. By confining attention to displacements, speeds, and accelerations along a path, the motion is effectively confined to one dimension, even though it is in fact three-dimensional. By definition there can be no speed normal to the path of motion, but there must be accelerations if the path is not straight.

5.2. Differentiation of vectors

The position of a particle in space can be defined relative to a point of reference by a vector. The velocity of the particle is likewise defined by a vector. As velocity is defined as the rate of change of position, a velocity vector must be obtained formally by differentiation with respect to time of a position vector. In the earlier chapters this formal step was avoided by reasoning in physical rather than mathematical terms. We will now consider direct differentiation of vectors.

5.2.1. The derivative of a vector

In time δt a particle moves from P to P′, Fig. 5.1, so that its position vector $\mathbf{a}$ becomes $\mathbf{a}+\delta\mathbf{a}$. The derivative of $\mathbf{a}$ with respect to time is defined as

$$\frac{\mathrm{d}\mathbf{a}}{\mathrm{d}t} \equiv \lim_{\delta t\to 0}\frac{\delta\mathbf{a}}{\delta t}. \tag{5.3}$$

The change $\delta\mathbf{a}$, and hence the rate of change, are both vectors. It is to be noted particularly that $\mathbf{a}$, being a vector, can change both its magnitude and its direction so that a being constant does not mean that $\mathrm{d}\mathbf{a}/\mathrm{d}t = 0$.

5.2.2. The derivative of a unit vector

A unit vector, being of fixed magnitude, can change only its direction. Such a change is effected by a rotation about the origin, Fig. 5.2. Hence

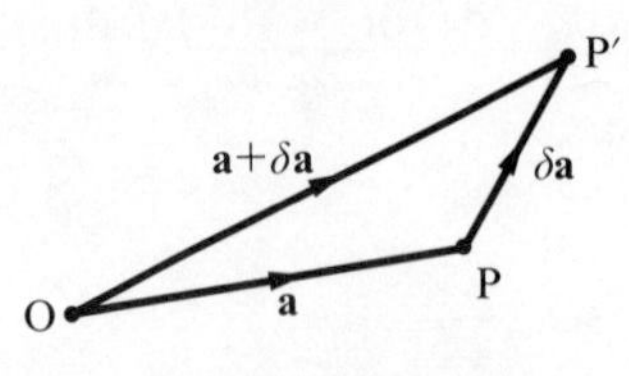

FIG. 5.1

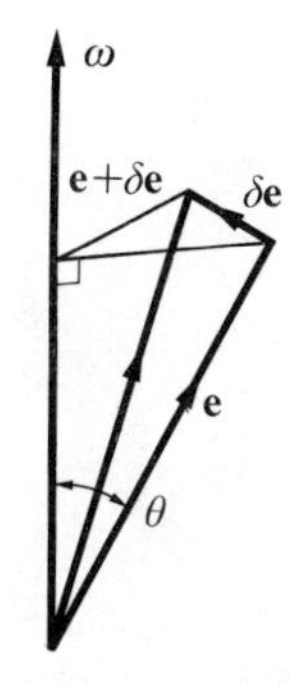

FIG. 5.2

the rate of change of a unit vector is related to its angular velocity, $\boldsymbol{\omega}$ say. From Fig. 5.2 $\delta\mathbf{e} = \boldsymbol{\omega} \times \mathbf{e}\,\delta t$. Hence

$$\frac{\mathrm{d}\mathbf{e}}{\mathrm{d}t} = \boldsymbol{\omega} \times \mathbf{e} \tag{5.4}$$

This is, of course, consistent with observing that the velocity of the tip of the vector is $\boldsymbol{\omega} \times \mathbf{e}$.

A very important special case is the rotation of a pair of orthogonal unit vectors about an axis perpendicular to their plane, Fig. 5.3.

$$\left.\begin{aligned} \frac{\mathrm{d}\mathbf{e}_1}{\mathrm{d}t} &= \boldsymbol{\omega} \times \mathbf{e}_1 = \omega\mathbf{e}_2 \\ \text{and} \qquad \frac{\mathrm{d}\mathbf{e}_2}{\mathrm{d}t} &= \boldsymbol{\omega} \times \mathbf{e}_2 = -\omega\mathbf{e}_1. \end{aligned}\right\} \tag{5.5}$$

The change of sign here is important. Its existence must be remembered, but in any particular case the association of the negative sign with one or other derivative is best determined by inspection.

5.2.3. The general expression for the derivative of a vector

It is self-evident that if λ is a scalar constant then

$$\frac{\mathrm{d}}{\mathrm{d}t}(\lambda\mathbf{a}) = \lambda\frac{\mathrm{d}\mathbf{a}}{\mathrm{d}t}.$$

A general vector $\mathbf{a} = a\mathbf{e}$ changes because of changes in both the

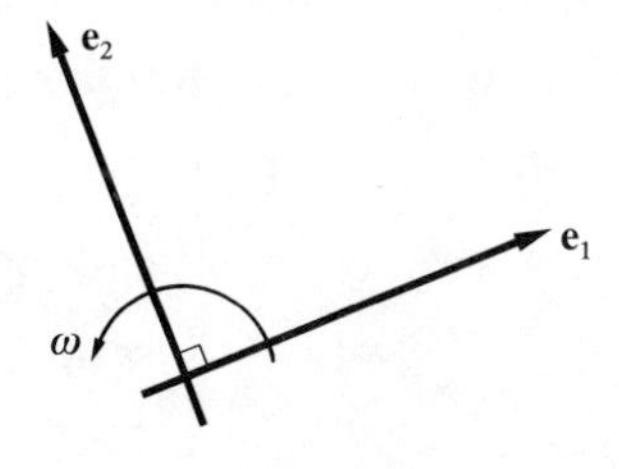

FIG. 5.3

magnitude and direction so that

$$\begin{aligned}
\frac{\mathrm{d}\mathbf{a}}{\mathrm{d}t} &= \lim_{\delta t \to 0} \frac{\delta \mathbf{a}}{\delta t} \\
&= \lim_{\delta t \to 0} \frac{(a+\delta a)(\mathbf{e}+\delta \mathbf{e}) - a\mathbf{e}}{\delta t} \\
&= \lim_{\delta t \to 0} \left(\frac{\delta a}{\delta t} \mathbf{e} + a \frac{\delta \mathbf{e}}{\delta t} \right) \\
&= \frac{\mathrm{d}a}{\mathrm{d}t} \mathbf{e} + a \frac{\mathrm{d}\mathbf{e}}{\mathrm{d}t} \\
&= \frac{\mathrm{d}a}{\mathrm{d}t} \mathbf{e} + a \boldsymbol{\omega} \times \mathbf{e} \\
&= \frac{\mathrm{d}a}{\mathrm{d}t} \mathbf{e} + \boldsymbol{\omega} \times \mathbf{a}. \qquad (5.6)
\end{aligned}$$

If **a** is a vector of constant length, as with a vector drawn between two points fixed in a rigid body, then $\mathrm{d}\mathbf{a}/\mathrm{d}t = \boldsymbol{\omega} \times \mathbf{a}$.

5.2.4. Derivative of the sum of two vectors

It is clear that if

$$\mathbf{c} = \mathbf{a} + \mathbf{b}$$

then

$$\mathbf{c} + \delta \mathbf{c} = (\mathbf{a} + \delta \mathbf{a}) + (\mathbf{b} + \delta \mathbf{b})$$

so that

$$\delta \mathbf{c} = \delta \mathbf{a} + \delta \mathbf{b}$$

and

$$\frac{\mathrm{d}\mathbf{c}}{\mathrm{d}t} = \frac{\mathrm{d}\mathbf{a}}{\mathrm{d}t} + \frac{\mathrm{d}\mathbf{b}}{\mathrm{d}t}. \qquad (5.7)$$

5.2.5. Derivative of a vector product

$$\begin{aligned}
\frac{\mathrm{d}}{\mathrm{d}t}(\mathbf{a} \times \mathbf{b}) &= \lim_{\delta t \to 0} \frac{\delta(\mathbf{a} \times \mathbf{b})}{\delta t} \\
&= \lim_{\delta t \to 0} \frac{(\mathbf{a} + \delta \mathbf{a}) \times (\mathbf{b} + \delta \mathbf{b}) - (\mathbf{a} \times \mathbf{b})}{\delta t}.
\end{aligned}$$

Now

$$\begin{aligned}
&(\mathbf{a} + \delta \mathbf{a}) \times (\mathbf{b} + \delta \mathbf{b}) - \mathbf{a} \times \mathbf{b} \\
&= \mathbf{a} \times \mathbf{b} + \mathbf{a} \times \delta \mathbf{b} + \delta \mathbf{a} \times \mathbf{b} + \delta \mathbf{a} \times \delta \mathbf{b} - \mathbf{a} \times \mathbf{b} \\
&= \mathbf{a} \times \delta \mathbf{b} + \delta \mathbf{a} \times \mathbf{b} + \delta \mathbf{a} \times \delta \mathbf{b}.
\end{aligned}$$

Ignoring $\delta \mathbf{a} \times \delta \mathbf{b}$ as being a second order small quantity it follows that

$$\frac{\mathrm{d}}{\mathrm{d}t}(\mathbf{a} \times \mathbf{b}) = \frac{\mathrm{d}\mathbf{a}}{\mathrm{d}t} \times \mathbf{b} + \mathbf{a} \times \frac{\mathrm{d}\mathbf{b}}{\mathrm{d}t}. \qquad (5.8)$$

Except that the order of the terms must be preserved, the differentiation of a vector product is directly analogous to the differentiation of a product of scalar functions.

5.2.6. Derivative of a scalar product

$$\frac{\mathrm{d}}{\mathrm{d}t}(\mathbf{a}\,.\,\mathbf{b}) = \lim_{\delta t\to 0}\frac{(\mathbf{a}+\delta\mathbf{a})\,.\,(\mathbf{b}+\delta\mathbf{b})-\mathbf{a}\,.\,\mathbf{b}}{\delta t}$$

$$= \mathbf{a}\,.\,\frac{\mathrm{d}\mathbf{b}}{\mathrm{d}t}+\mathbf{b}\,.\,\frac{\mathrm{d}\mathbf{a}}{\mathrm{d}t} \tag{5.9}$$

It can be concluded that the differentiation of products of vectors, including compound products, follows the same rules as for the differentiation of scalar functions except that where vector products are involved the order of terms must be preserved.

5.3. Applications of vector differentiation in planar motion

5.3.1. Motion in a circle

If a vector is expressed in terms of fixed unit vectors there is no difficulty in differentiating it, for if

$$\mathbf{r} = x\mathbf{i}+y\mathbf{j}+z\mathbf{k},$$

$$\frac{\mathrm{d}\mathbf{r}}{\mathrm{d}t} = \frac{\mathrm{d}x}{\mathrm{d}t}\mathbf{i}+\frac{\mathrm{d}y}{\mathrm{d}t}\mathbf{j}+\frac{\mathrm{d}z}{\mathrm{d}t}\mathbf{k}.$$

Because $\mathbf{i}$, $\mathbf{j}$, and $\mathbf{k}$ are fixed unit vectors $\mathrm{d}\mathbf{i}/\mathrm{d}t$, $\mathrm{d}\mathbf{j}/\mathrm{d}t$, and $\mathrm{d}\mathbf{k}/\mathrm{d}t$ are all zero.

Similarly,

$$\frac{\mathrm{d}^2\mathbf{r}}{\mathrm{d}t^2} = \frac{\mathrm{d}^2x}{\mathrm{d}t^2}\mathbf{i}+\frac{\mathrm{d}^2y}{\mathrm{d}t^2}\mathbf{j}+\frac{\mathrm{d}^2z}{\mathrm{d}t^2}\mathbf{k}.$$

If a particle moves with constant speed round a circle of radius a centred on the origin and in the plane $z=0$, then from Fig. 5.4

$$\mathbf{r} = a(\mathbf{i}\cos\omega t+\mathbf{j}\sin\omega t).$$

This assumes that the particle starts from the point $(a, 0)$ at time $t=0$. On differentiating once and then again we obtain

$$\frac{\mathrm{d}\mathbf{r}}{\mathrm{d}t} = a\omega(-\mathbf{i}\sin\omega t+\mathbf{j}\cos\omega t)$$

$$\frac{\mathrm{d}^2\mathbf{r}}{\mathrm{d}t^2} = -a\omega^2(\mathbf{i}\cos\omega t+\mathbf{j}\sin\omega t).$$

This is a clumsy way of describing the velocity and acceleration of the particle. It is easy enough to spot that $\mathrm{d}^2\mathbf{r}/\mathrm{d}t^2 = -\omega^2\mathbf{r}$, and that in itself

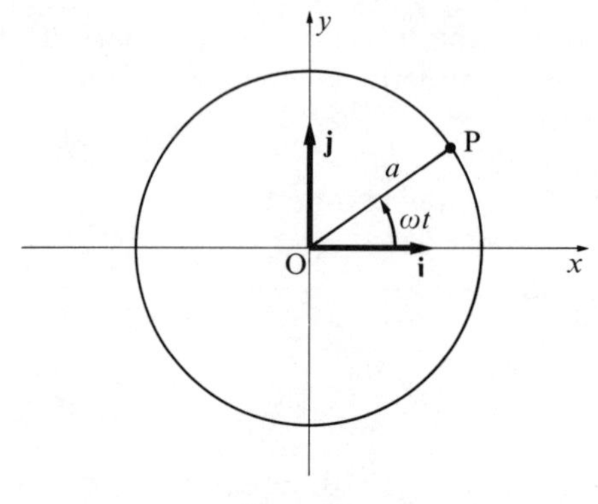

FIG. 5.4

suggests that there ought to be a more direct way of doing things, but it is not easy to see what the analogous expression for the velocity is.

A much tidier and more revealing analysis follows if $\mathbf{r}$ is expressed as $a\mathbf{e}_1$, Fig. 5.5

If

$$\mathbf{r} = a\mathbf{e}_1$$

then

$$\frac{\mathrm{d}\mathbf{r}}{\mathrm{d}t} = a\frac{\mathrm{d}\mathbf{e}_1}{\mathrm{d}t} = a\omega\mathbf{e}_2$$

and

$$\frac{\mathrm{d}^2\mathbf{r}}{\mathrm{d}t^2} = -a\omega^2\mathbf{e}_1.$$

If ω is not constant $\mathbf{e}_1$ and $\mathbf{e}_2$ still provide the most convenient basis for expressing the motion.

$$\frac{\mathrm{d}\mathbf{r}}{\mathrm{d}t} = a\omega\mathbf{e}_2 \quad \text{as for constant } \omega \text{, and}$$

$$\frac{\mathrm{d}^2\mathbf{r}}{\mathrm{d}t^2} = -a\left(\omega^2\mathbf{e}_1 + \frac{\mathrm{d}\omega}{\mathrm{d}t}\mathbf{e}_2\right).$$

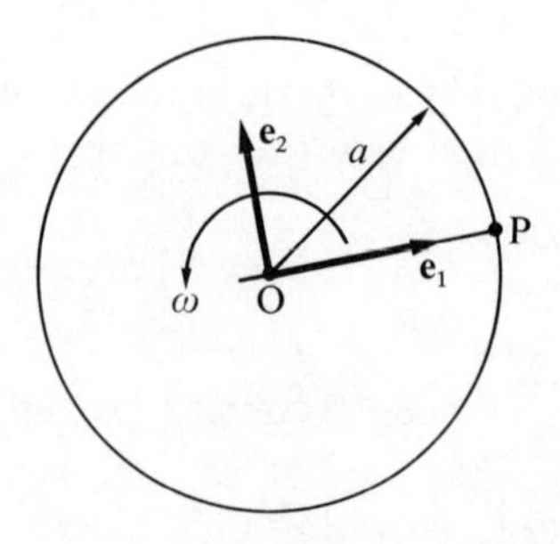

FIG. 5.5

An alternative derivation of the same results, though one that may be less acceptable on first reading is as follows: If $\mathbf{r}=\mathbf{a}$, then $\mathrm{d}\mathbf{r}/\mathrm{d}t=\boldsymbol{\omega}\times\mathbf{a}$, and $\mathrm{d}^2\mathbf{r}/\mathrm{d}t^2=(\mathrm{d}\boldsymbol{\omega}/\mathrm{d}t)\times\mathbf{a}+\boldsymbol{\omega}\times(\mathrm{d}\mathbf{a}/\mathrm{d}t)=(\mathrm{d}\boldsymbol{\omega}/\mathrm{d}t)\times\mathbf{a}+\boldsymbol{\omega}\times(\boldsymbol{\omega}\times\mathbf{a})$. For planar motion $\boldsymbol{\omega}\times(\boldsymbol{\omega}\times\mathbf{a})=-\omega^2\mathbf{a}$.

5.3.2. Planar motion expressed in polar co-ordinates

A particle P moves along a curved fixed path in a plane. At any instant its position is defined by the distance r from a fixed point O and the angle θ between OP and a fixed datum Ox. Unit vectors $\mathbf{e}_1$ and $\mathbf{e}_2$ are introduced as in Fig. 5.6(a). Then

$$\mathbf{r}=r\mathbf{e}_1$$

$$\frac{\mathrm{d}\mathbf{r}}{\mathrm{d}t}=\frac{\mathrm{d}r}{\mathrm{d}t}\mathbf{e}_1+r\frac{\mathrm{d}\mathbf{e}_1}{\mathrm{d}t}$$

$$=\frac{\mathrm{d}r}{\mathrm{d}t}\mathbf{e}_1+r\frac{\mathrm{d}\theta}{\mathrm{d}t}\mathbf{e}_2.$$

To abbreviate the writing and to clarify expressions we shall henceforth, where it is convenient to do so, write $\dot{r}$ for $(\mathrm{d}r/\mathrm{d}t)$, and $\dot{\theta}$ for $(\mathrm{d}\theta/\mathrm{d}t)$, etc. A dot over a variable denotes differentiation with respect to time. A second differentiation is denoted by a second dot, e.g. $\mathrm{d}^2r/\mathrm{d}t^2=\ddot{r}$, and so on, though clearly, after three differentiations this convention ceases to be advantageous. With this convention

$$\mathbf{e}_1=\dot{\theta}\mathbf{e}_2 \quad \text{and} \quad \mathbf{e}_2=-\dot{\theta}\mathbf{e}_1.$$

The expression for $\mathrm{d}\mathbf{r}/\mathrm{d}t$ becomes

$$\dot{\mathbf{r}}=\dot{r}\mathbf{e}_1+r\dot{\theta}\mathbf{e}_2. \tag{5.10}$$

On differentiating again

$$\ddot{\mathbf{r}}=(\ddot{r}\mathbf{e}_1+\dot{r}\dot{\mathbf{e}}_1)+(\dot{r}\dot{\theta}\mathbf{e}_2+r\ddot{\theta}\mathbf{e}_2+r\dot{\theta}\dot{\mathbf{e}}_2)$$
$$=(\ddot{r}-r\dot{\theta}^2)\mathbf{e}_1+(r\ddot{\theta}+2\dot{r}\dot{\theta})\mathbf{e}_2. \tag{5.11}$$

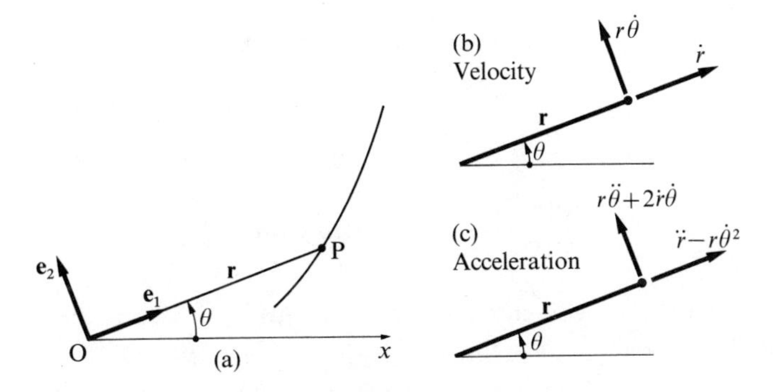

FIG. 5.6

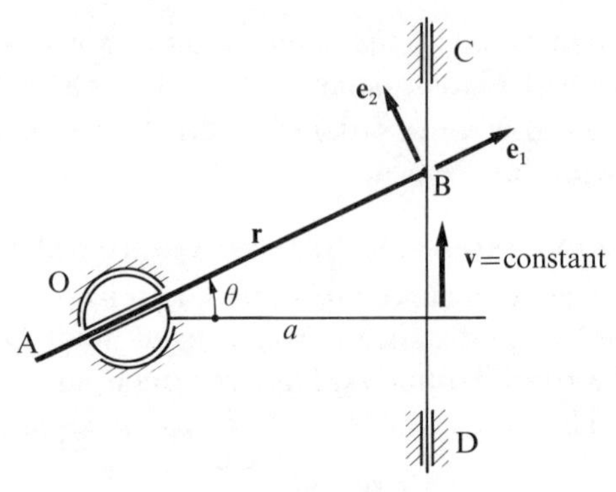

FIG. 5.7

The first term in the brackets on the right-hand side of this equation is referred to as the radial acceleration, and the second term as the transverse acceleration.

Example 5.1: In Fig. 5.7, CD is moving with constant velocity through fixed guides. Member AB is hinged to CD at B and slides through a swivel block at O. Determine the angular velocity and acceleration of AB.

Let OB $= r$ and let θ be the angle between r and the normal from O to CD. Let $\mathbf{e}_1$ and $\mathbf{e}_2$ be unit vectors as shown.

$$\frac{\mathrm{d}\mathbf{r}}{\mathrm{d}t}=\mathbf{v}=\dot{r}\mathbf{e}_1+r\dot{\theta}\mathbf{e}_2.$$

Taking the scalar product with $\mathbf{e}_2$ to eliminate the unknown $\dot{r}$ we find

$$\mathbf{v}\,.\,\mathbf{e}_2=\dot{r}\mathbf{e}_1\,.\,\mathbf{e}_2+r\dot{\theta}\mathbf{e}_2\,.\,\mathbf{e}_2$$

$$\Rightarrow \qquad \mathbf{v}\cos\theta=\mathbf{r}\dot{\theta}$$

$$\Rightarrow \qquad \dot{\theta}=\mathbf{v}\cos\theta/r=\mathbf{v}\cos^2\theta/a.$$

The scalar product of $\mathbf{v}$ with $\mathbf{e}_1$ yields r:

$$\mathbf{v}\,.\,\mathbf{e}_1=\mathbf{r}\mathbf{e}_1\,.\,\mathbf{e}_1+r\dot{\theta}\mathbf{e}_2\,.\,\mathbf{e}_1$$

$$\Rightarrow \qquad \dot{r}=v\sin\theta.$$

The acceleration of B is given by

$$\frac{\mathrm{d}^2\mathbf{r}}{\mathrm{d}t^2}=(\ddot{r}-r\dot{\theta}^2)\mathbf{e}_1+(r\ddot{\theta}+2\dot{r}\dot{\theta})\mathbf{e}_2.$$

But B has constant velocity and hence $\mathrm{d}^2\mathbf{r}/\mathrm{d}t^2=0$. As $\mathbf{e}_1$ and $\mathbf{e}_2$ are at right angles to each other the two terms on the right-hand side of the expression for $\mathrm{d}^2\mathbf{r}/\mathrm{d}t^2$ cannot cancel each other out. It follows that the coefficients of $\mathbf{e}_1$ and $\mathbf{e}_2$ must be zero individually.

Equating the coefficient of $\mathbf{e}_2$ to zero yields

$$\begin{aligned}\ddot{\theta} &= -2\dot{r}\dot{\theta}/r \\ &= -2v^2 \sin\theta \cos\theta/r^2 \\ &= -2v^2 \sin\theta \cos^3\theta/a^2.\end{aligned}$$

5.3.3. Components of the velocity and acceleration of a particle along and normal to its path

In a time interval δt a particle moves from P to P′, Fig. 5.8, a distance δs along its path. $\mathbf{e}_t$ and $\mathbf{e}_n$ are unit vectors directed tangential and normal to the path so that the small displacement is

$$\begin{aligned}\delta\mathbf{r} &= \delta s\mathbf{e}_t \\ &= \dot{s}\,\delta t\mathbf{e}_t.\end{aligned}$$

Hence the velocity of the particle is

$$\mathbf{v} = \frac{\mathrm{d}\mathbf{r}}{\mathrm{d}t} = \dot{s}\mathbf{e}_t. \tag{5.12}$$

Its acceleration is

$$\begin{aligned}\mathbf{a} = \frac{\mathrm{d}^2\mathbf{r}}{\mathrm{d}t^2} &= \ddot{s}\mathbf{e}_t + \dot{s}\frac{\mathrm{d}\mathbf{e}_t}{\mathrm{d}t} \\ &= \ddot{s}\mathbf{e}_t + \dot{s}\frac{\mathrm{d}\psi}{\mathrm{d}t}\mathbf{e}_n\end{aligned}$$

where ψ is the angle between $\mathbf{e}_t$ and a fixed datum. So

$$\mathbf{a} = \ddot{s}\mathbf{e}_t + \dot{s}\frac{\mathrm{d}s}{\mathrm{d}t}\cdot\frac{\mathrm{d}\psi}{\mathrm{d}s}\mathbf{e}_n.$$

But $\mathrm{d}\psi/\mathrm{d}s = 1/R$, where R is the radius of curvature CP, so

$$\mathbf{a} = \ddot{s}\mathbf{e}_t + \frac{v^2}{R}\mathbf{e}_n. \tag{5.13}$$

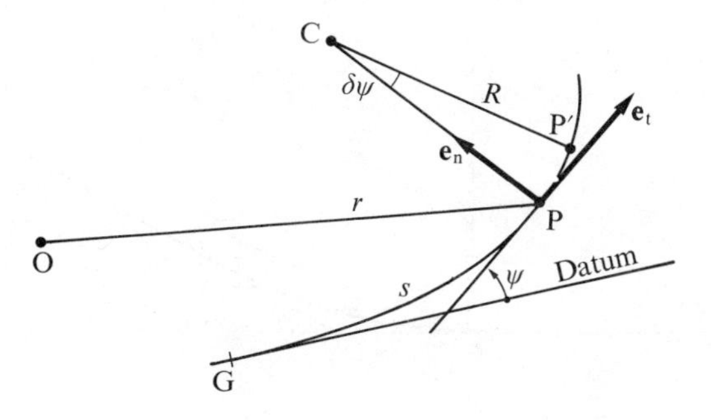

FIG. 5.8

It may be noted that the co-ordinates of P, its distance s round the curve from a datum G and the angle ψ between $\mathbf{e}_t$ and a given direction, which might conveniently be the direction of the tangent at G, do not feature in the expressions for the velocity and acceleration.

Example 5.2: Two straight lengths of road are to be joined by a bend. In order to minimize discomfort to drivers and passengers in vehicles the transverse acceleration is to be kept as small as possible, and so is the rate of change of transverse acceleration. The length of curve between the two straight sections of road is S and the total change of direction ϕ is small. Determine the curve.

A circular arc would give the least value for v^2/R but is ruled out by the requirement that the rate of change of transverse acceleration is to be kept as small as possible. A sudden change from a straight run to a circular path would entail an instantaneous change in the transverse acceleration from zero to v^2/R, with an infinite rate of change. This is troublesome. The rate of change will be minimized if it has a constant absolute value with a change of sign midway round the curve. Let us assume a set of co-ordinate axes at the end of one of the straights, as in Fig. 5.9. The smallness of the change of direction ϕ permits two approximations to be made: first, the distance from the origin measured round the curve to any point is approximately equal to the x co-ordinate of the point; second, the curvature $1/R$ of the curve is given approximately by $\mathrm{d}^2y/\mathrm{d}x^2$. We would expect the curve to be symmetrical about its centre point C. After all, traffic may be assumed to flow in both directions. So let us assume that the curvature increases at a constant rate until $x = S/2$. The differential equation to the curve is then

$$\frac{\mathrm{d}^2y}{\mathrm{d}x^2} = bx$$

$$\Rightarrow \qquad \frac{\mathrm{d}y}{\mathrm{d}x} = \tfrac{1}{2}bx^2 + \mathrm{A}.$$

But $\mathrm{d}y/\mathrm{d}y = 0$ when $x = 0$, so that $\mathrm{A} = 0$. Also $\mathrm{d}y/\mathrm{d}x = \phi/2$ when $x = S/2$,

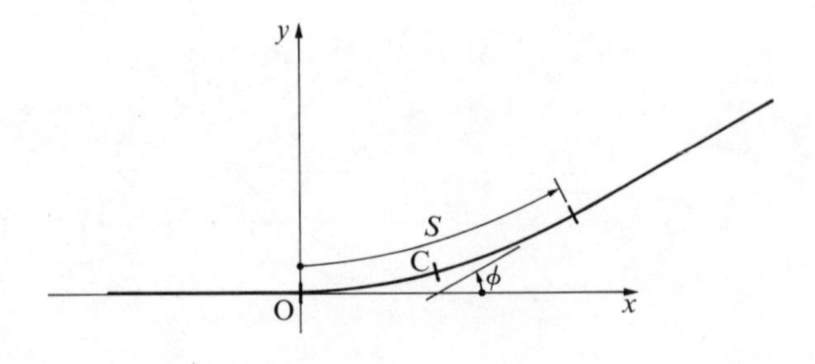

FIG. 5.9

hence $b = 4\phi/S^2$

$$\Rightarrow \qquad \frac{dy}{dx} = 4\phi x^2/S^2$$

$$\Rightarrow \qquad y = \tfrac{4}{3}\phi x^3/S^2 + D.$$

But $y = 0$ when $x = 0$, so that $D = 0$ and

$$y = \tfrac{4}{3}\phi x^3/S^2.$$

This is the equation to the curve between O and C. The other half of the curve is a mirror image in the normal at C. The minimum radius of curvature is $2\phi/S$ and is obtained by substituting for b and x in the expression for d^2y/dx^2. For a circle, $1/R = \phi/S$. Hence the maximum transverse acceleration at a given speed for the curve we have derived is double that which would be obtained with a circular path. In practice there can be a compromise, trading a decrease in the maximum transverse acceleration for an increase in the rate of change.

5.3.4. The determination of path curvature

The derivation of the expression for the acceleration of a particle just given implies that the radius of curvature of the path is known. It can happen that knowledge of the acceleration of a particle is a convenient way of finding the radius of curvature of its path.

Assuming that the velocity and acceleration of the particle are known, we can extract $\ddot{s}$ from

$$\mathbf{a} = \ddot{s}\mathbf{e}_t + \frac{v^2}{R}\mathbf{e}_n$$

by taking the scalar product with $\mathbf{v} = v\mathbf{e}_t$. This gives

$$\mathbf{a} \cdot \mathbf{v} = \ddot{s}v$$

$$\Rightarrow \qquad \ddot{s} = \mathbf{a} \cdot \mathbf{v}/v. \tag{5.14}$$

To find R we can eliminate $\ddot{s}$ from the equation for $\mathbf{a}$ by taking the vector product of both sides with $\mathbf{v}$:

$$\mathbf{a} \times \mathbf{v} = \frac{v^3}{R}\mathbf{e}_n \times \mathbf{e}_t.$$

$\mathbf{e}_n \times \mathbf{e}_t$ is a unit vector so that

$$\frac{1}{R} = \frac{|\mathbf{a} \times \mathbf{v}|}{v^3}. \tag{5.15}$$

Example 5.3: A Cardan drag-link mechanism, Fig. 5.10, consists of a fixed gear wheel (sun) with n teeth meshing with a planetary gear which has n/2 teeth through an idler, which may be of any convenient size. The gears are kept in mesh by being mounted in a triangular frame OAC which

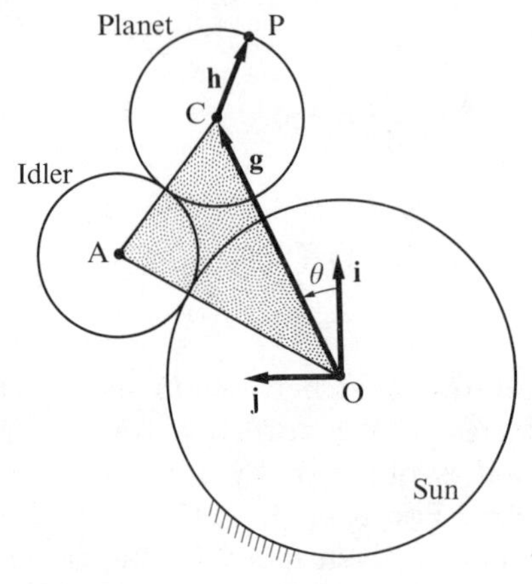

FIG. 5.10

rotates about the centre of the fixed gear. What is the path followed by a point P on the pitch circle of the planetary gear? What is the radius of curvature of the path?

Let the datum configuration be when P is collinear with the centres of the fixed and planetary gears. Let **g** be the vector drawn from the centre O of the fixed gear to the centre C of the planetary gear, and let **h** be the vector drawn from C to P. The location of P relative to the centre of the fixed gear is

$$\mathbf{r}=\mathbf{g}+\mathbf{h}.$$

Before this equation can be differentiated the angular velocities of **g** and **h** must be known. $\boldsymbol{\omega}_\mathbf{g}$ is the angular velocity with which the frame is driven. To determine $\boldsymbol{\omega}_\mathbf{h}$ imagine for a moment that the frame is fixed in position and that the fixed gear is free to rotate. Allow the 'fixed' gear to rotate θ clockwise. The resulting rotation of the planet gear is 2θ clockwise. Now let the gear wheels all be locked to the frame and rotate everything through an angle θ anticlockwise about O. The net result is no ratation of the fixed gear, a rotation θ anticlockwise for **g** and a rotation θ clockwise for **h**. Hence, $\boldsymbol{\omega}_\mathbf{g}=-\boldsymbol{\omega}_\mathbf{h}=\boldsymbol{\omega}$, say. Now

$$\begin{aligned}\mathbf{v}=\mathrm{d}\mathbf{r}/\mathrm{d}t&=\boldsymbol{\omega}\times\mathbf{g}-\boldsymbol{\omega}\times\mathbf{h}\\&=\boldsymbol{\omega}\times(\mathbf{g}-\mathbf{h})\\\mathbf{a}&=\boldsymbol{\omega}\times(\boldsymbol{\omega}\times\mathbf{g})+(-\boldsymbol{\omega})\times\{(-\boldsymbol{\omega})\times\mathbf{h}\}\\&=-\omega^2(\mathbf{g}+\mathbf{h}).\end{aligned}$$

With a view to calculating $1/R$ we now evaluate

$$\begin{aligned}\mathbf{a}\times\mathbf{v}&=-\omega^2(\mathbf{g}+\mathbf{h})\times\{\boldsymbol{\omega}\times(\mathbf{g}-\mathbf{h})\}\\&=-\omega^2[\boldsymbol{\omega}\{(\mathbf{g}+\mathbf{h})\cdot(\mathbf{g}-\mathbf{h})\}-(\mathbf{g}-\mathbf{h})\{\boldsymbol{\omega}\cdot(\mathbf{g}+\mathbf{h})\}]\\&=-\omega^2\boldsymbol{\omega}(g^2-h^2),\end{aligned}$$

so that

$$|\mathbf{a}\times\mathbf{v}| = \omega^3(g^2-h^2).$$

Also

$$\begin{aligned} v^3 &= (\mathbf{v}\cdot\mathbf{v})^{\frac{3}{2}} \\ &= [\{\boldsymbol{\omega}\times(\mathbf{g}-\mathbf{h})\}\cdot\{\boldsymbol{\omega}\times(\mathbf{g}-\mathbf{h})\}]^{\frac{3}{2}} \\ &= \{\omega^2(\mathbf{g}-\mathbf{h})\cdot(\mathbf{g}-\mathbf{h})\}^{\frac{3}{2}} \\ &= \omega^3(g^2-2gh\cos 2\theta+h^2)^{\frac{3}{2}}. \end{aligned}$$

Hence

$$\frac{1}{R} = \frac{g^2-h^2}{(g^2-2gh\cos 2\theta+h^2)^{3/2}}.$$

To determine the path of P unit vectors **i** and **j** are introduced, **i** being along the line from which θ is measured. Then

$$\begin{aligned} \mathbf{r} &= \mathbf{g}+\mathbf{h} \\ &= g(\mathbf{i}\cos\theta+\mathbf{j}\sin\theta)+h(\mathbf{i}\cos\theta-\mathbf{j}\sin\theta) \\ &= (g+h)\cos\theta\,\mathbf{i}+(g-h)\sin\theta\,\mathbf{j} \\ &= x\mathbf{i}+y\mathbf{j}. \end{aligned}$$

Thus $x=(g+h)\cos\theta$ and $y=(g-h)\sin\theta$. On eliminating θ the equation of the path of P is found to be

$$\frac{x^2}{(g+h)^2}+\frac{y^2}{(g-h)^2}=1,$$

which is the equation to an ellipse.

There is, of course, nothing special about the particular point P that we have considered: all points on the circumference of the planet gear describe similar ellipses about a common centre. In every case the major axis coincides with the line along which **g** and **h** are collinear.

An application which has been suggested for this mechanism is to provide a dwell action. In Fig. 5.11 CP is made equal to the radius of curvature of the ellipse traced by the Cardan drag-link mechanism for the position where $\theta=90°$, when P is at P_0. As P moves round the ellipse the

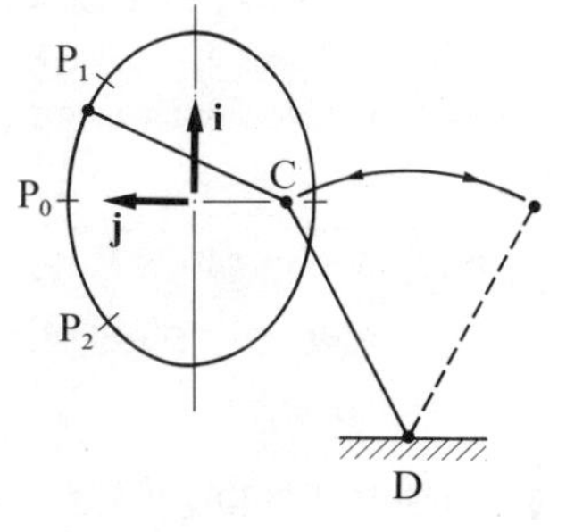

FIG. 5.11

output lever CD oscillates. However, as P moves between P_1 and P_2, the movement of CD is very small so that for a finite period of time it effectively 'dwells' in the extreme position. Such a dwelling action is frequently required in process machinery where something is required to be kept stationary whilst it is filled or has some other operation carried out on it.

5.4. Spatial motion of a rigid body

Planar motion is such an important special case of general motion that it merits the special attention that we have given it. The results for planar motion can be generalized to make them apply to spatial motion, a point that will now be pursued. In doing so it will be seen how vector differentiation provides an alternative derivation of eqn (3.2) presented in Chapter 3.

If two points A and B of a rigid body have position vectors $\mathbf{r}_A$ and $\mathbf{r}_B$, and $\boldsymbol{\rho}$ is the vector from A to B, then

$$\mathbf{r}_B = \mathbf{r}_A + \boldsymbol{\rho}.$$

Differentiating with respect to time we have

$$\frac{d\mathbf{r}_B}{dt} = \frac{d\mathbf{r}_A}{dt} + \frac{d\boldsymbol{\rho}}{dt}$$

and from eqn (5.6), noting that $d\boldsymbol{\rho}/dt = 0$,

$$\frac{dr_B}{dt} = \frac{dr_A}{dt} + \boldsymbol{\omega} \times \boldsymbol{\rho}$$

or

$$\mathbf{v}_B = \mathbf{v}_A + \boldsymbol{\omega} \times \boldsymbol{\rho} \tag{5.16}$$

where $\boldsymbol{\omega}$ is the angular velocity of the body.

The acceleration of B is obtained by a further differentiation with respect to time

$$\begin{aligned} \mathbf{a}_B &= \mathbf{a}_A + \frac{d\boldsymbol{\omega}}{dt} \times \boldsymbol{\rho} + \boldsymbol{\omega} \times \frac{d\boldsymbol{\rho}}{dt} \\ &= \mathbf{a}_A + \frac{d\boldsymbol{\omega}}{dt} \times \boldsymbol{\rho} + \boldsymbol{\omega} \times (\boldsymbol{\omega} \times \boldsymbol{\rho}). \end{aligned} \tag{5.17}$$

If A is a fixed point and AB rotates with angular velocity in a plane, so that $\boldsymbol{\omega} \cdot \boldsymbol{\rho} = 0$,

$$\begin{aligned} \mathbf{a}_B &= \boldsymbol{\omega} \times (\boldsymbol{\omega} \times \boldsymbol{\rho}) \\ &= \boldsymbol{\omega}(\boldsymbol{\omega} \cdot \boldsymbol{\rho}) - \boldsymbol{\rho}(\boldsymbol{\omega} \cdot \boldsymbol{\omega}) \\ &= -\boldsymbol{\rho}\omega^2, \end{aligned}$$

which is the well-known standard result for the centripetal acceleration of a point moving round a circle.

5.5. Moving frames of reference

The examples in this chapter have used two techniques. One simply expresses vectors in terms of components in fixed directions so that differentiation involves merely the rates of change of the scalar multipliers of the fixed unit vectors. In the other, unit vectors are visualized as moving against a fixed background, and their rates of change are as would be seen by a fixed observer. Instead of choosing an arbitrary set of fixed unit vectors the observer takes as his reference directions the directions in the fixed plane with which the moving set of vectors are momentarily coincident. In this way the expression for the acceleration of a particle, eqn (5.13),

$$\mathbf{a} = \ddot{s}\mathbf{e}_t + \frac{v^2}{R}\mathbf{e}_n$$

is the acceleration relative to a fixed background, and not relative to a moving plane which is fixed to the moving $\mathbf{e}_n$ and $\mathbf{e}_t$ vectors.

The problem of determining the absolute motion of a particle, or a body, from the known motion relative to a second moving body was encountered earlier, in the velocity analysis of mechanisms. A particular example to which reference may be made is the quick-return mechanism on page 24 where a block slides on a swinging link. The same sort of considerations apply in computing the take-off trajectory of a rocket relative to the earth to achieve a particular path in space.

In Fig. 5.12 O is a fixed origin, A is a reference point fixed to a moving body, and P is a particle which has a velocity $\partial\boldsymbol{\rho}/\partial t$ relative to the moving body. The position of P relative to the fixed origin is given by

$$\mathbf{r} = \mathbf{r}_A + \boldsymbol{\rho}.$$

If the particle were stationary relative to the body its velocity would be

$$\frac{d\mathbf{r}}{dt} = \frac{d\mathbf{r}_A}{dt} + \boldsymbol{\omega} \times \boldsymbol{\rho},$$

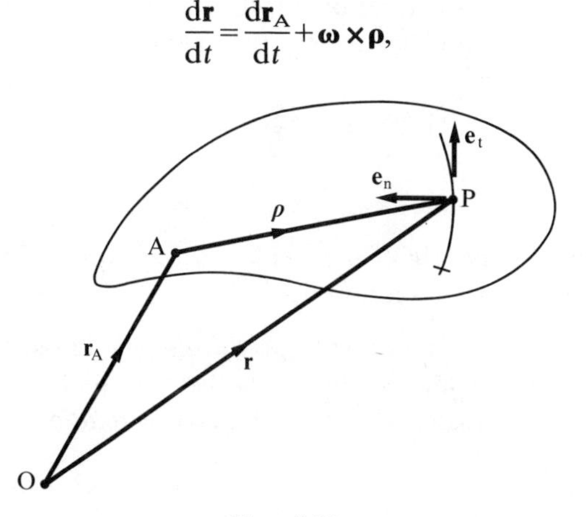

FIG. 5.12

the only possible change in $\boldsymbol{\rho}$ being a change in direction. To allow for the motion of the particle relative to the body we must add the relative velocity $\partial\boldsymbol{\rho}/\partial t$ to give

$$\frac{\mathrm{d}\mathbf{r}}{\mathrm{d}t}=\frac{\mathrm{d}\mathbf{r}_\mathrm{A}}{\mathrm{d}t}+\boldsymbol{\omega}\times\boldsymbol{\rho}+\partial\boldsymbol{\rho}/\partial t \tag{5.18}$$

or

$$\mathbf{v}_\mathrm{P}=\mathbf{v}_\mathrm{A}+\boldsymbol{\omega}\times\boldsymbol{\rho}+\mathbf{v}_\mathrm{rel}. \tag{5.18a}$$

Eqn (5.18a) is identical to eqn (1.4). Whether its present derivation is more acceptable than that given in Chapter 1 is a matter of taste. The virtue of the present analysis lies in its easy extension to determine the acceleration of P and, if necessary, higher order derivatives. In differentiating eqn (5.18a) to find the absolute acceleration of P account must be taken of the change in $\boldsymbol{\rho}$ caused by the rotation of the moving body, thus

$$\begin{aligned}\frac{\mathrm{d}\boldsymbol{\rho}}{\mathrm{d}t}&=\boldsymbol{\omega}\times\boldsymbol{\rho}+\frac{\partial\boldsymbol{\rho}}{\partial t}\\&=\boldsymbol{\omega}\times\boldsymbol{\rho}+\mathbf{v}_\mathrm{rel}.\end{aligned}$$

Differentiation of eqn (5.18a) gives

$$\begin{aligned}\mathbf{a}_\mathrm{P}&=\mathbf{a}_\mathrm{A}+\frac{\mathrm{d}\boldsymbol{\omega}}{\mathrm{d}t}\times\boldsymbol{\rho}+\boldsymbol{\omega}\times\frac{\mathrm{d}\boldsymbol{\rho}}{\mathrm{d}t}+\frac{\mathrm{d}}{\mathrm{d}t}\left(\frac{\partial\boldsymbol{\rho}}{\partial t}\right)\\&=\mathbf{a}_\mathrm{A}+\dot{\boldsymbol{\omega}}\times\boldsymbol{\rho}+\boldsymbol{\omega}\times\left(\boldsymbol{\omega}\times\boldsymbol{\rho}+\frac{\partial\boldsymbol{\rho}}{\mathrm{d}t}\right)+\left(\frac{\partial^2\boldsymbol{\rho}}{\partial t^2}+\boldsymbol{\omega}\times\frac{\partial\boldsymbol{\rho}}{\partial t}\right)\\&=\mathbf{a}_\mathrm{A}+\dot{\boldsymbol{\omega}}\times\boldsymbol{\rho}+\boldsymbol{\omega}\times(\boldsymbol{\omega}\times\boldsymbol{\rho})+\frac{\partial^2\boldsymbol{\rho}}{\partial t^2}+2\boldsymbol{\omega}\times\frac{\partial\boldsymbol{\rho}}{\partial t}.\end{aligned} \tag{5.19}$$

Eqns (5.18a) and (5.19) become less cumbersome when it is recognized that $(\mathbf{v}_\mathrm{A}+\boldsymbol{\omega}\times\boldsymbol{\rho})$ is the velocity of point P in the body, and that $(\mathbf{a}_\mathrm{A}+\dot{\boldsymbol{\omega}}\times\boldsymbol{\rho}+\boldsymbol{\omega}\times(\boldsymbol{\omega}\times\boldsymbol{\rho}))$ is the acceleration of the body at P, for these are the velocity and acceleration that the particle would have if it were fixed to the body. So

$$\mathbf{v}_\mathrm{P}=\mathbf{v}_\mathrm{body}+\mathbf{v}_\mathrm{rel} \tag{5.20}$$

and

$$\mathbf{a}_\mathrm{P}=\mathbf{a}_\mathrm{body}+\mathbf{a}_\mathrm{rel}+2\boldsymbol{\omega}\times\mathbf{v}_\mathrm{rel}. \tag{5.21}$$

The third term on the right-hand side of the eqn (5.21) is called the Coriolis component of acceleration.

Example 5.4: A block P slides with speed $\dot{r}$ along a straight rod that rotates in a plane about one end. Determine the velocity and acceleration of the block when it is at a distance r from the axis of rotation, Fig. 5.13.

From eqn (5.20)

$$\mathbf{v}_\mathrm{P}=r\omega\mathbf{e}_2+\dot{r}\mathbf{e}_1$$

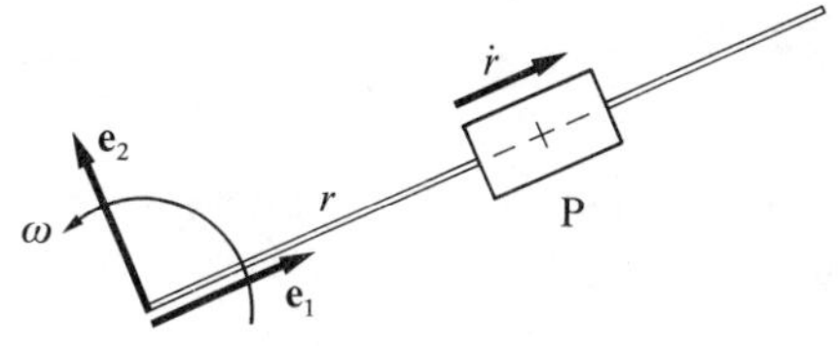

FIG. 5.13

and from eqn (5.21)

$$\mathbf{a}_P = \dot{r}\omega\mathbf{e}_2 + r\dot{\omega}\mathbf{e}_2 - r\omega^2\mathbf{e}_1 + \ddot{r}\mathbf{e}_1 + \omega\dot{r}\mathbf{e}_2$$
$$= (\ddot{r} - r\omega^2)\mathbf{e}_1 + (r\dot{\omega} + 2\omega\dot{r})\mathbf{e}_2.$$

These results are, of course, identical to the expressions for the velocity and acceleration of a particle in plane polar co-ordinates, eqns (5.10) and (5.11).

Example 5.5: A vehicle is travelling due north at latitude $\lambda = 55°$N at a speed of 80 km h^{-1}. Treating the world as a sphere rotating about a fixed axis through its centre determine the components of the absolute acceleration of the vehicle.

The radius of the world is $R = 6{\cdot}4 \times 10^6$ m.
Its angular speed is $\omega = 2\pi \div (24 \times 3600) = 72{\cdot}8 \times 10^{-6}$ rad s^{-1}.
The speed of the vehicle is $v = 22{\cdot}2$ m s^{-1}.
The body (i.e. world) acceleration at the vehicle is

$$(R \cos \lambda)\omega^2 \approx 0{\cdot}02 \text{ m s}^{-2},$$

the direction being towards the earth's axis.

The acceleration of the vehicle relative to the ground due to the curvature of the earth's surface is $v^2/R \approx 0{\cdot}08 \times 10^{-3}$ m s^{-2} towards the centre of the earth.

The Coriolis component of acceleration is

$$2\omega v \sin \lambda \approx 2 \times 10^{-3} \text{ m s}^{-2}$$

in a direction tangential to the earth's surface and perpendicular to the velocity of the vehicle relative to the earth, in accordance with the rule for the direction of a vector product.

The smallness of the v^2/R component is perhaps not surprising. The magnitude of the Coriolis component is also so small that it seems to be negligible. In this case, it is. It must not be assumed that it will always be so for its effect is extremely important on the flow of ocean currents, and of air-streams over the surface of the earth. In the more obvious of engineering applications, as for example in the motion of machinery, the Coriolis component of acceleration can be of over-riding importance.

5.6. Acceleration analysis of planar mechanisms

When it is applied to planar motion, eqn (5.17) for the acceleration of a general point B in a lamina given the acceleration of another point A reduces to

$$\mathbf{a}_B = \mathbf{a}_A + \dot{\boldsymbol{\omega}} \times \boldsymbol{\rho} - \omega^2 \boldsymbol{\rho} \tag{5.21}$$

and eqn (5.19) for the acceleration of a point P that is moving relative to the body at B becomes

$$\mathbf{a}_P = \mathbf{a}_A + \dot{\boldsymbol{\omega}} \times \boldsymbol{\rho} - \omega^2 \boldsymbol{\rho} + \frac{\partial^2 \boldsymbol{\rho}}{\partial t^2} + 2\boldsymbol{\omega} \times \frac{\partial \boldsymbol{\rho}}{\partial t}. \tag{5.22}$$

These equations are the basis for acceleration analysis of planar mechanism in a way that is analogous to the application of eqns (1.2) and (1.4) in velocity analysis. When written in terms that are consistent with those used in eqns (5.21) and (5.22) the velocity equations are

$$\mathbf{v}_B = \mathbf{v}_A + \boldsymbol{\omega} \times \boldsymbol{\rho} \tag{5.23}$$

and

$$\mathbf{v}_P = \mathbf{v}_A + \boldsymbol{\omega} \times \boldsymbol{\rho} + \partial \boldsymbol{\rho}/\partial t. \tag{5.24}$$

The traditional procedure is to draw vector polygons to represent eqns (5.21) and (5.22) as appropriate. The polygons representing individual links are combined together into an acceleration diagram for the mechanism. Apart from the special case of impending motion it is inevitably necessary to precede acceleration analysis by velocity analysis, so that the drawing of acceleration diagrams is nearly always preceded by the drawing of a velocity diagram.

The method is best demonstrated by examples.

Example 5.6: The slider-crank chain. In the crank/coupler/piston mechanism shown in Fig. 5.14 the crank AB turns with constant angular velocity $\boldsymbol{\omega}_{AB}$. Determine the velocity and acceleration of the piston C for the given configuration.

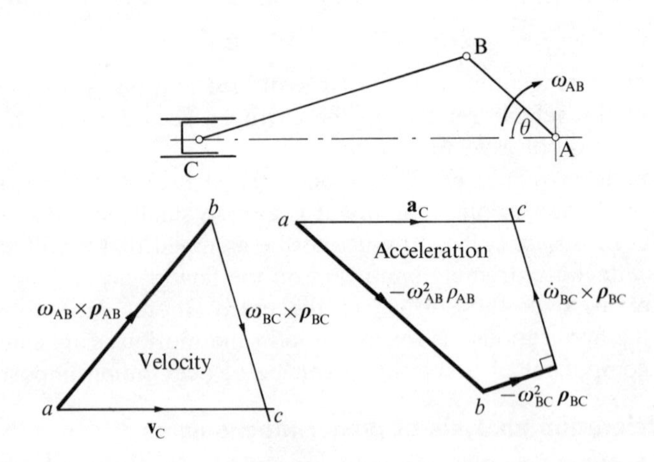

FIG. 5.14

The construction of the velocity diagram calls for no special comment. $\mathbf{v}_B = \boldsymbol{\omega}_{AB} \times \boldsymbol{\rho}_{AB}$ is known both in magnitude and direction. For the coupler

$$\mathbf{v}_C = \mathbf{v}_B + \boldsymbol{\omega}_{BC} \times \boldsymbol{\rho}_{BC}.$$

The unknown quantities in this equation are the magnitudes of $\mathbf{v}_C$ and $\boldsymbol{\omega}_{BC} \times \boldsymbol{\rho}_{BC}$. However, knowing their directions, *ac* and *bc* can be drawn to complete the velocity diagram. The magnitude of $\mathbf{v}_C$ is then obtained by measuring *ac* on a velocity diagram drawn to scale. Although it is not specifically called for, $\boldsymbol{\omega}_{BC}$ is needed in the construction of the acceleration diagram and must be determined by measuring *bc* and dividing by the length BC.

The acceleration of B is known and is represented to scale in the acceleration diagram by *ab*. Note that $\mathbf{a}_B = -\omega_{AB}^2 \boldsymbol{\rho}_{AB}$ so *ab* is drawn parallel to but in the opposite direction from AB. For the coupler BC

$$\mathbf{a}_C = \mathbf{a}_B - \boldsymbol{\omega}_{BC}^2 \boldsymbol{\rho}_{BC} + \boldsymbol{\omega}_{BC} \times \boldsymbol{\rho}_{BC}.$$

The unknown quantities in this equation are the magnitudes of $\mathbf{a}_C$ and $\dot{\boldsymbol{\omega}}_{BC} \times \boldsymbol{\rho}_{BC}$. We can make a start by drawing the known vector $-\omega_{BC}^2 \boldsymbol{\rho}_{BC}$ from point *b*. A line is then drawn at the end of, and perpendicular to this vector. This line is in the direction of $\dot{\boldsymbol{\omega}}_{BC} \times \boldsymbol{\rho}_{BC}$.

A second line of indefinite length is now drawn through *a* in the direction of a_C. The two lines just drawn intersect at *c*, and so a_C is determined.

In Fig. 5.14, and in the others to follow, vectors that are known *ab initio* are drawn bold whilst those that are known initially only in direction are drawn light. Another convention which is adopted is that labels *a*, *b*, *c*, etc., are always associated with corresponding points in the diagram of the mechanism. No label is written at the end of $-\boldsymbol{\omega}_{BC}^2 \boldsymbol{\rho}_{BC}$ remote from *b* because that point is not associated with a particular point in the actual mechanism and so requires no label.

Example 5.7: Four-bar mechanism.

The construction of the acceleration diagram for the four-bar mechanism in Fig. 5.15 follows that for the slider-crank mechanism through $-\omega_{AB}^2 \boldsymbol{\rho}_{AB}$, $-\omega_{BC}^2 \boldsymbol{\rho}_{BC}$ and a line of indefinite length in the direction of $\dot{\boldsymbol{\omega}}_{BC} \times \boldsymbol{\rho}_{BC}$. To locate *c* we must add the graphical representation of

$$\mathbf{a}_C = \mathbf{a}_D - \omega_{DC}^2 \boldsymbol{\rho}_{DC} + \dot{\boldsymbol{\omega}}_{DC} \times \boldsymbol{\rho}_{DC}.$$

$\mathbf{a}_D$ is, of course, zero so that *a* and *d* coincide in the acceleration diagram. Hence $-\omega_{DC}^2 \boldsymbol{\rho}_{DC}$ is drawn and then followed by a line in the direction of $\boldsymbol{\omega}_{DC} \times \boldsymbol{\rho}_{DC}$ (i.e. perpendicular to DC). The polygon closes at the intersection *c*. By measuring the length of the vector $\dot{\boldsymbol{\omega}}_{DC} \times \boldsymbol{\rho}_{DC}$ and dividing by DC we determine the angular acceleration $\boldsymbol{\omega}_{DC}$ of the rocker.

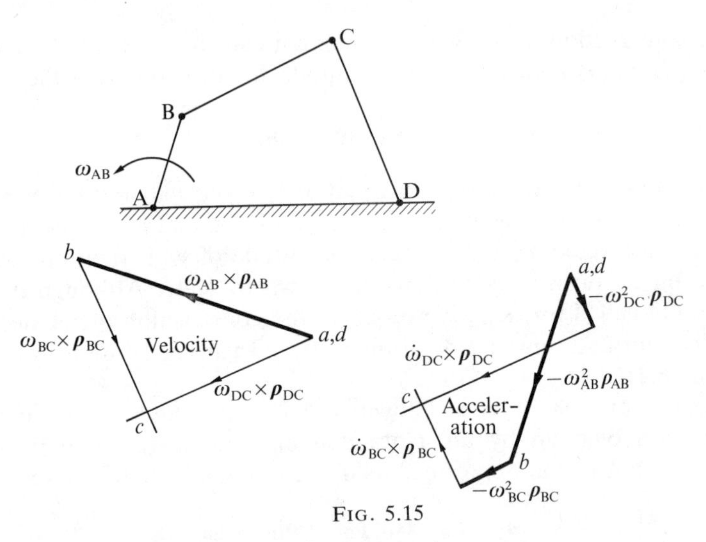

FIG. 5.15

Example 5.8: Quick-return mechanism.

The determination of the angular acceleration of link CD and the acceleration of D for a given motion of the crank AB of the quick-return mechanism in Fig. 5.16 is rather more difficult than was finding the angular acceleration of the rocker in the crank and rocker mechanism in Fig. 5.15. The sliding of block B on the swinging link CD requires the use of eqn (5.22) whereas so far eqn (5.21) has been sufficient.

The velocity diagram is identical to that in Fig. 1.26 with the numerical subscript being used to distinguish between block B and the coincident point on CD. The main problem in applying eqn (5.22) is in deciding

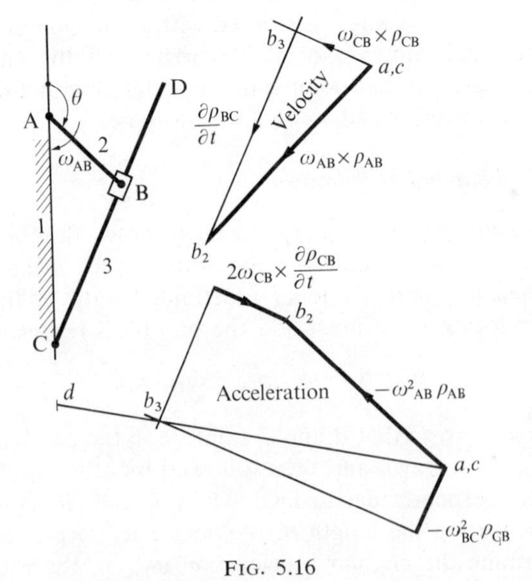

FIG. 5.16

what is sliding on what. Are we to think of the block sliding on the rod, or the rod sliding through the block? The answer is, of course, that it does not matter provided that we properly follow through the logical consequences of our choice. Because the author finds it easier, we will regard the block as sliding on the rod so that B is identified with P in eqn (5.22). This choice is already implied by the direction of the arrow head associated with $\partial\boldsymbol{\rho}_{CB}/\partial t$ in Fig. 1.26 and repeated in Fig. 5.16. This is the direction of the velocity of B_2, the block, relative to the link CD. In the acceleration diagram, as in the velocity diagram, a and c are coincident. The first vector to be drawn is

$$\mathbf{a}_B = -\omega_{AB}^2\boldsymbol{\rho}_{AB}.$$

Eqn (5.22) becomes

$$\mathbf{a}_B = \mathbf{a}_C + \dot{\boldsymbol{\omega}}_{CB} \times \boldsymbol{\rho}_{CB} - \omega_{CB}^2\boldsymbol{\rho}_{CB} + \frac{\partial^2\boldsymbol{\rho}_{CB}}{\partial t^2} + 2\boldsymbol{\omega}_{CB} \times \frac{\partial\boldsymbol{\rho}_{CB}}{\partial t}$$

where $\mathbf{a}_C = 0$. In spite of its length there are only two unknown scalar quantities in this equation: $\mathbf{a}_B$ has already been determined by the preceding equation; $\boldsymbol{\rho}_{CB}$ is known; $\boldsymbol{\omega}_{CB}$ and $\partial\boldsymbol{\rho}_{CB}/\partial t$ are both determined from the velocity diagram. The directions of $\boldsymbol{\omega}_{CB} \times \boldsymbol{\rho}_{CB}$ and $\partial^2\boldsymbol{\rho}_{CB}/\partial t^2$ are both known, leaving only their magnitudes as the two unknown scalar quantities.

In principle, the vectors on the right-hand side of eqn (5.22) can be added in any order, but it proves to be advantageous to have a set procedure. This is directed towards ensuring that b_3 corresponding to B_3 on the link CD ends up as an identifiable point in the completed acceleration diagram. The temptation to be resisted is to lump $-\omega_{CB}^2\boldsymbol{\rho}_{CB}$ and $2\boldsymbol{\omega}_{CB} \times \partial\boldsymbol{\rho}_{CB}/\partial t$ together on the grounds that they are both fully known vectors.

Let us consider the terms one by one: $\mathbf{a}_B$, represented by ab_2, is already drawn. $-\omega_{CB}^2\boldsymbol{\rho}_{CB}$ is fully determined and can be drawn. It can be followed by a line of indefinite length drawn in the direction of $\boldsymbol{\omega}_{CB} \times \boldsymbol{\rho}_{CB}$. The sum of these two vectors leads to b_3 which represents the acceleration of B on link CD. All that is known at the moment is that b_3 lies somewhere on the line just drawn. As we are unable to continue with the two final terms on the right-hand side of the equation we will transfer them to the other side of the equation.

$\partial\boldsymbol{\rho}_{CB}/\partial t$ is obtained in magnitude and direction from the velocity diagram, and so is $\boldsymbol{\omega}_{CB}$ $(=\boldsymbol{\omega}_{CD})$. We see that $\boldsymbol{\omega}_{CB}$ is clockwise so that the direction of the Coriolis term $2\boldsymbol{\omega}_{CB} \times \partial\boldsymbol{\rho}_{CB}/\partial t$ is obtained by rotating 90° clockwise from the positive direction of $\partial\boldsymbol{\rho}_{CB}/\partial t$. Taking account of the change of sign due to its transposition to the left-hand side of the equation, the vector is drawn as shown from b_2. The polygon is now closed by a line in the direction of $\partial^2\boldsymbol{\rho}_{CB}/\partial t^2$, that is parallel to CD. b_3 is located at the point of closure.

The length of $\boldsymbol{\omega}_{CB} \times \boldsymbol{\rho}_{CB}$ can now be measured and ω_{CB} is determined.

$$\mathbf{a}_D = -\omega_{CD}^2 \boldsymbol{\rho}_{CD} + \dot{\boldsymbol{\omega}}_{CD} \times \boldsymbol{\rho}_{CD}$$
$$= (-\omega_{CD}^2 \boldsymbol{\rho}_{CB} + \dot{\boldsymbol{\omega}}_{CD} \times \boldsymbol{\rho}_{CB}) \times CD/CB.$$

It follows that d is located in the acceleration diagram by joining cb_3 and extending it in the ratio CD/CB.

The crux of this particular problem lies in deducing the correct direction for the Coriolis term. The reader is advised to follow the steps in detail and to verify that his or her diagram would look the same had the choice been made to regard the block as the rotating plane on which the link is sliding.

5.6.1. The acceleration image

The final step in the last example whereby the acceleration of a third point on a rigid link is determined when the accelerations of two other points are known is an example of a particular application of the image theorem. A similar theorem has already been met in relation to velocities.

In Fig. 5.17 a lamina has angular velocity $\boldsymbol{\omega}$ and angular acceleration $\dot{\boldsymbol{\omega}}$. Let us assume that $\mathbf{a}_A$ and $\mathbf{a}_B$ are both known, or have been found. In the acceleration diagram a and b are connected by the vectors $-\omega^2 \boldsymbol{\rho}_{AB}$ and $\boldsymbol{\omega} \times \boldsymbol{\rho}_{AB}$. The angle γ between ab and $-\boldsymbol{\rho}_{AB}$ is $\tan^{-1} \dot{\omega}/\omega^2$, and for given ω and $\dot{\omega}$ would be the same whatever the direction of AB.

It follows that there will be the same angle γ between bc in the acceleration diagram and $-\boldsymbol{\rho}_{BC}$, and between ca and $-\boldsymbol{\rho}_{CA}$. Hence, triangle abc in the acceleration diagram, the acceleration image of ABC on the lamina, is similar to triangle ABC.

The same can be said for any geometric figure drawn on the lamina, for such a figure can be divided into a set of triangles. It is important to note that if the order ABC is clockwise on the lamina, it is also clockwise in the acceleration diagram and vice versa. If C happens to be collinear with A and B then the location of C is determined by dividing db in the same ratio that C divides AB.

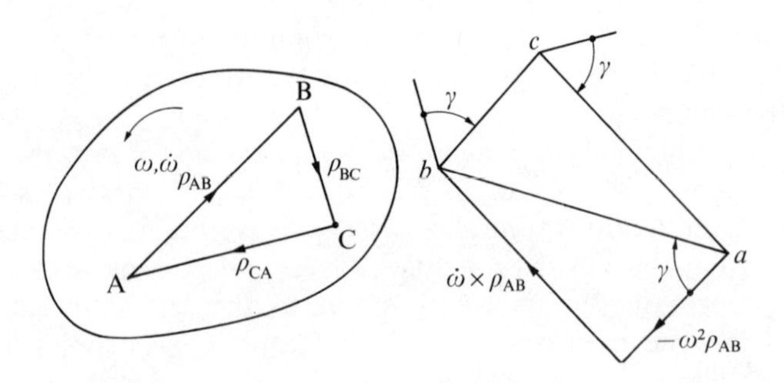

FIG. 5.17

5.6.2. Acceleration analysis of cam mechanisms

An understanding of how accelerations are calculated for cam mechanisms is important because, unless steps are taken to avoid the situation, it can easily happen that the acceleration pattern will make such mechanisms unworkable at high speeds. The same is true, of course, for other mechanisms but the problem is particularly acute in cam design.

Example 5.9: Fig. 5.18 shows a cam mechanism with a reciprocating roller follower. The cam profile consists of arcs of circles as indicated. Determine the change in acceleration as the point of contact between the cam and follower moves from the flank to the nose when the cam rotates with constant angular velocity.

The figure shows the point of contact between the roller and the cam to be on the flank of the cam. As the cam turns, so x increases. At first sight the calculation of the velocity and acceleration of the follower appears to be rather complicated, and to require consideration of the way in which the roller moves in contact with the cam. It becomes apparent that this is not necessary on realizing that the distance between the centre B of the cam profile and the centre C of the roller is constant during the period that the point of contact P is on the flank of the cam. This means that, as long as P is between Q and S, the motion of the follower is precisely the same as it would be if CB were connected by a rigid link. For this period,

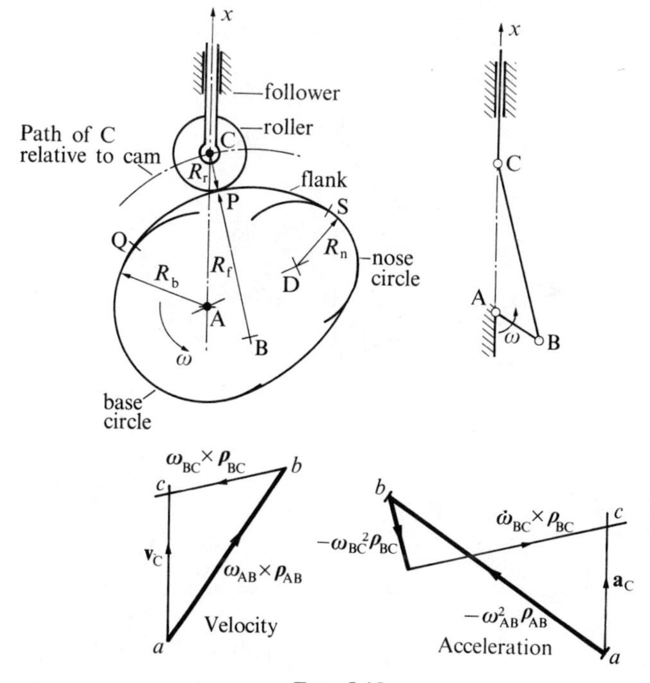

FIG. 5.18

the slider-crank mechanism drawn to the right of the cam is said to be the equivalent mechanism for the cam and follower. The velocity and acceleration diagrams for the equivalent mechanisms are constructed in the way already described, and are as shown.

Now consider what happens as P passes from one side to the other of S. Up until this instant the equivalent mechanism is ABC. After it the equivalent mechanism will be ADC, where D is the centre of the nose circle of the cam. Both mechanisms give the same velocity for C, as shown in Fig. 5.19. By chance AB has turned out to be perpendicular to AC with the effect that b and c coincide in the velocity diagram. The coincidence of c for the two mechanisms ABC and ADC is, however, not a matter of chance and follows from the geometry of the mechanism and velocity diagrams, as can be seen if the construction is thought through.

The situation is different with the acceleration diagrams. The diagrams for the two equivalent mechanisms are superimposed in the figure and it can be seen that there must be a sudden change in the acceleration of the

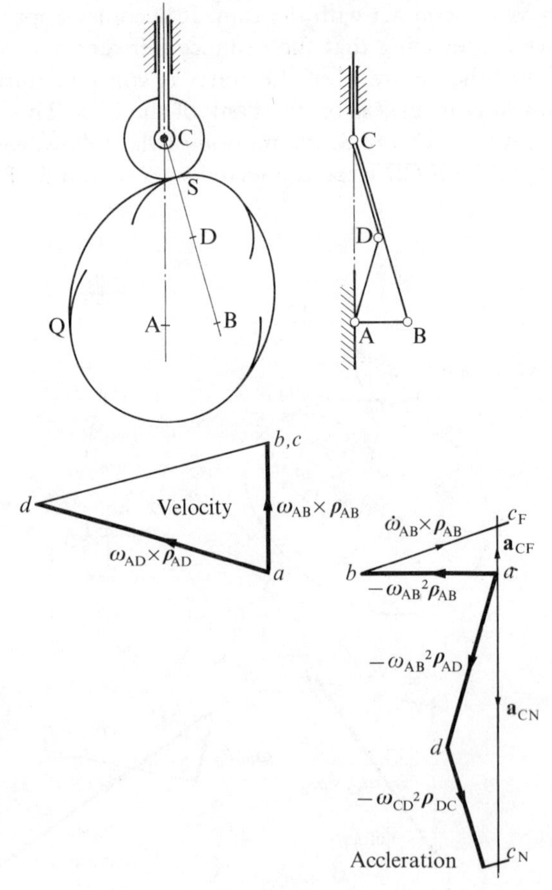

FIG. 5.19

cam follower as contact passes S and from the flank to the nose. This sudden change in acceleration implies a sudden change in the forces in the mechanism. This may not matter very much at low speeds, but at high speeds the effect is to induce vibrations and the system ceases to operate satisfactorily. If the cam is required to operate at high speed it is necessary to join the flank and nose circles by a transition curve in order to avoid the instantaneous changes in acceleration.

5.6.3. Alternative methods of velocity and acceleration analysis of linkage mechanisms

It will have been noted that for velocity and acceleration analysis of linkage mechanisms we have relied so far on actually drawing velocity and acceleration diagrams to scale, except perhaps for very simple situations where it is easy to apply trigonometry to sketches of the vector diagrams. The reason for this almost complete reliance on graphical methods is that until recently there has been no satisfactory alternative. It is rarely possible to derive a simple algebraic expression for the velocity and acceleration output of a mechanism. The quick-return mechanism in Fig. 5.16 is an exception. In this case it can be shown that

$$\omega_{CD} = \frac{1 + p\cos\theta}{(p + \cos\theta)^2}\,\omega_{AB}$$

and

$$\dot{\omega}_{CD} = \frac{-p\sin\theta(p^2 - 1)}{(1 + 2p\cos\theta + p^2)^2}\,\omega_{AB}^2$$

where $p = AC/AB$. But in the case of the apparently simpler example of the slider-crank mechanism in Fig. 5.14 the expressions for the velocity and acceleration of the piston are compound infinite series in terms of the rotation θ of the crank and $n = BC/AB$. For most mechanisms it is not even worth considering an algebraic solution.

With the four-bar mechanism cumbersome calculations are needed simply to determine the configuration. In order to calculate the crank angle at D, Fig. 5.15, when the angle at A is given, it is necessary to solve the triangles ABD and BCD in turn. With only a slide-rule available as a computational aid it is very much quicker to draw the mechanism to scale. It is then natural to follow with graphical evaluation of velocities and accelerations.

The development of electronic pocket calculators has made velocity and acceleration analysis by purely numerical calculation more attractive. Certainly the configuration of a four-bar mechanism can be calculated in a period of time comparable with that required to draw the mechanism and with much less inconvenience. The succeeding calculation of velocity and acceleration is then best based on the vector equations. We start with

$$\mathbf{v}_C = \mathbf{v}_B + \boldsymbol{\omega}_{BC} \times \boldsymbol{\rho}_{BC}$$

$$\Rightarrow \qquad \boldsymbol{\omega}_{CD} \times \boldsymbol{\rho}_{CD} = \boldsymbol{\omega}_{AB} \times \boldsymbol{\rho}_{AB} + \boldsymbol{\omega}_{BC} \times \boldsymbol{\rho}_{BC}.$$

If we wish to determine the angular velocity of the rocker we can take the scalar product with $\boldsymbol{\rho}_{BC}$ to eliminate the unknown $\boldsymbol{\omega}_{BC}$:

$$(\boldsymbol{\omega}_{CD}\times\boldsymbol{\rho}_{DC})\cdot\boldsymbol{\rho}_{BC}=(\boldsymbol{\omega}_{AB}\times\boldsymbol{\rho}_{AB})\cdot\boldsymbol{\rho}_{BC}+0$$

$$\Rightarrow\quad(\boldsymbol{\rho}_{DC}\times\boldsymbol{\rho}_{BC})\cdot\boldsymbol{\omega}_{CD}=(\boldsymbol{\rho}_{AB}\times\boldsymbol{\rho}_{BC})\cdot\boldsymbol{\omega}_{AB}$$

$$\Rightarrow\quad\omega_{CD}=\frac{(\boldsymbol{\rho}_{AB}\times\boldsymbol{\rho}_{BC})\cdot\mathbf{k}}{(\boldsymbol{\rho}_{DC}\times\boldsymbol{\rho}_{BC})\cdot\mathbf{k}}\omega_{AB}$$

where $\mathbf{k}$ is a unit vector perpendicular to the plane of the mechanism. If we want to calculate $\dot{\boldsymbol{\omega}}_{CD}$ we also have to evaluate

$$\omega_{BC}=-\frac{(\boldsymbol{\rho}_{AB}\times\boldsymbol{\rho}_{DC})\cdot\mathbf{k}}{(\boldsymbol{\rho}_{BC}\times\boldsymbol{\rho}_{DC})\cdot\mathbf{k}}\omega_{AB}.$$

The unit vector $\mathbf{k}$ has been introduced to make the final expressions for ω_{CD} and ω_{BC} mathematically correct, but in fact $\mathbf{k}$ affects the numerical calculations not at all, for $(\boldsymbol{\rho}_{AB}\times\boldsymbol{\rho}_{BC})\cdot\mathbf{k}=\text{AB}\,.\,\text{BC}\sin\theta_B$ where θ_B is the angle between the crank and the coupler.

The acceleration equation for the mechanism, assuming $\dot{\omega}_{AB}=0$, is

$$-\omega_{AB}^2\boldsymbol{\rho}_{AB}-\omega_{BC}^2\boldsymbol{\rho}_{BC}+\dot{\boldsymbol{\omega}}_{BC}\times\boldsymbol{\rho}_{BC}=-\omega_{DC}^2\boldsymbol{\rho}_{DC}+\dot{\boldsymbol{\omega}}_{DC}\times\boldsymbol{\rho}_{DC}.$$

The unknown $\dot{\boldsymbol{\omega}}_{BC}$ is eliminated by taking the scalar product of all the terms in the equation with $\boldsymbol{\rho}_{BC}$ to give

$$-\omega_{AB}^2\boldsymbol{\rho}_{AB}\cdot\boldsymbol{\rho}_{BC}-\omega_{BC}^2\rho_{BC}^2=-\omega_{DC}^2\boldsymbol{\rho}_{DC}\cdot\boldsymbol{\rho}_{BC}+(\boldsymbol{\rho}_{DC}\times\boldsymbol{\rho}_{BC})\cdot\dot{\boldsymbol{\omega}}_{DC}$$

$$\Rightarrow\quad\dot{\omega}_{DC}=(\omega_{DC}^2\boldsymbol{\rho}_{DC}\cdot\boldsymbol{\rho}_{BC}-\omega_{AB}^2\boldsymbol{\rho}_{AB}\cdot\boldsymbol{\rho}_{BC}-\omega_{BC}^2\rho_{BC}^2)\div(\boldsymbol{\rho}_{DC}\times\boldsymbol{\rho}_{BC})\cdot\mathbf{k}$$

In spite of its cumbersome appearance the numerical evaluation of this equation is not difficult on a calculator. If a programable calculator is available it is a simple matter to evaluate the scalar and vector products by subroutines. A program for the complete calculation outlined above for determining the configuration, velocities, and accelerations of the mechanism need not be very long, and can be used to calculate results for a whole cycle of operations.

5.7. Problems

1. A circular disc of radius r rolls at constant speed without slip along Ox. A point P on the circumference is initially at O. Express the velocity of P in terms of θ and

(a) The fixed unit vectors $\mathbf{i}$ and $\mathbf{j}$.

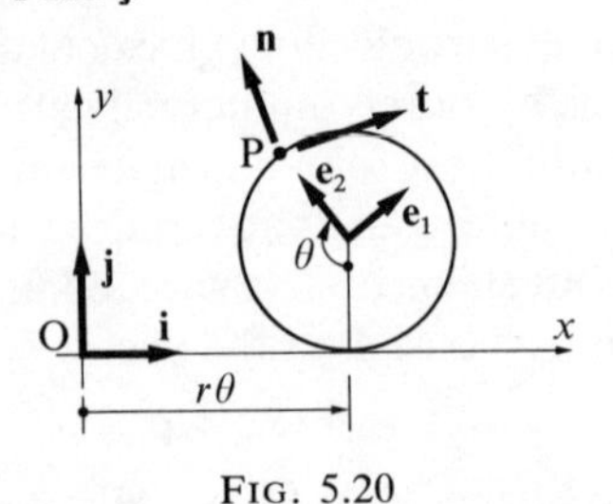

FIG. 5.20

(b) The rotating unit vectors $\mathbf{e}_1$ and $\mathbf{e}_2$.
(c) The moving unit vectors **n** and **t** which are normal and parallel to the path of P at P.
Derive the acceleration of P for each case by differentiation.

2. A unit vector **e** rotates in a plane with angular velocity $\boldsymbol{\omega}$. Show that $\mathbf{e} = \boldsymbol{\omega} \times \mathbf{e}$.
The position vector **r** of a point P in plane motion rotates with angular velocity $\boldsymbol{\omega}$. The radial and transverse unit vectors, $\mathbf{e}_1$ and $\mathbf{e}_2$, are positive in the directions shown in Fig. 5.21. Prove that

$$\ddot{\mathbf{r}} = (\ddot{r} - r\omega^2)\mathbf{e}_1 + (r\dot{\omega} + 2\dot{r}\omega)\mathbf{e}_2.$$

A radar station, tracking a sea-going hovercraft, uses a rotating aerial which is kept pointing at the target. As the hovercraft passes due South of the aerial, its range is 2 km and is increasing at a *constant* rate of 11 m s^{-1}. The aerial is then turning at $\frac{1}{2}^\circ$ s^{-1} towards the East. Find the speed and course of the hovercraft. What is the angular acceleration of the aerial if the hovercraft is making a turn at constant speed?

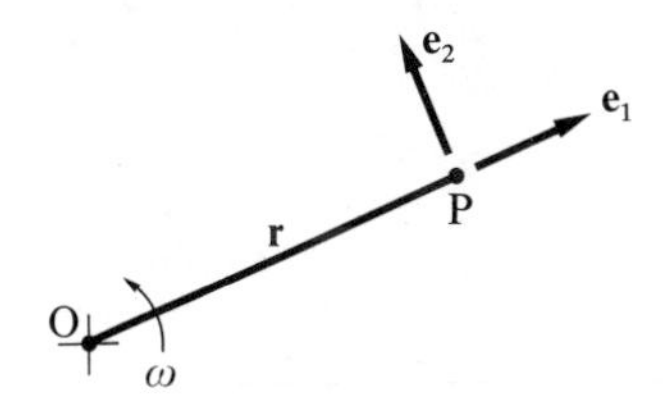

FIG. 5.21

3. In an elliptic trammel the ends A and B of a straight rigid bar move at right angles to each other in slides that intersect at O. The velocity $\mathbf{v}_A$ is constant. Determine the acceleration of B when OAB $= \theta$. What is then the acceleration of the centre point C of AB?

4. A point P moves along a straight line, whose perpendicular distance OQ from a fixed point O is 100 mm, such that the angular velocity of OP is constant and equal to $\pi/36$ rad s^{-1}.
Express the velocity and acceleration of P in terms of (a) the time after passing Q, and (b) its displacement x from Q. Find values for the velocity and acceleration (i) when $t = 6$ s, and (ii) when $x = 200$ mm.

5. A thin, inextensible cord is attached to the rim of a fixed circular disc of centre Q and radius a, and is wrapped round it. A point P of the cord is initially in contact with the disc at a point O. The cord is unwrapped from the disc and during the process the free part PZ remains straight and is tangential to the disc at the point of contact Z. PZ rotates with constant angular velocity ω.
(a) Taking axes Oxy in the plane of the disc, Ox coinciding with OQ produced, find the Cartesian co-ordinates of P as a function of time.
Find P's velocity and acceleration as functions of time, and show that the results are consistent with Z being the centre of curvature of the path of P.
(b) Taking unit vectors $\mathbf{e}_1$ and $\mathbf{e}_2$ along and perpendicular to PZ, differentiate the position vector $\overrightarrow{QZ}$ to obtain the velocity and acceleration of P, and again confirm that Z is the centre of curvature of the path of P.
(c) Show that the velocity and acceleration obtained in part (a) are consistent with those obtained in answer to part (b).

6. Confirm from the results of questions 1 and 3 that although the instantaneous centre of a body has zero velocity it does not necessarily have zero acceleration.
Given that the acceleration of the instantaneous centre of a lamina is $\mathbf{a}_I$, what is the acceleration of a point P distance r from I such that IP is at an angle ϕ to $\mathbf{a}_I$? Hence show

that the locus of points for which the path curvature is instantaneously zero (i.e. $\rho = \infty$) is a circle of diameter $\mathbf{a}_I/\omega^2$.

Note. This is known as the inflection circle. The inflection circle is important in advanced kinematics and provides a valuable tool in the generation of approximate straight-line motion.

7. At the particular instant depicted, crank AB of the four-bar mechanism in Fig. 5.22 is rotating with uniform angular velocity 5 rad s^{-1}. Find the velocity and acceleration of the mid-point of BC.

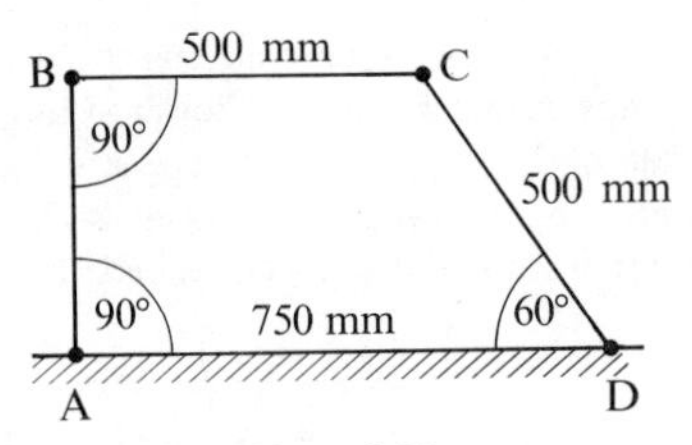

FIG. 5.22

8. In the mechanism shown in Fig. 5.23 arm OA rotates anticlockwise at 300 rev min^{-1}. At the instant shown, with OA 30° below the line OE, find the angular acceleration of the triangular member ABC and the acceleration of the slider E.

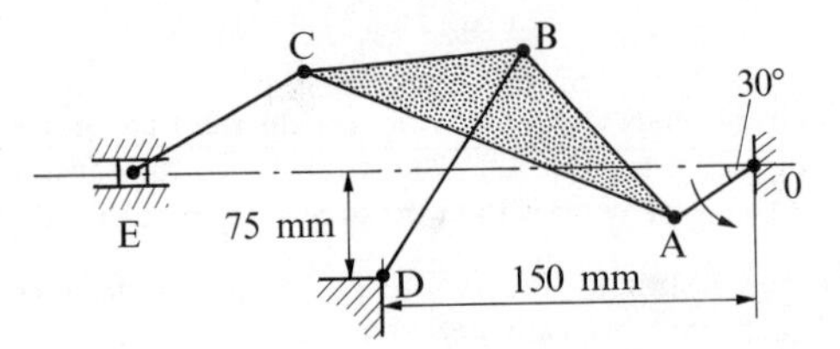

OA=25 mm, AB=BC=100 mm, AC=175 min, BD=150 mm, CE=75 mm

FIG. 5.23

9. In the mechanism shown in Fig. 5.24, A and B are fixed points, AC $= a$, BD $= 3a$, AB $= a\sqrt{3}$. If AC is rotating with constant angular velocity ω, find the acceleration of D in magnitude and direction when $\angle$CAB $= 30°$.

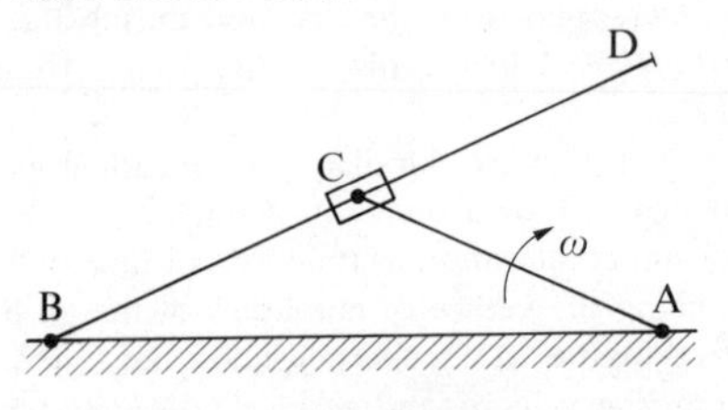

FIG. 5.24

10. Determine the angular acceleration of BC and the acceleration of C for the mechanism of question 14, Chapter 1 (p. 29) in its given configuration.

11. Determine the acceleration of E in question 15, Chapter 1 (p. 30).

12. Sketch an equivalent mechanism for the cam mechanism in question 12, Chapter 1 (p. 29), and determine the acceleration of the follower for the given position.

13. The cam shown in Fig. 5.25 rotates about O with constant angular velocity ω and drives a roller follower whose radius is $r/4$. Determine in terms of r and ω, the maximum acceleration of the follower whilst it is in contact with the straight portions of the cam profile.

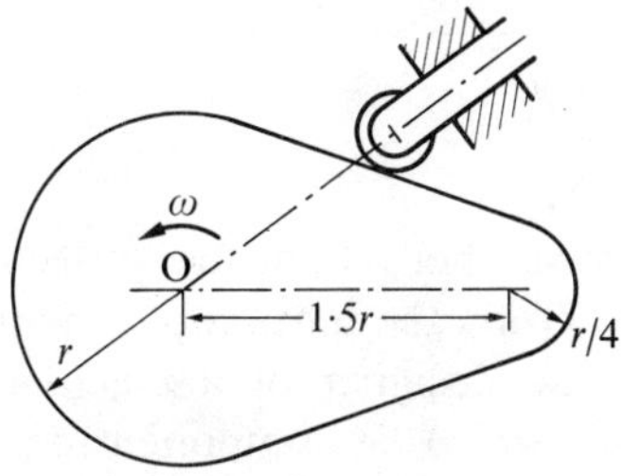

FIG. 5.25

14. What is the acceleration of the follower in question 13 when it is in the position of maximum lift?

6. Newton's laws for particle motion

It is appropriate to recall here that the objective of dynamics is to develop the methods whereby the motion of a body can be related to the forces that act on it. The designer of a machine needs to be able to calculate the forces to which a component is subjected when it is constrained to move in a particular way. Likewise it is necessary to be able to predict the path taken by a spacecraft under the action of the forces that act upon it. The study of dynamics, the interrelationship between force and motion, starts with a statement of Newton's three laws of motion.

6.1. The First Law

Newton's First Law is axiomatic. It states that if the forces on a particle are in equilibrium the velocity of the particle is constant: if $\mathbf{P}=0$, $\mathbf{v}$ is constant.

This law defines what is meant by equilibrium. It will be noted that the velocity may be zero.

6.2. The Second Law

Newton's Second Law relates the resultant force acting on a particle to the rate of change of velocity. It introduces the concept of momentum, which is defined as the product of the mass m of a particle and it velocity $\mathbf{v}$.

The Second Law states that the rate of change of momentum of a particle is proportional to the resultant force $\mathbf{P}$ that acts on it. That is

$$\mathbf{P}=k\frac{\mathrm{d}}{\mathrm{d}t}(m\mathbf{v}),$$

where k is a constant of proportionality. We will always contrive to keep m constant so that the equation becomes

$$\mathbf{P}=km\mathbf{a}$$

where $\mathbf{a}=\mathrm{d}\mathbf{v}/\mathrm{d}t$ is the acceleration of the particle.

The value of k depends on the choice of units in which $\mathbf{P}$, m, and $\mathbf{a}$ are measured. It is a matter of convenience to choose these units in such a manner that $k=1$. This is done by introducing a unit of force, the newton, and defining 1 newton (N) as the force that accelerates a mass of one kilogram (kg) at a rate of 1 metre per second per second ($\mathrm{m\,s^{-2}}$).

The metre and the second are themselves defined by reference to the wavelength and frequency of a specific naturally occurring radiation. In principle, anyone, anywhere, could independently devise experiments to determine the standard metre and the standard second.

The unit of mass is defined by reference to an international prototype that is in the custody of the International Bureau for Weights and Measures. All masses are measured in principle by comparison with this prototype. The means of comparison is provided by Newton's Third Law.

6.3. The Third Law

The Third Law states that action and reaction are equal and opposite. This means that if one particle B exerts a force $\mathbf{P}_{AB}$ on another particle A (action) there must consequently be a force $\mathbf{P}_{BA}$ on a particle B (reaction) such that

$$\mathbf{P}_{AB} = -\mathbf{P}_{BA}.$$

In the absence of any other forces the particles accelerate so that from the second law

$$m_A \mathbf{a}_A = -m_B \mathbf{a}_B$$

so that

$$\frac{m_B}{m_A} = \frac{\mathbf{a}_A}{\mathbf{a}_B}.$$

In principle we can therefore compare masses with each other, and ultimately with the international prototype, by measuring accelerations.

It might be thought that the measurement of acceleration is a simple process involving merely the use of a ruler and a stopwatch. For most purposes this is so. If, however, great precision is required there is a complication. This arises because Newton's Second Law is based on the concept of absolute acceleration, which implies the existence, somewhere in space, of a set of fixed axes relative to which all accelerations are measured. It is only with reference to such a set of axes that a particle in equilibrium would continue to move in a state of uniform motion or to remain at rest. We refer to such a frame of reference as Newtonian.

For most engineering purposes it is sufficiently accurate to regard the earth as providing a Newtonian frame of reference. It is evident that measurements relative to the earth cannot give absolute accelerations because the earth itself is in motion. In particular, it is rotating. The effect of the earth's rotation on acceleration was demonstrated in an example on page 119, where it was seen to be very small. Even so this effect can be detected by simple experiments and in some circumstances is of major significance.

A further complication to be noted in passing but not studied here in detail is the effect of high velocities. When the speed of a particle approaches the speed of light c Newton's Second Law is modified. The special theory of relativity then tells us that

$$\mathbf{P}=\frac{\mathrm{d}}{\mathrm{d}t}\left\{\frac{m\mathbf{v}}{\sqrt{(1-v^2/c^2)}}\right\}.$$

In most engineering applications this reduces to

$$\mathbf{P}=m\mathbf{a} \tag{6.1}$$

6.4. The weight of a body

The mass of a body is a measure of the amount of material that it contains and is a constant property of the body. The usual evidence of the mass of an object is its weight, that is the earth's gravitational force. At any location the weight is proportional to the mass, but the weight of a given mass varies with its location on the earth's surface and the height above the surface.

A particle falling freely close to the earth's surface does so with a constant acceleration $\mathbf{g}$. It follows from Newton's Second Law that the gravitational force acting on the particle, its weight, is $m\mathbf{g}$. The particle is presumed to be falling in a vacuum so that there is no air resistance. The weight, being a force, is measured in newtons.

The weight of a body is so often a factor to be considered that it is tempting to represent it by a single symbol, $\mathbf{W}$ say, instead of $m\mathbf{g}$. Whilst it would be unduly pedantic to object to this in equilibrium problems where there is no motion, and where the weight may be the only significant 'external' force, it is wise always to use $m\mathbf{g}$ to represent the weight in dynamics. Otherwise it is possible to arrive at expressions that involve both $\mathbf{W}$ and $\mathbf{g}$ in circumstances where gravity is irrelevant. Although the value of $\mathbf{g}$ varies with location on the earth's surface it is usually sufficient to assume the constant value of $9.81\ \mathrm{m\,s^{-1}}$.

6.5. Solution of the equation of motion $\mathbf{P}=m\mathbf{a}$

Calculation of $\mathbf{P}$ from the known acceleration of a single particle is a trivial exercise. Calculation of the acceleration when $\mathbf{P}$ is given is equally trivial. Our main concern is to solve the equation

$$\mathbf{P}=m\mathbf{a} \tag{6.1}$$

to determine the motion of a particle over an extended period of time. That is to say we seek the solution of

$$m\frac{\mathrm{d}^2\mathbf{r}}{\mathrm{d}t^2}=\mathbf{P} \tag{6.2}$$

where $\mathbf{P}$ may be a function of t, $\mathbf{r}$, or $\mathbf{v}$ either separately or in combination.

The procedure for solving eqn (6.2) depends mainly on the form in which **P** is expressed, but it can also depend on the form in which the solution is required. The solution does not present any special difficulties on account of its vectorial nature, for there is no objection in splitting it up into three component scalar equations:

$$\begin{aligned} m\ddot{x} &= p_x \\ m\ddot{y} &= p_y \\ m\ddot{z} &= p_z. \end{aligned} \tag{6.3}$$

Taking the first of these equations as typical, alternative methods of solution will be presented. Up to a point, this repeats Section 1 of Chapter 5, except that we will now attach physical significance to what was previously a purely mathematical exercise.

6.5.1. Solution of $m\ddot{x} = p(t)$

When the force is a function only of time, integration is quite straightforward. If $p(t)$ is readily integrable an analytical expression can be found for x. Otherwise it is necessary to use graphical or numerical integration.

In many instances we will be satisfied with the first integral

$$[m\dot{x}]_1^2 = \int_1^2 p(t)\,\mathrm{d}t$$

$$\Rightarrow \qquad mv_2 - mv_1 = \int_1^2 p(t)\,\mathrm{d}t \tag{6.4a}$$

The product of the mass and velocity was earlier defined as the momentum of a particle. The integral on the right-hand side of this equation is defined as the *impulse* of the force. The use of the word impulse suggests that the force is applied over a small time interval. Although this is very often the case in the circumstances in which the equation is most often applied, it does not have to be so. In words this equation is expressed as:

$$(\textit{Change of momentum}) = (\textit{Impulse}) \tag{6.4b}$$

The consequences of this equation will be studied later.

6.5.2. Solution of $m\ddot{x} = p(x)$

It is less obvious how to proceed when the force is a function of position, but some progress is always possible.

The acceleration can be manipulated to express it as a derivative with respect to x by the following argument:

$$\begin{aligned} \frac{\mathrm{d}^2x}{\mathrm{d}t^2} &= \frac{\mathrm{d}}{\mathrm{d}t}\left(\frac{\mathrm{d}x}{\mathrm{d}t}\right) = \frac{\mathrm{d}x}{\mathrm{d}t}\frac{\mathrm{d}}{\mathrm{d}x}\left(\frac{\mathrm{d}x}{\mathrm{d}t}\right) \\ &= v\frac{\mathrm{d}v}{\mathrm{d}x} = \frac{\mathrm{d}}{\mathrm{d}x}\left(\tfrac{1}{2}v^2\right). \end{aligned}$$

A single integration with respect to x yields

$$[\tfrac{1}{2}mv^2]_1^2 = \int_1^2 p(x)\,\mathrm{d}x$$

$$\Rightarrow \qquad \tfrac{1}{2}mv_2^2 - \tfrac{1}{2}mv_1^2 = \int_1^2 p(x)\,\mathrm{d}x. \tag{6.5a}$$

The physical significance of this equation will be studied more closely in later sections. The reader may well recognize, however, that $\frac{1}{2}mv^2$ is known as the *kinetic energy* of a particle and that the integral on the right-hand side of the equation is the *work* done by the force as the particle moves from one position to the other. In words this equation is expressed as:

(*Change of kinetic energy*) = (*Work done by the applied force*). (6.5b)

The integration of the equations of motion has already been referred to briefly in Chapter 5. In this reappraisal the integrations have been set in contexts both of which prove to be very important in the development of our subject.

It must be emphasized that neither $p(t)$ nor $p(x)$ need to be readily integrable mathematical functions, as the integrals in eqns (6.4a) and (6.5a) can always be evaluated numerically. It should be noted also that if p is constant the two solutions are equally applicable.

Example 6.1: Fig. 6.1 shows a simplified form of cathode-ray tube. Determine the speed v_a of an electron as it reaches the aperture in the anode, and its deflection δ at the screen when voltages $\pm V$ are applied to the plates.

The progress of an electron from the cathode to the screen must be studied in stages.

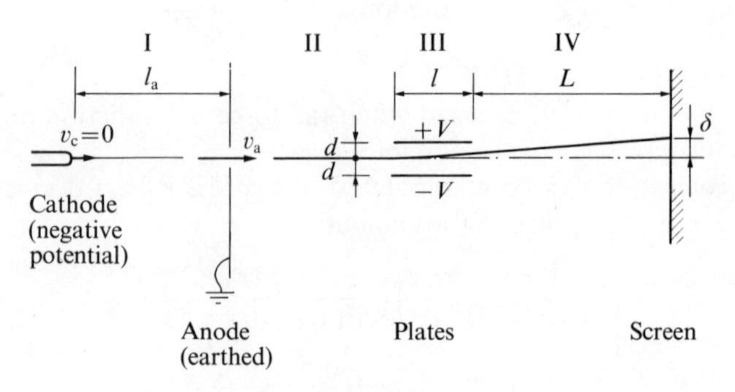

FIG. 6.1

Region I. The electron is here in a potential gradient V_a/l so that its equation of motion is

$$m\frac{d^2x}{dt^2}=eV_a/l$$

where e is the electron charge.

The right hand side of this equation is a constant so it appears that the integration can be with respect to either x or t. As the time of transit is unknown, integration with respect to t is impractical. The distance travelled l_a is known so eqn (6.5a) is used to give

$$\tfrac{1}{2}mv_a^2-\tfrac{1}{2}mv_c^2=\int_0^{l_a} eV_a/l_a\,dx.$$

For practical purposes $v_c=0$ so that

$$v_a^2=2eV_a/m.$$

Region II. There is no force on the electron in this region so that Newton's First Law applies and the velocity is constant at $\mathbf{v}_a$.

Region III. Here advantage is taken if the vectorial nature of Newton's Second Law by considering motions in the x and y directions separately, Fig. 6.2.

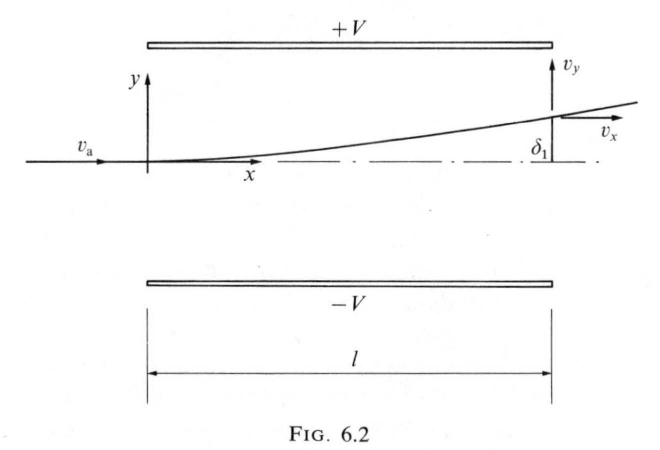

FIG. 6.2

As there is no potential gradient in the x-direction this component of velocity remains constant and on emerging from the plates $v_x=v_a$. As this velocity is constant, the time of transit though the plates is $\tau_1=l/v_a$. (Fringing effects are neglected.)

The potential gradient in the y-direction is V/d, so that the equation for the y-component of motion is

$$m\frac{d^2y}{dt^2}=eV/d.$$

Knowing the time interval for which the equation holds we can integrate it with respect to t

$$m\frac{\mathrm{d}y}{\mathrm{d}t}=(eV/d)t+A.$$

As $\mathrm{d}y/\mathrm{d}t=0$ at $t=0$, $A=0$. On integrating again

$$my=(eV/2d)t^2+B.$$

As $y=0$ at $t=0$, $B=0$. Also at $t=\tau_1$, $y=\delta_1$, hence

$$\frac{eV}{2md}\left(\frac{l}{v_\mathrm{a}}\right)^2=\frac{V}{V_\mathrm{a}}\frac{l^2}{4d}.$$

Also at $t=\tau_1$

$$\frac{\mathrm{d}y}{\mathrm{d}t}=v_\mathrm{y}=\frac{eV}{md}\frac{l}{v_\mathrm{a}}.$$

Region IV. Again the electron is in a region where it is free from applied forces and Newton's First Law applies. The velocity has two components, v_a axially and v_y transversely. The time of transit is $\tau_2=L/v_\mathrm{a}$.

$$\begin{aligned}\delta&=\delta_1+v_\mathrm{y}\tau_2\\&=\frac{V}{V_\mathrm{a}}\frac{l^2}{4d}+\frac{eV}{md}\frac{l}{v_\mathrm{a}}\frac{L}{v_\mathrm{a}}\\&=\frac{V}{V_\mathrm{a}}\frac{l}{2d}\left(\tfrac{1}{2}l+L\right).\end{aligned}$$

As the geometry of a cathode-ray tube is fixed, the deflection is directly proportional to the ratio of the applied voltage to the anode voltage.

Fig. 6.1 implies that the electrons move in a straight line from the cathode to the hole in the anode. In practice it is necessary to focus the stream of electrons on to the aperture. This may be done electromagnetically.

An electrical coil wound round the tube produces an electromagnetic field $\mathbf{B}$ that is parallel to the axis of the tube. The force on an electron moving in this field is $-e\mathbf{v}\times\mathbf{B}$, so that its equation of motion is

$$\begin{aligned}m\mathbf{a}=m\frac{\mathrm{d}\mathbf{v}}{\mathrm{d}t}&=e\mathbf{B}\times\mathbf{v}\\\frac{\mathrm{d}\mathbf{v}}{\mathrm{d}t}&=\left(\frac{e}{m}\mathbf{B}\right)\times\mathbf{v}\\&=\boldsymbol{\omega}\times\mathbf{v}.\end{aligned}$$

From this equation we can infer that the vector $\mathbf{v}$ rotates with constant angular velocity $\boldsymbol{\omega}=(e/m)\mathbf{B}$. Any component of $\mathbf{v}$ that is parallel to $\mathbf{B}$ is unaffected by this rotation, the cross product of such a component with $\mathbf{B}$

being zero. The component of **v** that is perpendicular to **B** rotates steadily without change of magnitude. When viewed in the direction of **B** the path appears to be circular. In fact it is a helix due to the component of velocity parallel to **B**. The beam is focused when the time of transit from cathode to anode just matches the time taken by an electron to complete one orbit as seen axially, so that $t = l_a/(v_a/2) = 2\pi/\omega$.

Electromagnetic focusing provides our first example of **P** as a function of **v**. It is a somewhat special example in that it is possible to proceed directly to the solution of the vector differential equation.

6.5.3. Solution of $m\ddot{x} = p(v)$

The equation can be written in the form

$$m\frac{dv}{dt} = p(v)$$

so that

$$t = \int_{v_1}^{v_2} \frac{m\,dv}{p(v)}. \tag{6.6}$$

Alternatively, use can again be made of

$$\begin{aligned} m\frac{d^2x}{dt^2} &= mv\frac{dv}{dx} = p(v) \\ \Rightarrow \qquad x_2 - x_1 &= \int_{v_1}^{v_2} \frac{mv}{p(v)}\,dv. \end{aligned} \tag{6.7}$$

Again, both integrals can be evaluated graphically.

Example 6.2: The motion of a particle of a particle falling freely under gravity through the atmosphere.

The terminal velocity is achieved when the air resistance kv^2 just balances the weight mg. So that the terminal velocity is

$$v_T = \sqrt{mg/k}.$$

The time taken by a particle to attain a velocity v having started from rest is given by eqn (6.6),

$$t^2 \int_0^v \frac{m}{mg - kv^2}\,dv = \frac{1}{g}\int_0^v \frac{v_T^2}{v_T^2 - v^2}\,dv.$$

If $v = \beta v_T$ this integral becomes

$$\begin{aligned} t &= \frac{v_T}{g}\int_0^\beta \frac{d\beta}{1-\beta^2} \\ &= \frac{v_T}{2g}\log_e\frac{1+\beta}{1-\beta}. \end{aligned}$$

The distance travelled in this time is

$$x = \frac{v_T^2}{g}\int_0^\beta \frac{\beta}{1-\beta^2}\,\mathrm{d}\beta$$
$$= \frac{v_T^2}{2g}\log_e \frac{1}{(1-\beta^2)}.$$

It frequently occurs that **P** is a function of both position and velocity. In Fig. 6.3 the mass rests on a horizontal plane. Its movement is constrained by a spring of stiffness s. The force exerted by the spring is sx in

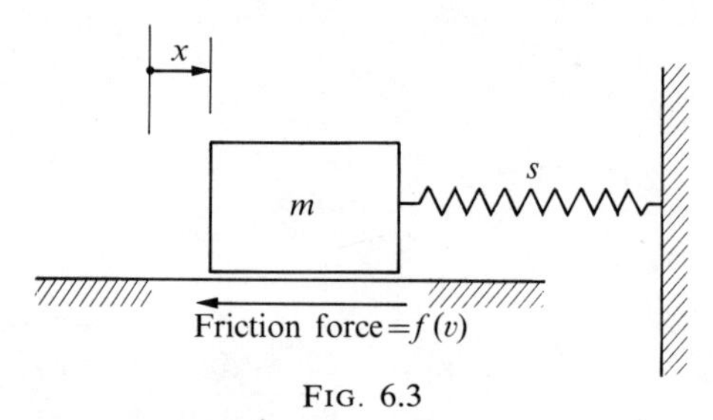

FIG. 6.3

opposition to a compression (or extension) of an amount x. Motion is opposed by friction of an amount $f(v)$ between the mass and the plane on which it rests. The equation of motion is

$$m\frac{\mathrm{d}^2x}{\mathrm{d}t^2} = -(sx + f(v))$$

or

$$m\frac{\mathrm{d}^2x}{\mathrm{d}t^2} - f(v) + sx = 0.$$

The solution of this equation presents difficulties unless $f(v) = \lambda\,\mathrm{d}x/\mathrm{d}t$, when the equation becomes an ordinary differential equation and is solved by the standard methods.

6.6. Momentum and moment of momentum

It has been shown in deducing eqns (6.4a) and (6.5a) that when the force applied to a particle is a function of position only, a function of time only, or a constant, the integration of the equation of motion as far as the first integral is always possible and is quite standard. It follows that it is possible to proceed straight to an integrated form of the equation of motion, either to eqn (6.4a) or eqn (6.5a), without starting with $\mathbf{P} = m\mathbf{a}$. This proves to have advantages far beyond the mere saving of a small amount of labour.

6.6.1. Conservation of linear momentum

Integration of

$$\mathbf{P} = m\mathbf{a} = \mathrm{d}(m\mathbf{v})/\mathrm{d}t$$

with respect to time yields immediately

$$\int_1^2 \mathbf{P}\,\mathrm{d}t = m\mathbf{v}_2 - m\mathbf{v}_1. \tag{6.8}$$

As was pointed out in relation to eqn (6.4a) this can be expressed verbally as

$$(\textit{Impulse}) = (\textit{Change of momentum}).$$

If $\mathbf{P}=0$, then $m\mathbf{v}$ is constant. This result is known as *the principle of the conservation of linear momentum*, which states that *in the absence of an applied force the momentum of a particle is constant.* It is no more than a restatement of Newton's First Law. Problems which can be solved by its direct application are usually fairly trivial. But when applied to systems of two or more particles the principle proves to be very powerful. For the present we will study the dynamics of a single particle further, but from a slightly different standpoint.

6.6.2. Conservation of moment of momentum

In Chapter 3 the concept of the moment of a vector was introduced. Following the definition given there, the moment $\mathbf{h}$ of momentum of a particle about a point O, Fig. 6.4, is

$$\mathbf{h} = \mathbf{r} \times (m\mathbf{v}).$$

Now consider the basic equation of motion

$$\mathbf{P} = \frac{\mathrm{d}(m\mathbf{v})}{\mathrm{d}t}$$

and let us take moments of both sides about O:

$$\mathbf{r} \times \mathbf{P} = \mathbf{r} \times \frac{\mathrm{d}(m\mathbf{v})}{\mathrm{d}t}.$$

The right-hand side of this equation can be manipulated into a different form. We start by noting that

$$\frac{\mathrm{d}}{\mathrm{d}t}(\mathbf{r}\times\mathbf{v}) = \frac{\mathrm{d}\mathbf{r}}{\mathrm{d}t}\times\mathbf{v} + \mathbf{r}\times\frac{\mathrm{d}\mathbf{v}}{\mathrm{d}t}.$$

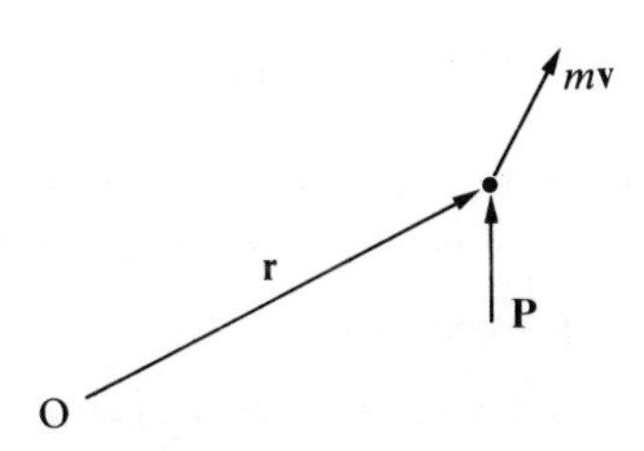

FIG. 6.4

As $\mathbf{v}=\mathrm{d}\mathbf{r}/\mathrm{d}t$ the first term on the right-hand side of this equation is zero so

$$\mathbf{r}\times\frac{\mathrm{d}\mathbf{v}}{\mathrm{d}t}=\frac{\mathrm{d}}{\mathrm{d}t}(\mathbf{r}\times\mathbf{v}).$$

Hence

$$\mathbf{r}\times\mathbf{P}=\frac{\mathrm{d}}{\mathrm{d}t}(\mathbf{r}\times m\mathbf{v}).$$

Denoting the moment $\mathbf{r}\times\mathbf{P}$ by $\mathbf{Q}$ this equation attains the standard form

$$\mathbf{Q}=\frac{\mathrm{d}\mathbf{h}}{\mathrm{d}t}. \tag{6.9}$$

This is a very important equation. It states that the moment about any point of the resultant force acting on a particle equals the rate of change of moment of momentum about the point. In particular, if $\mathbf{Q}$ is zero then $\mathbf{h}$ is constant. This is *the principle of conservation of moment of momentum,* and may be stated formally as follows:
if the forces that act on a particle have zero moment about a point O, the moment of momentum about O is constant.

Satellite motion provides a classical example of the principle.

6.6.3. Orbital motion of a single particle

Consider a planet moving freely through space under the influence only of the gravitational attraction of its sun, which we will take to be fixed, Fig. 6.5.

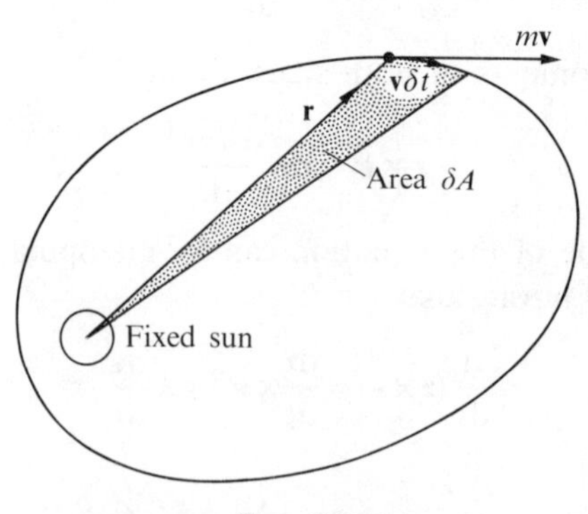

FIG. 6.5

As the force on the planet acts through the centre of the sun the moment of the planet's momentum about the sun's centre is constant,

$$\mathbf{h}=\mathbf{r}\times(m\mathbf{v})=\text{constant}.$$

Taking the scalar product with $\mathbf{r}$

$$\Rightarrow \qquad \mathbf{r}\cdot\mathbf{h}=\mathbf{r}\cdot[\mathbf{r}\times(m\mathbf{v})]=0.$$

It follows that $\mathbf{r}$ is always perpendicular to the constant vector $\mathbf{h}$ and hence that the motion is in a fixed plane through O. An impulse applied to knock the satellite out of its plane results in the motion being in a new plane through O.

A further conclusion follows. The area swept out by the vector $\mathbf{r}$ in a small time interval δt is $\delta A = \frac{1}{2}|\mathbf{r} \times (\mathbf{v}\,\delta t)|$.
Hence

$$\frac{\mathrm{d}A}{\mathrm{d}t} = \frac{1}{2}|\mathbf{r} \times \mathbf{v}| = \frac{h}{2m} = \text{constant}.$$

This is Kepler's Second Law for planetary motion. It states that the vector drawn from the centre of the sun to a planet sweeps out area at a constant rate.

We may note in passing that Kepler's First Law states that the path of a planet is an ellipse with the sun centred on one of the foci. The proof of this law requires the law of gravitational attraction to be known. The Second Law holds whatever the law of gravitational attraction might be. The present proof of the Second Law demonstrates how the momentum principles can be used to arrive at a significant conclusion even though there may be insufficient information to deduce a fully detailed solution.

6.7. Energy

The concept of momentum has been introduced as a consequence of the need to solve the equation of motion for a particle when the applied force is a function of time. A parallel study in which force is a function of position leads to the consideration of energy.

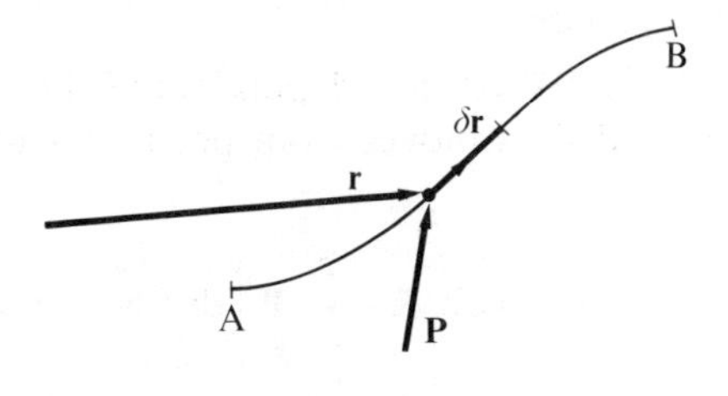

FIG. 6.6

The work done by a force $\mathbf{P}$ when its point of application moves through space an amount $\delta\mathbf{r}$ is $\mathbf{P}\,.\,\delta\mathbf{r}$. If the force is applied to a particle of mass m, the work done on the particle in moving from A to B is

$$\int_A^B \mathbf{P}\,.\,\mathrm{d}\mathbf{r} = \int_A^B m\mathbf{a}\,.\,\mathrm{d}\mathbf{r}.$$

The integrand on the right-hand side of this equation can be simplified in the following way:

$$\mathbf{a}\,.\,\mathrm{d}\mathbf{r}=\frac{\mathrm{d}\mathbf{v}}{\mathrm{d}t}\frac{\mathrm{d}\mathbf{r}}{\mathrm{d}t}\,\mathrm{d}t$$

$$=\left(\frac{\mathrm{d}\mathbf{v}}{\mathrm{d}t}\,.\,\mathbf{v}\right)\mathrm{d}t$$

$$=\frac{1}{2}\frac{\mathrm{d}}{\mathrm{d}t}(\mathbf{v}\,.\,\mathbf{v})\,\mathrm{d}t$$

$$=\tfrac{1}{2}\mathrm{d}(v^2).$$

Hence

$$\int_A^B \mathbf{P}\,.\,\mathrm{d}\mathbf{r}=\int_A^B \tfrac{1}{2}m\,\mathrm{d}(v^2)$$

$$=\tfrac{1}{2}mv_B^2-\tfrac{1}{2}mv_A^2=\Delta T. \qquad (6.10)$$

The quantity $\frac{1}{2}mv^2$ is defined as the *kinetic energy* (energy of motion) of the particle. Expressed verbally the equation is,

(*The work done on a particle in a displacement*) = (*Its change in kinetic energy*).

Viewed as a mathematical exercise, the energy equation is simply an alternative to the momentum equation as a first integral of the equation of motion. Its practical value is that, just as with the momentum equation, there are circumstances in which it is particularly useful. This happens when it is permissible to assume that the total energy of the system is constant, and the energy is said to be conserved. It is the nature of **P** that determines whether energy is conserved.

6.7.1. Conservative forces

A force is *conservative* if $\int_1^2 \mathbf{P}\,.\,\mathrm{d}\mathbf{r}$ is independent of the path traced by the point of application of **P** as it moves from position 1 to position 2. This means, Fig. 6.7

$$\int_1^2 \mathbf{P}\,.\,\mathrm{d}\mathbf{r} \text{ for path A}=\int_1^2 \mathbf{P}\,.\,\mathrm{d}\mathbf{r} \text{ for path B}$$

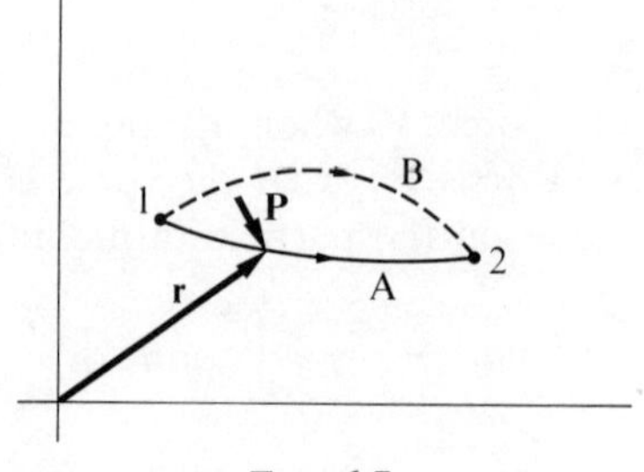

FIG. 6.7

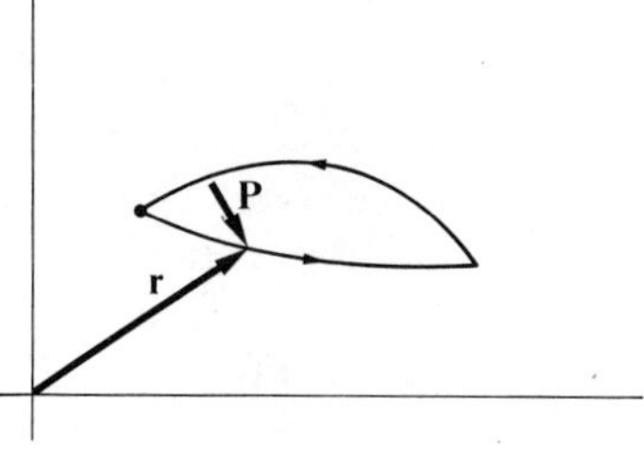

FIG. 6.8

for all possible choices of path. It also means, Fig. 6.8

$$\oint \mathbf{P} \,.\, d\mathbf{r} = 0 \qquad (6.11)$$

where $\oint$ denotes that whatever path is followed it always finishes at the same point as where it started.

Example 6.3: Evaluate $\int_1^2 \mathbf{P} \,.\, d\mathbf{r}$ *for the force due to gravitational attraction.*

(a) *Constant gravity.*

The force on a particle moving freely in a constant gravitational field is $m\mathbf{g}$ where $\mathbf{g}$ is the acceleration due to gravity. The work done on the particle by gravity is, Fig. 6.9

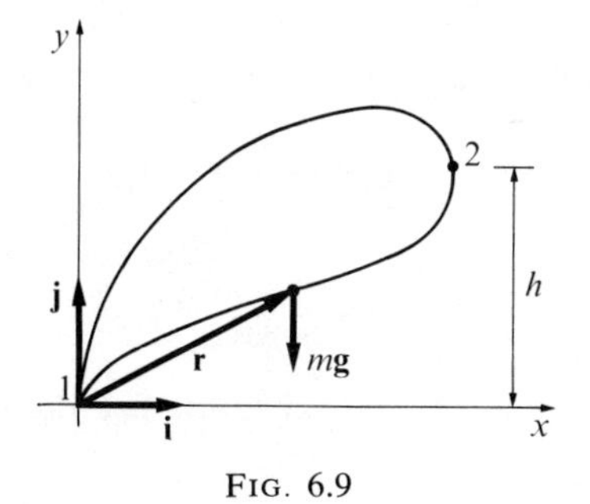

FIG. 6.9

$$\begin{aligned} &\int_1^2 m\mathbf{g} \,.\, d\mathbf{r} \\ &\quad = \int_1^2 m\mathbf{g} \,.\, \mathbf{i}\, dx + \int_1^2 m\mathbf{g} \,.\, \mathbf{j}\, dy \\ &\quad = -\int_1^2 mg\, dy \\ &\quad = -mgh \qquad (6.12a) \end{aligned}$$

irrespective of the path actually followed.

(b) Varying gravity

The assumption of constant gravity is sufficiently accurate for normal purposes only if h is very much less than the radius of the earth. In satellite motion and in the mechanics of space travel it is necessary to allow for a variation in gravitational attraction in accordance with the inverse-square law.

The work done by gravity is then, Fig. 6.10

$$\int_1^2 m\mathbf{g}\,.\,d\mathbf{r}$$

where $\mathbf{g}$ is now a variable vector whose magnitude is $g(R_0/r)^2$, R_0 being the radius of the earth and g the gravitational constant at the surface of

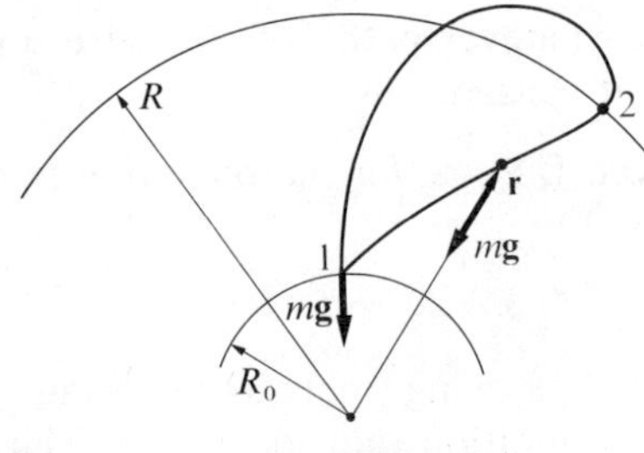

FIG. 6.10

the earth. As $\mathbf{g}$ is always directed towards the earth's centre it is always collinear with $\mathbf{r}$ and in the opposite direction so the work done is

$$-\int_1^2 mg\left(\frac{R_0}{r}\right)^2 dr$$
$$= -mgR_0^2\left(\frac{1}{R_0}-\frac{1}{R}\right). \tag{6.13b}$$

If $(R-R_0)=h$ is small compared with R_0, then $V=-mgh$ as in a constant gravity field.

Example 6.4: Evaluate $\int_1^2 \mathbf{P}\,.\,d\mathbf{r}$ *for the force exerted by an elastic spring.*

For simplicity we will consider a linear spring of stiffness s for which the force required to cause an extension x is sx. The force which the

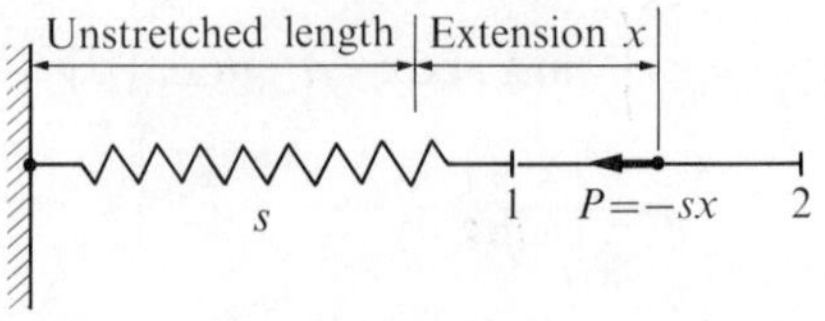

FIG. 6.11

spring applies to whatever is stretching it is $-sx$. The total work done in stretching it from an initial extension x_1 to a final one x_2 is

$$\int_1^2 (-sx)\,.\,\mathrm{d}x = -\frac{s}{2}(x_2^2 - x_1^2). \tag{6.14a}$$

This result is not affected if the spring is stretched beyond its final state and then allowed to contract, provided that the maximum extension is not sufficient to cause permanent damage, thereby invalidating the relationship between force and extension.

Example 6.5: Show that the force due to friction is non-conservative.

Friction is an example of a non-conservative force. Consider a block resting on a rough horizontal plane. A force is applied to move it slowly from one position to another. Then the force is reversed to restore the block to its original position. As the force is always in opposition to the incremental displacement $\mathbf{P}\,.\,\delta\mathbf{r}$ is always negative. It follows that $\oint \mathbf{P}\,.\,\mathrm{d}\mathbf{r} \neq 0$.

6.7.2. Potential energy

When a system, of which a particle is a member, exerts a force on the particle thereby causing it to move from an initial state of rest, the system does work on the particle and as a result the particle acquires kinetic energy. For example, the force of gravity acting on a free particle causes it to fall. The system is the source of this energy and so we can say that it has the potential to impart kinetic energy to the particle. As a result of energy being transferred to the particle the energy stored in the system, apart from the kinetic energy imparted to the particle, has dropped. The *change* in *potential energy* of the system is defined to be

$$\Delta V = -\int_A^B \mathbf{P}\,.\,\mathrm{d}\mathbf{r} \tag{6.15}$$

where $\mathbf{P}$ is the force applied to the particle as it moves from A to B.

Example 6.6: Derive an expression for the potential energy due to gravity assuming:

(*a*) *Constant gravity*
From eqns (6.12a) and (6.15)

$$\Delta V = mgh. \tag{6.12b}$$

(*b*) *Varying gravity*
From eqns (6.13a) and (6.15)

$$V = mgR_0^2 \frac{1}{R_0} - \frac{1}{R} \tag{6.13b}$$

Example 6.7: Derive an expression for the potential energy stored in an elastic spring.

From eqns (6.14a) and (6.15)

$$\Delta V = \tfrac{1}{2}s(x_2^2 - x_1^2). \tag{6.14b}$$

It is usually convenient to take $x_1 = 0$ as datum. Then the potential energy at extension x is

$$V = \tfrac{1}{2}sx^2. \tag{6.14c}$$

6.7.3. Conservation of energy

The principle of conservation of energy states that in a conservative system the sum of the potential and kinetic energy is constant. This follows from eqns (6.10) and (6.15) which combine together to give

$$\Delta V + \Delta T = 0$$

or

$$V + T = \text{constant}. \tag{6.16}$$

T is always zero for a particle at rest. V is defined relative to whatever datum is most convenient, e.g. $\mathbf{r} = 0$.

Considerable care must be exercised in using eqn (6.16) for it applies only to conservative systems. Systems in which friction is present are non-conservative, for energy is dissipated in heat. Similarly, there is usually a 'loss' of energy in impact between masses. When colliding bodies separate they have often suffered permanent deformation at the point of contact and/or they are vibrating. In either case energy has been converted into a form from which it is irrecoverable.

This does not mean that it is incorrect to write $\int \mathbf{P} \cdot d\mathbf{r} = \Delta T$ in such circumstances, but instead of eqn (6.16) we must write

$$\Delta V + \Delta T + \Delta D = 0$$

where ΔD is the work done by non-conservative forces. Provided that enough is known about the system, ΔD can be calculated. The use of the concept of energy is, however, most helpful when the system being considered is conservative.

6.7.4. Equilibrium and stability

In eqn (6.15) the finite change ΔV in potential energy associated with a force $\mathbf{P}$ on a particle is defined as

$$\Delta V = -\int \mathbf{P} \cdot d\mathbf{r}.$$

For an infinitesimal displacement

$$\delta \mathbf{V} = -\mathbf{P} \cdot \delta \mathbf{r}.$$

If the motion of the particle is so constrained that it has only one degree of freedom allowing movement in, say, the x direction $\delta V = -P_x \delta x$, and in the limit

$$\frac{dV}{dx} = -P_x. \tag{6.17}$$

If the potential energy happens to be a known function of x this relationship can be used to find P_x. When the particle is in equilibrium P_x is zero, and so is $\mathrm{d}V/\mathrm{d}x$. This relationship can be used to find the equilibrium configuration for a system.

Example 6.8: A concentrated mass is attached at B to a light rod AB of length a. It is held in equilibrium at an angle θ to the upward vertical by a spring, one end of which is attached to B, and the other to a fixed point vertically above A as in Fig. 6.12. Determine the value of θ for equilibrium if the unstretched length of the spring is a and its stiffness is s.

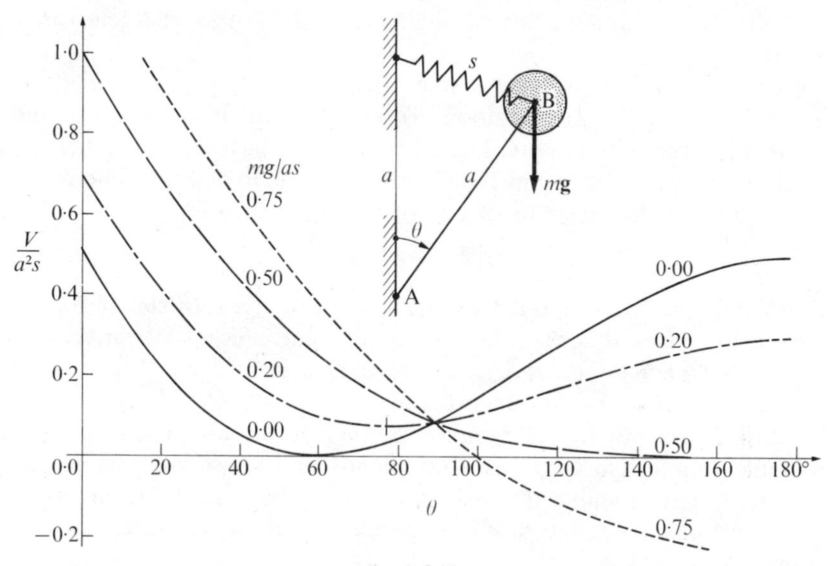

FIG. 6.12

We can arbitrarily take the level of A to be the datum for gravitational potential energy and the unstretched condition of the spring as the datum for elastic potential energy. These two contributions to the potential energy V of the whole system do not need to be measured from datums that correspond to one particular configuration of the system. Bringing the components into line with each other would merely involve the addition of constant terms which would disappear when V is differentiated in accordance with eqn (6.17). Hence

$$V = mga\cos\theta + \tfrac{1}{2}s(2a\sin\theta/2 - a)^2$$

$$P_\theta = \frac{\mathrm{d}V}{\mathrm{d}\theta} = -mga\sin\theta + a^2s(2\sin\theta/2 - 1)\cos\theta/2$$

$$= 0 \text{ for equilibrium.}$$

On rearranging the terms we find that $\mathrm{d}V/\mathrm{d}\theta = 0$ when

$$\sin\theta/2 = 0{\cdot}5(1 - mg/as)^{-1}$$

and when

$$\cos \theta/2 = 0.$$

The second solution, $\theta = 180°$, is an obvious one from consideration of symmetry, and holds whatever the relative values of the parameters a, m, and s. The existence of the first solution does depend on a, m, and s. It can be seen that for $\sin \theta/2 \leqslant 1$, $mg/as \leqslant 0{\cdot}5$. This is illustrated in Fig. 6.12 where V/a^2s is plotted against θ for a range of values of mg/as. For zero mass, $\mathrm{d}V/\mathrm{d}\theta = 0$ at $\theta = 60°$ and $\theta = 180°$. At $mg/as = 0{\cdot}2$, $\mathrm{d}V/\mathrm{d}\theta = 0$ at $\theta = 77{\cdot}4°$ and $\theta = 180°$. For $mg/as = 0{\cdot}5$ the slope is zero only for $\theta = 180°$. For greater values of mg/as, e.g. $0{\cdot}75$, this characteristic is emphasized.

It cannot be assumed that a system such as that shown in Fig. 6.12 will remain in an equilibrium position. To see why this is so we must study eqn (6.17) a bit more closely. Indeed we must go back to eqn (6.15) from which it was derived and which defines the change in potential energy of a system in a displacement $\mathbf{dr}$ of one of its particles to be

$$\mathrm{d}V = -\mathbf{P} \cdot \mathbf{dr},$$

where $\mathbf{P}$ is the force applied by the system to the particle. If $\mathbf{P} \cdot \mathbf{dr}$ is positive it means that $\mathbf{P}$ is in a direction that causes $\mathbf{r}$ to increase. It follows that when $\mathrm{d}V/\mathrm{d}x$ is negative it means that the force $\mathbf{P}_x$ causes x to tend to increase in the positive direction.

The practical way of determining whether or not a system is in a state of stable equilibrium is to disturb it slightly and to see what happens. If the equilibrium is stable the system will move back to the configuration from which it was displaced. If it is unstable the initial displacement tends to increase.

An individual particle will be pushed in the direction of x increasing if $\mathrm{d}V/\mathrm{d}x$ is negative, and in the opposite direction if $\mathrm{d}V/\mathrm{d}x$ is positive.

In the example in Fig. 6.12 the system is in equilibrium for the case $m = 0$ when $\theta = 60°$. We see that $P_\theta = \mathrm{d}V/\mathrm{d}\theta$ is negative for $\theta < 60°$ and positive for $\theta > 60°$. So, whichever way the particle is disturbed from $\theta = 60°$, forces will operate to return it to the 60° position. For the same case, $m = 0$, the situation is reversed when $\theta = 180°$. There $\mathrm{d}V/\mathrm{d}\theta$ changes from positive to negative as θ increases. This means that if the mass is displaced to either side of this position and then released, it will move farther away from $\theta = 180°$. Hence, for $m = 0$ the equilibrium is stable when $\theta = 60°$ and unstable when $\theta = 180°$.

As m is increased the value of θ for stable equilibrium increases. For $mg/as = 0{\cdot}5$ the two equilibrium positions coalesce, and to a first order approximation $\mathrm{d}V/\mathrm{d}\theta$ is zero on either side of $\theta = 180°$. For $mg/as = 0{\cdot}75$, $\theta = 180°$ is clearly a position of stable equilibrium.

It can be seen from this example that stable equilibrium is associated with a minimum value of V, that is $\mathrm{d}^2V/\mathrm{d}\theta^2$ is negative, and that unstable equilibrium is associated with a maximum for V, for which $\mathrm{d}^2V/\mathrm{d}\theta^2$. Left to itself a system always moves to a position for which the potential energy is a minimum.

For the particular example considered here

$$\frac{\mathrm{d}^2V}{\mathrm{d}\theta^2} = a^2s(1 - mg/as)\cos\theta + -\sin\theta/2.$$

For $\theta = 180°$, $\mathrm{d}^2V/\mathrm{d}\theta^2 = a^2s(mg/as - 0{\cdot}5)$. This is positive, and the equilibrium is stable, when $mg/as > 0{\cdot}5$.

When $m = 0$ the equilibrium value $\theta = 60°$ gives a positive value to $\mathrm{d}^2V/\mathrm{d}\theta^2$.

The concept of potential energy leads, as our example shows, to a well defined method for investigating the stability of the equilibrium states of a system. Its importance grows when the ideas are developed and extended to deal with systems with several degrees of freedom. Because the method is well defined there is a temptation to think always in terms of potential energy when stability is investigated. It may be a waste of time to do this when the equilibrium states are already known. With simple systems it is usually possible to determine very quickly whether the forces that operate when it is displaced will cause it to return to the equilibrium state or to move away from it.

6.8. Some single particle systems

This chapter concludes with a number of examples illustrating application of Newton's laws in the various ways that we have developed.

6.8.1. Simple gravity pendulum

A particle of mass m is suspended in a gravity field g from a fixed point by an inextensible string of length l. The particle oscillates in a fixed vertical plane.

It will be left as an exercise for the reader to write down the equation of motion by direct application of Newton's laws. Here the energy equation, eqn (6.16) will be used.

Let θ be the angle of the string to the vertical, Fig. 6.13. The velocity is $l\dot{\theta}$ so that the kinetic energy is

$$T = -ml^2\dot{\theta}^2.$$

Taking the lowest point in the path of the particle as the datum for potential energy

$$\begin{aligned} V &= mgh \\ &= mgl(1 - \cos\theta). \end{aligned}$$

The sum of the kinetic and potential energies is constant so

$$\tfrac{1}{2}ml^2\dot{\theta}^2 + mgl(1 - \cos\theta) = \text{constant}.$$

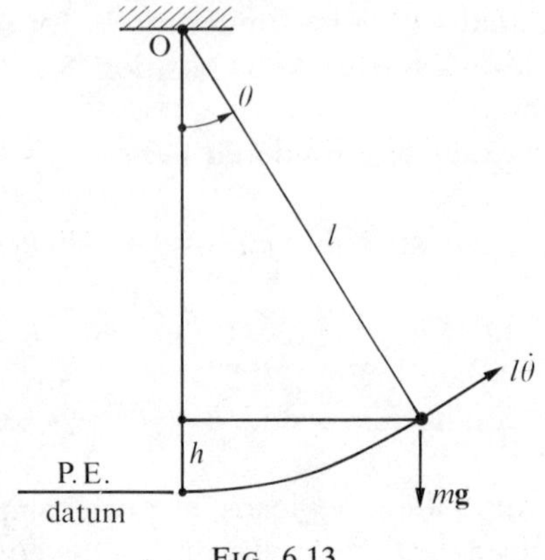

FIG. 6.13

This equation cannot be solved to give θ as a function of time in terms of elementary mathematical functions so we will confine attention to the situation in which θ never becomes large, say greater than 15°. Then

$$(1-\cos\theta)\approx\tfrac{1}{2}\theta^2$$

and the energy equation, which is a form of the equation of motion, becomes

$$\frac{l}{g}\dot{\theta}^2+\theta^2=\text{constant}.$$

When θ is a maximum, θ_0 say, $\dot{\theta}$ is zero so that the constant on the right-hand side of this equation is θ_0^2, giving

$$\frac{l}{g}\dot{\theta}^2+\theta^2=\theta_0^2.$$

The equation is a standard form

$$(\dot{\theta}/\omega_n)^2+\theta^2=\theta_0^2 \tag{6.18}$$

where in this case $\omega_n=\sqrt{(g/l)}$. The motion that it represents is simple harmonic of amplitude θ_0 and at a circular frequency ω_n. The solution is

$$\theta=\theta_0\cos\omega_n t, \tag{6.19}$$

which can be verified by substitution in eqn (6.18) or can be obtained by direct integration as follows:

$$\dot{\theta}/\omega_n=(\theta_0-\theta)^{\frac{1}{2}}$$

$$\Rightarrow \qquad \omega_n t=\int\frac{d\theta}{(\theta_0^2-\theta^2)^{\frac{1}{2}}}+\text{constant}$$

$$=\sin^{-1}(\theta/\theta_0)+C.$$

If $\theta = \theta_0$ at $t = 0$, $C = -\pi/2$

$\Rightarrow$
$$\theta = \theta_0 \cos \omega_n t.$$

6.8.2. Spring-mounted mass

A mass m is supported by a light spring of stiffness s in a gravity field g. The obvious datum from which to measure potential energy is when the mass is at the level at which the spring is unstrained. Then, on extending the spring x the energy stored therein is $-sx^2$, from eqn (6.14c). Taking the same datum for potential energy due to gravity

$$V = \tfrac{1}{2}sx^2 - mgx.$$

It is to be noted that the component due to gravity is here $-mgx$ whereas with the pendulum it was $+mgh$. This is because the positive direction for x has been taken to be in the same direction as g so that $V = -\int F\,dx$ is negative. With the pendulum an increase in θ caused a displacement of the mass against the force of gravity.

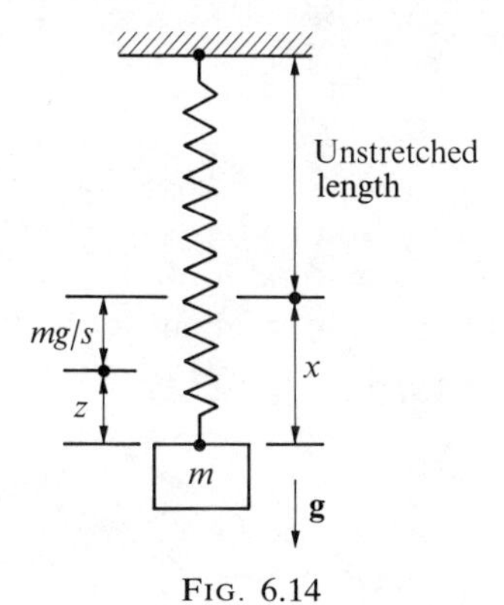

FIG. 6.14

The kinetic energy is this example is $\frac{1}{2}m\dot{x}^2$ so that, provided there is no damping

$$\tfrac{1}{2}m\dot{x}^2 + \tfrac{1}{2}sx^2 - mgx = constant$$

$\Rightarrow$
$$\frac{m}{s}x^2 + (x - mg/s)^2 = constant.$$

If $x = x_0$ when $\dot{x} = 0$ the constant is $(x_0 - mg/s)^2$ so that

$$\frac{m}{s}x^2 + (x - mg/s)^2 = (x_0) - mg/s)^2.$$

A change of variable putting $(x - mg/s) = z$ transforms this equation to the standard form of eqn (6.18):

$$\frac{m}{s}\dot{z}^2 + z^2 = z_0^2$$

so that the motion is simple harmonic with amplitude z_0 and circular frequency $\omega_n = \sqrt{(s/m)}$.

On closer inspection it is realized that mg/s is the static extension of the spring due to the dead weight $m\mathbf{g}$, so that z is the displacement of the mass from its equilibrium position. The equation of motion in terms of z is the equation that would have been derived had we simply ignored the dead weight $m\mathbf{g}$ and the static extension due to it. This is because the weight is at all times supported by a constant component of the spring force. It is very helpful to know that in problems such as this it is permissible to ignore gravity and treat the springs as if they were unstressed in the equilibrium position. Of course, if there is any doubt it is safer to include all components of potential energy as we have done here.

6.8.3. Satellite motion

The equation of motion for a satellite with a fixed sun is, Fig. 6.15,

$$m\frac{\mathrm{d}^2\mathbf{r}}{\mathrm{d}t^2} = -mg\frac{R_0^2\mathbf{r}}{\mathbf{r}^3}$$

or

$$\ddot{\mathbf{r}} = -k\mathbf{r}/r^3 \tag{6.20}$$

where R_0 is the radius of the sun and g is the acceleration due to gravity at the surface of the sun. The method for a full solution of this equation is not immediately obvious. For the moment let us be content with partial solutions that reveal the salient features of the motion.

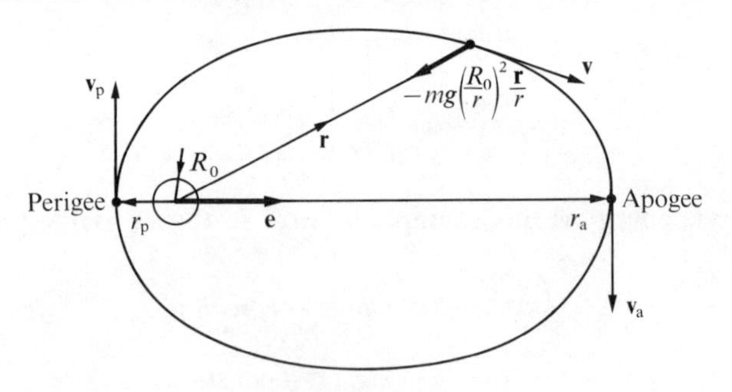

FIG. 6.15

Forming the vector product with $\mathbf{r}$ on both sides of eqn (6.20) we have

$$\mathbf{r} \times \ddot{\mathbf{r}} = 0.$$

This can be integrated immediately to give

$$\mathbf{r} \times \dot{\mathbf{r}} = \text{constant} = \mathbf{h}_0, \text{ say.} \tag{6.21}$$

This equation can be recognized as a statement that the momentum of the particle about the sun is constant with $\mathbf{h}_0 = \mathbf{h}/m$. In Section 6.6.3 we concluded from this result that the path is confined to a plane.

The energy counterpart of eqn (6.21) can be obtained by taking the scalar product of both sides of eqn (6.20) with $\dot{\mathbf{r}}$, but it is simpler to proceed directly.

Relative to the surface of the sun the potential energy of the satellite is given by eqn (6.13b) as

$$V = mgR_0^2\left(\frac{1}{R_0} - \frac{1}{r}\right).$$

The sum of the kinetic and potential energies is constant so

$$\tfrac{1}{2}mv^2 - mgR_0^2/r = \text{constant}$$

where the term mgR_0^2/R_0 has been absorbed into the constant. Dividing through by m and substituting k for gR_0^2 we have

$$\tfrac{1}{2}v^2 - k/r = \text{constant}. \tag{6.22}$$

6.8.3.1. Perigee, apogee, and velocity of escape. If r_p and v_p are the radius and velocity of the satellite at perigee, the lowest point in the orbit, eqn (6.22) becomes

$$\tfrac{1}{2}v^2 - k/r = \tfrac{1}{2}v_p^2 - k/r_p.$$

At apogee, the highest point in the orbit, the radius is r_a and the velocity is v_a so that

$$\tfrac{1}{2}v_p^2 - k/r_p = \tfrac{1}{2}v_a^2 - k/r_a.$$

Eqn (6.21) gives

$$r_p v_p = r_a v_a = h_0.$$

Taken together these two equations give

$$\tfrac{1}{2}(v_p^2 - v_a^2) = \frac{k}{h_0}(v_p - v_a)$$

$$\Rightarrow \qquad v_p + v_a = 2k/h_0 \tag{6.23}$$

$$\Rightarrow \qquad v_a/v_p = \frac{2k}{h_0 v_p} - 1.$$

But

$$r_a v_a = r_p v_p$$

so that

$$r_a = r_p\left[\frac{2k}{h_0 v_p} - 1\right]^{-1}$$

$$= r_p\left[\frac{2gR_0^2}{r_p v_p^2} - 1\right]^{-1}. \tag{6.24}$$

If v_p is made sufficiently large r_a tends to infinity. The 'velocity of escape' v_e is $\sqrt{(2gR_0^2/r_p)}$. The velocity for a circular orbit of radius r_p is $v_c = \sqrt{(gR_0^2/r_p)}$ so that $v_e = v_c\sqrt{2}$.

6.8.3.2. Satellite orbits. There seems to be no obvious direct way of solving eqn (6.20) to obtain the path of the satellite. Indeed, but for the knowledge of previous work (Kepler), there is no reason to assume that a solution exist in terms of simple mathematical functions. In these circumstances we have to resort to mathematical experiment, the aim being in this case, to manipulate the right-hand side of

$$\ddot{\mathbf{r}} = -k\mathbf{r}/r^3$$

into a form that can readily be integrated. Intuition suggests a study of

$$\frac{\mathrm{d}}{\mathrm{d}t}\left(\frac{\mathbf{r}}{r}\right) = \frac{r\dot{\mathbf{r}} - \dot{r}\mathbf{r}}{r^2}$$

$$= \frac{r^2\dot{\mathbf{r}} - r\dot{r}\mathbf{r}}{r^3}.$$

Now

$$r^2 = \mathbf{r} \cdot \mathbf{r}$$

and

$$r\dot{r} = \frac{1}{2}\frac{\mathrm{d}}{\mathrm{d}t}(r^2)$$

$$= \frac{1}{2}\frac{\mathrm{d}}{\mathrm{d}t}(\mathbf{r} \cdot \mathbf{r})$$

$$= \mathbf{r} \cdot \dot{\mathbf{r}}.$$

So

$$r^2\dot{\mathbf{r}} - r\dot{r}\mathbf{r} = (\mathbf{r} \cdot \mathbf{r})\dot{\mathbf{r}} - (\mathbf{r} \cdot \dot{\mathbf{r}})\mathbf{r}$$

$$= \mathbf{r} \times (\dot{\mathbf{r}} \times \mathbf{r})$$

$$= \mathbf{r} \times \mathbf{h}_0,$$

and

$$\frac{\mathrm{d}}{\mathrm{d}t}\left(\frac{\mathbf{r}}{r}\right) = \frac{-\mathbf{r} \times \mathbf{h}_0}{r^3}.$$

On taking the vector product of both sides of eqn (6.20) with $\mathbf{h}_0$ and using this last result we find

$$\ddot{\mathbf{r}} \times \mathbf{h}_0 = k\frac{\mathrm{d}}{\mathrm{d}t}\left(\frac{\mathbf{r}}{r}\right).$$

As $\mathbf{h}_0$ is a constant, this equation can be integrated to give

$$\dot{\mathbf{r}} \times \mathbf{h}_0 = k(\mathbf{r}/r) + \mathbf{B} \qquad (6.25)$$

where $\mathbf{B}$ is a constant (vector) of integration.

Noting that $\mathbf{r} \cdot (\dot{\mathbf{r}} \times \mathbf{h}_0) = \mathbf{h}_0 \cdot (\mathbf{r} \times \dot{\mathbf{r}}) = h_0^2$, we now take the scalar product of both sides of eqn (6.25) with $\mathbf{r}$ to obtain

$$h_0^2 = kr + \mathbf{B} \cdot \mathbf{r}. \tag{6.26}$$

Referring to Fig. 6.15, at perigee v is perpendicular to r, so $\dot{\mathbf{r}} \times \mathbf{h}_0 = -r_p v_p^2 \mathbf{e}$, where $\mathbf{e}$ is a unit vector directed along the line joining perigee and apogee. Substitution for $\dot{\mathbf{r}} \times \mathbf{h}_0$ in eqn (6.25) yields

$$\mathbf{B} = (k - r_p v_p^2)\mathbf{e}$$
$$= -(v_p^2 - v_c^2) r_p \mathbf{e}.$$

On substituting this in eqn (6.26) and replacing h_0^2 by $r_p v_p^2$ we find

$$\frac{r}{\mu^2 - 1} = \frac{r_p \mu^2}{\mu^2 - 1} + \mathbf{e} \cdot \mathbf{r} \tag{6.27}$$

where $\mu = v_p/v_c$. If $\mu = 1$, r is constant at r_p and the path is circular. In general $v_p > v_c$ and $\mu > 1$.

Eqn (6.27) is the equation to a conic. It expresses the distance from a point on the orbit to a fixed point, the focus, as a constant times the distance from the point to a fixed line, the directrix, Fig. 6.16.

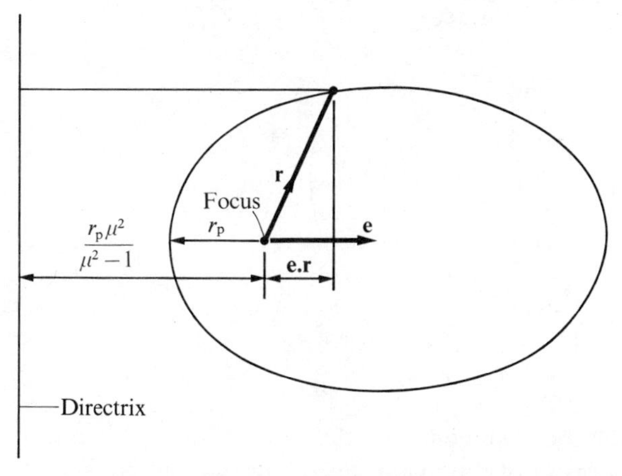

FIG. 6.16

If the constant ε, equal here to $(\mu^2 - 1)^{\frac{1}{2}}$, is less than unity the path is an ellipse. If $\varepsilon > 1$ the path is a hyperbola. The intermediate case where $\varepsilon = 1$ occurs when $\mu = \sqrt{2}$ and the velocity at perigee equals the velocity of escape. In this case the path is a parabola.

The method given here of integrating eqn (6.20) is an exercise in vector manipulation that does little to add to an understanding of mechanics. It is given for the sake of completeness. A much more elegant proof that the path is an ellipse uses geometry and follows directly from the energy relationship, eqn (6.22):

$$-v^2 - k/r = \text{constant}.$$

If p is the perpendicular distance from the centre of the sun to the tangent of the satellite's path then, from eqn (6.21)

$$pv = h_0.$$

Hence

$$\frac{h_0^2}{2p^2} - \frac{k}{r} = \text{constant.}$$

This is one of the standard forms of the equation to an ellipse†. Unfortunately, only a skilled geometer is likely to recognize it as such.

6.8.4. Spring-controlled orbit

A mass moves round a circular path under the constraint of an elastic spring, Fig. 6.17, the spring being coplanar with the path. The angular velocity is Ω so that the force P in the spring is $ma\Omega^2$, and its extension is P/s where s is the stiffness of the spring. If the mass is now given a small radial impulse its motion is a combination of the original steady rotation and a small amplitude vibration.

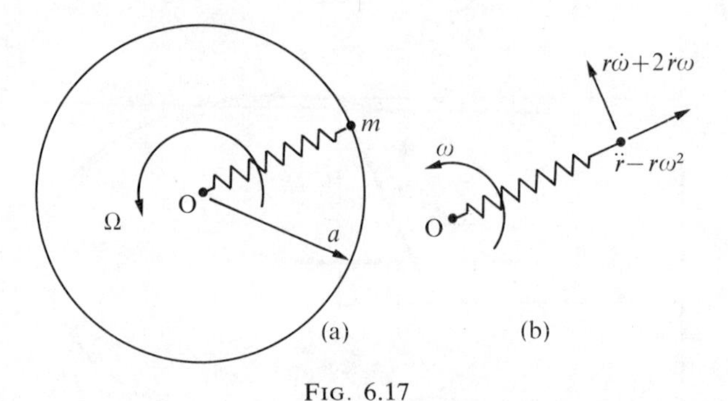

FIG. 6.17

It must not be assumed that the circumferential motion will be unaffected as a result of the disturbance having been applied radially. After the disturbance the only force on the mass is that exerted by the spring so that the moment of its momentum about O must remain constant at its original value. As the radius varies, the angular velocity must also vary. Taking the instantaneous angular velocity to be ω, the accelerations are as shown in Fig. 6.17(b). If r_0 is the unstretched length of the spring the equation of motion in the radial direction is

$$m(r\omega^2 - \ddot{r}) = s(r - r_0)$$

and in the circumferential direction it is

$$m(r\dot{\omega} + 2\dot{r}\omega) = 0.$$

† See Lockwood, E. H. (1961). *A book of curves*. Cambridge University Press.

This is a pair of simultaneous equations with ω and r as unknown functions of time. Fortunately the second equation can be integrated on sight after multiplying through by r

$$m(r^2\dot{\omega}+2r\dot{r}\omega)=0$$

$$\Rightarrow \qquad \frac{\mathrm{d}}{\mathrm{d}t}(mr^2\omega)=0$$

$$\Rightarrow \qquad mr^2\omega=\text{constant.}$$

This result amounts to no more than a statement that the moment of momentum is constant. The constant is the original value $ma^2\Omega$. It follows that $\omega=\Omega(a/r)^2$.

On substituting for ω in the first equation of motion we find

$$ma^4\Omega^2/r^3-m\ddot{r}=s(r-r_0).$$

The substitution $\ddot{r}=\dot{r}\,\mathrm{d}\dot{r}/\mathrm{d}r=\frac{1}{2}(\dot{r}^2)/\mathrm{d}r$ allows integration to yield an equation for $\dot{r}^2$. It can be recognized as a statement that the total energy of the system is constant. There is then some difficulty in completing the solution. The problem is simplified if we take account now of the fact that the perturbation is small. Let

$$r=a+e$$

where $e \ll a$. Then

$$ma^4\Omega^2/r^3 \approx ma\Omega^2(1-3e/a)$$

and

$$m\ddot{r}=m\ddot{e}$$

so that the equation of motion becomes

$$ma\Omega^2(1-3e/a)-m\ddot{e}=s(a-r_0+e).$$

In the initial steady-state rotation $s(a-r_0)=P=ma\Omega^2$. On cancelling these terms the equation of motion reduces to

$$m\ddot{e}+(s+3m\Omega^2)e=0.$$

This can be recognized as the equation for simple harmonic motion at a natural circular frequency of $\sqrt{(s/m+3\Omega^2)}$.

The question might well be asked: Why choose a purely algebraic method of solution to this problem rather than a vector method similar to that used to study satellite motion? In part the answer is that we are trying here to illustrate different methods of approach. The main reason is, however, that the methods chosen appear to the writer to be the most appropriate for each of the problems. What is most appropriate in any particular case is often a matter of taste, and the reader must develop his own taste. What needs to be recognized is that there is no one method that is 'best' in all circumstances.

6.8.5. Motion of a spherical pendulum

A particle of mass m is suspended in a gravity field from a fixed point by an inextensible string of length l so that it moves in a spherical surface. There are two special cases for the motion of the mass. In one it moves in a single vertical plane and the motion is that of the simple pendulum studied in Section 6.8.1. In the other case the mass moves in a horizontal circular path with the string sweeping out a conical surface. We will study the general motion.

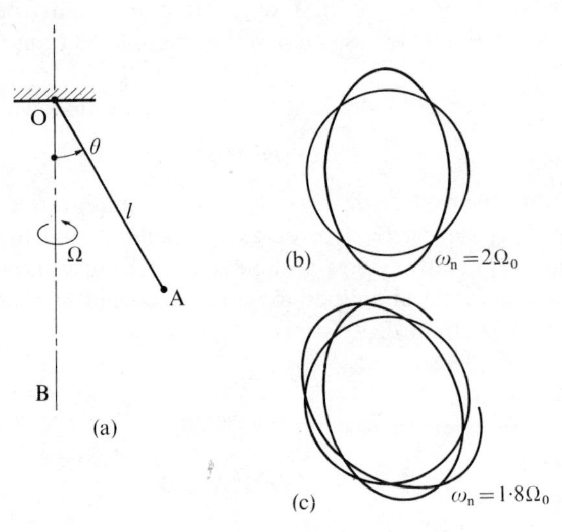

FIG. 6.18

At any instant the string AO is at an angle θ with the downward vertical OB, Fig. 6.18. At the same instant the plane AOB turns about OB with an angular velocity Ω. The mass has several components of acceleration. In the view depicted in Fig. 6.18(a) the components are $l\dot{\theta}^2$ and $l\ddot{\theta}$ along and perpendicular to AO, and $r\Omega^2$ directed inwards from A. Perpendicular to plane AOB there is a further component of acceleration $(r\dot{\Omega}+2\dot{r}\Omega)$. It should be noted that the mass has only two degrees of freedom so that the variable $r=l\sin\theta$ is redundant though its introduction facilitates the presentation of the analysis. Due to the complexity of the motion it is very easy to make an error in writing down the components of acceleration and setting up the equations of motion. A direct appeal to the momentum and energy equations allows this difficulty to be circumvented.

The only forces on the mass are its weight and the tension in the string. The latter passes through point O and the former is parallel to OB. Hence the moment of momentum of the particle about OB is constant,

$$mr^2\Omega = \text{constant}$$

$$\Rightarrow \qquad r^2\Omega = r_0^2\Omega_0,$$

where r_0 and Ω_0 are the initial values of r and Ω.

There is no dissipation of energy so that

$$T+V=\text{constant}$$

$$\Rightarrow \qquad \tfrac{1}{2}m\{(l\dot{\theta})^2+(r\Omega)^2\}-mgl\cos\theta=\text{constant}.$$

On eliminating r between the two equations and putting $r_0=l\sin\theta_0$ we find

$$\tfrac{1}{2}\dot{\theta}^2-\frac{g}{l}\cos\theta=\frac{-\tfrac{1}{2}\Omega_0^2\sin^4\theta_0}{\sin^2\theta}+\text{constant}.$$

There seems to be little hope of obtaining a general solution to this equation so we will study some special cases.

First, if $\Omega_0=0$ the problem is that of the simple pendulum and the equation reduces to

$$\tfrac{1}{2}\dot{\theta}^2-\frac{g}{l}\cos\theta=\text{constant}$$

as derived in Section 6.8.1.

Second, if θ is constant at θ_0 with $\dot{\theta}=0$, then $\Omega_0^2\cos\theta_0=g/l$, which is the result for a conical pendulum.

If a conical pendulum is given a small impulse, the problem is similar to that for the quasi-circular motion of the spring-controlled mass studied in the previous sections. As before we split the motion up into two components putting $\theta=\theta_0+\psi$, where $\psi\ll\theta_0$. Substitution of this in the general equation of motion given above involves some difficulty because it will be found necessary to retain terms in ψ^2. It is better first to differentiate the equation with respect to time to give

$$\ddot{\theta}+\frac{g}{l}\sin\theta=\Omega_0^2\sin^4\theta_0\frac{\cos\theta}{\sin^3\theta}$$

before making the substitutions

$$\sin\theta=\sin\theta_0+\psi\cos\theta_0$$

$$\sin^4\theta=\sin^4\theta_0+4\psi\cos\theta_0\sin^3\theta_0$$

$$\cos\theta=\cos\theta_0-\psi\sin\theta_0$$

and yielding in turn

$$\ddot{\psi}+\Omega_0^2(4\cos^2\theta_0+\sin^2\theta_0)\psi=0.$$

This is the equation to simple harmonic motion with natural circular frequency

$$\omega_n=\Omega_0(1+3\cos^2\theta_0)^{\frac{1}{2}}.$$

If θ_0 is small, $\omega_n\approx 2\Omega_0$ and the path seen in plan is elliptical as shown in Fig. 8.16(b). If $\theta_0=30°$, $\omega_n=1.8\Omega_0$. This means that the perturbation completes only 1·8 cycles in the time that it takes the main motion to complete one cycle. The result is that the motion has the characteristic of a slowly rotating ellipse, Fig. 6.18(c).

6.8.6. Stability of a rotating simple pendulum

The simple pendulum in Fig. 6.19 has a light, rigid rod in place of an inextensible string. At O the rod is pivoted freely about a horizontal axis to a vertical shaft. If the shaft is stationary the rod and mass behave as a simple pendulum. We will consider what happens when it rotates at a steady speed Ω.

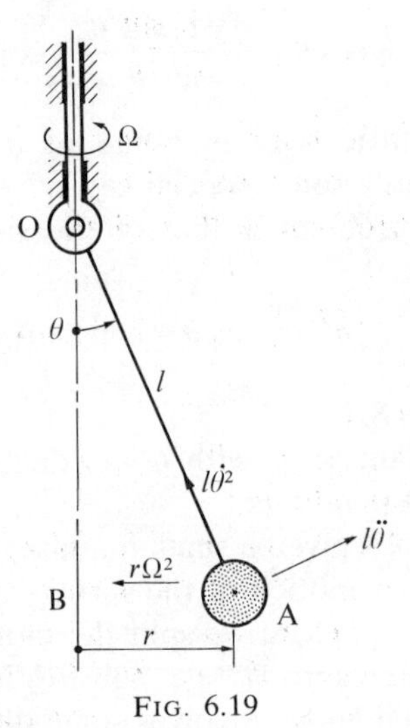

FIG. 6.19

It was emphasized above that as a consequence of the moment of momentum being conserved about the vertical axis through O the precessional speed Ω of a spherical pendulum must vary. It follows that in the present example a driving or restraining torque must be applied about the vertical axis to keep Ω constant. The system is consequently not a conservative one and caution must be exercised in using the energy equation. Indeed, the safest course is to eschew its use altogether in such cases, and to make direct use of Newton's laws.

The components of acceleration of the mass are precisely as they are for the spherical pendulum except that Ω is now constant. The components of acceleration in the plane OAB are as shown in Fig. 6.19. The forces acting on the mass in this plane are the tension in the rod and the weight mg. The equation of motion for the direction perpendicular to the rod is therefore

$$mg \sin\theta = m(r\Omega^2\cos\theta - l\ddot{\theta})$$

$$\Rightarrow \qquad \ddot{\theta} + (g/l - \Omega^2\cos\theta)\sin\theta = 0.$$

If θ is small this equation reduces to

$$\ddot{\theta} + (g/l + \Omega^2)\theta = 0,$$

the equation for simple harmonic motion at a natural circular frequency $\omega_n = (g/l - \Omega^2)^{\frac{1}{2}}$. When $\Omega = 0$ the natural frequency is that for a simple pendulum. As Ω is increased the frequency of the oscillations decreases and becomes zero when $\Omega = (g/l)^{\frac{1}{2}}$. At this stage the mass can be displaced slightly from its central position and it will simply stay there.

On increasing Ω still further $(g/l - \Omega^2)$ becomes negative, $-\beta^2$ say, giving the equation of motion

$$\frac{d^2\theta}{dt^2} - \beta^2\theta = 0.$$

The general solution of this equation is

$$\theta = Ae^{\beta t} + Be^{-\beta t}.$$

The constants A and B depend upon the initial velocity and displacement of the mass. In theory A and B could both be zero and the mass would remain in equilibrium with $\theta = 0$. The equilibrium would, however, be unstable, for the slightest disturbance would result in θ tending to grow indefinitely with time due to the term $Ae^{\beta t}$. The motion would not, in fact, be unlimited because the equation of motion has been simplified by making the assumption that θ is small. Once θ has grown sufficiently the simplified equation of motion ceases to apply.

The stable equilibrium position for $\Omega^2 > g/l$ is given by

$$\cos\theta = g/l\Omega^2,$$

as for a conical pendulum.

It can be verified as an exercise that the substitution $\theta = \theta_0 + \psi$, where $\theta_0 = \cos^{-1}(g/l\Omega^2)$ leads to

$$\ddot{\psi} + \Omega^2 \sin^2\theta_0 \psi = 0$$

as the equation of motion for small oscillations about the equilibrium position. Once more the motion is simple harmonic the natural circular frequency being on this occasion $\Omega \sin\theta_0$.

6.9. Problems

1. An 80 kg man stands on a spring-operated weighing machine in a lift. The machine registers 100 kg. What is the acceleration of the lift?

   There is in the lift a pendulum clock which keeps good time when unaccelerated. How much does it gain or lose now?

2. A ship moves in a straight line in still water. With its engines stopped, it is brought to rest by the water resistance. At one instant its speed is 3 m s^{-1} and 60 s later it has fallen to 1·8 m s^{-1}.

   Taking the resistance to vary as the speed squared for speeds over 0·6 m s^{-1} and as the speed for speeds below 0·6 m s^{-1}, find the distance travelled in coming to rest from a speed of 3 m s^{-1}.

3. The table shows the net thrust on a 7250 kg aircraft at various ground speeds up to the take-off speed of 192 km h^{-1}. Estimate the ground run before the aircraft is airbourne.

Speed (km h^{-1})	0	48	96	144	192
Thrust (kN)	21·5	20·7	19·8	18·5	16·7

4. The variation of resistance with speed of a ship of 80 000 Mg is given in the table below. For speeds between 25 and 50 km h^{-1} the propulsive force exerted by the propellors may be assumed to be constant at 10 MN. Find the time taken to increase the speed from 25 to 50 km h^{-1} and the distance travelled in that time.

Speed (km h^{-1})	25	30	35	40	45	50
Resistance (MN)	1·85	2·65	3·55	4·80	6·10	7·60

5. The resistance force to motion of a railway train is kv^2 per unit mass of the train. If the train starts from rest and the power is constant, show that the train would need to travel an infinite distance to attain full speed, but will attain half full speed in a distance $\frac{1}{3k}\log_e(8/7)$.

6. A railway train of mass 6×10^5 kg is being driven along a straight level track by an electromagnetic device which applies a force of 2 MN for the first 50 m of each kilometre travelled. The resistance to motion may be taken as $32{\cdot}4\,v^2$ N, where v is the speed of the train in m s^{-1}.

(i) What is the average speed of the train?

(ii) What is the maximum speed of the train?

(iii) Will the journey be comfortable?

It may be assumed that the fluctuations of speed are small.

7. An electrostatic cathode-ray tube is operated with its cathode at $-2{\cdot}00$ kV and its anode earthed. The beam of electrons passes between a pair of parallel deflecting plates 1·0 cm apart and 5·0 cm long in the axial direction. The beam then strikes a flat screen which is 20 cm from the centre of the plates and perpendicular to the axis of the tube.

The plates are held initially at +50 V and −50 V with respect to earth. Estimate the displacement of the spot on the screen when the leads to the deflection plates are interchanged. Neglect fringing effects.

8. An electron of mass m is accelerated through a potential difference V and projected perpendicularly into a magnetic field. The field is of uniform strength B within a circle of radius a, and there are no fringe effects. Determine B if the particle is to emerge radially from the field with its path deflected through an angle of 120° from its original direction. The force on an electron with charge $-e$ moving with velocity v in a magnetic field B is $-e\mathbf{v}\times\mathbf{B}$.

9. An aircraft of mass 5 Mg, on landing, is travelling horizontally at a speed 20 m s^{-1}. It is then brought to rest partly by the action of a constant frictional force of 50 kN and partly by the action of a system of wires which exert a restoring force $P(x)$ when displaced a distance x, as shown in Fig. 6.20.

Find the displacement when the aircraft comes to rest, and show that although it begins to move backwards it cannot regain the position of zero displacement.

10. A mass resting on the ground is attached to the lower end of a light vertical spring which is just taut. The upper end of the spring is then given a constant upward vertical velocity V and the mass begins to move when the upper end has risen h. Find an expression for the greatest extension of the spring in terms of V, h, and g.

11. A light rod AB of length $2l$ can rotate in a horizontal plane about a vertical axis AZ, and a bead of mass m can slide freely on the rod. Initially, the whole system is at rest with the bead situated midway between A and B. If a torque T is applied to the rod about the axis AZ for a very short time interval Δt, show that the bead reaches B with a sliding velocity $Q\sqrt{3}/2\,ml$, where $Q=\int_0^{\Delta t} T\,\mathrm{d}t$.

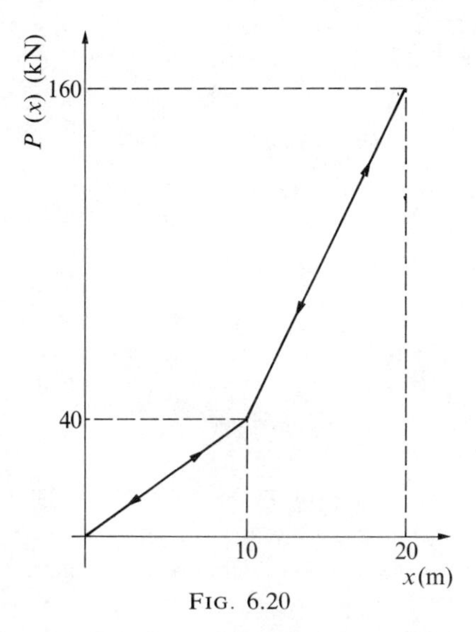

FIG. 6.20

12. A straight tube of length $4l$ rotates in a horizontal plane about a vertical axis through one end. The angular velocity is constant and equal to ω.

A smooth particle, which is free to slide along the tube, is released without radial velocity at a distance l from the axis of rotation. Show that the particle will leave the tube after a time $(\cosh^{-1}4)/\omega$, and find the absolute velocity and acceleration of the particle just before it reaches the end of the tube.

13. Calculate the height above the earth's surface of an artificial satellite in a 'stationary' circular orbit above the equator. Take the gravitational acceleration as $9{\cdot}8\ \mathrm{m\,s^{-2}}$ at the surface of the earth, and assume an inverse-square law of gravitational force with distance from the earth's centre. The circumference of the earth at the equator is 40 Mm.

14. Four equal rods are pinned together in the form of a rhombus which lies in a horizontal plane. Two opposite corners are connected by a spring. Across the other two corners there are applied equal and opposite inwardly directed forces.

Investigate the existence and stability of positions of equilibrium for this system by sketching suitable graphs.

15. A particle of mass m is free to move in a plane. It is subject to forces of attraction towards two fixed points of that plane $(\pm a, 0)$, the magnitude of each force being k times the distance from the particle to the point in question.

What is the potential energy V of the particle when it is at the point (x, y)? Sketch contours of V.

What is the resultant force on the particle and in which direction does it act? If the particle starts from rest describe its subsequent motion.

16. A particle lies in a two-dimensional potential field given by

$$V = 3x^2 + 2xy + y^3,$$

x and y being measured in metres and V in N m, and it is subjected to no forces other than those arising from this field.

(i) Find the magnitude and direction of the force at the point $(1, 2)$.

(ii) Given that the particle has a mass of 4 kg and moves freely between the point $(1, 2)$ where its speed is $6\ \mathrm{m\,s^{-1}}$, and the point $(2, 3)$, find its speed as it passes through the latter point.

17. A cylinder of radius R is fixed with its axis horizontal. Another cylinder of radius r and mass m is placed on top of the first cylinder so that their axes are parallel. The centre of gravity of the upper cylinder is h below its geometric centre.

Show that the position is stable provided $h > r^2/(R+r)$.

18. A uniform lamina ABCD has sides AB $= 2a$ and BC $= 2b$. A string of length $4a$ is connected to A and B and the lamina hangs in equilibrium on a small frictionless pulley with AB horizontal.

Show that the equilibrium is stable only if $b > /\sqrt{3}a$.

19. An artificial earth satellite is released from its launching vechicle at an altitude of 400 km with a speed of 9 km s^{-1} in a horizontal direction. If the mass of the satellite is 100 kg and the radius of the earth is 6·5 Mm, find how much energy is required.

(i) To raise the satellite to the given altitude.
(ii) To give it the necessary initial speed.

Neglect effects due to the rotation of the earth and air resistance.

20. An artificial satellite of mass m is transferred from one circular orbit at speed v_p, to another at speed $v_p/7$, by exerting a short-duration thrust tangentially at P, Fig. 6.21.

What is the required impulse at P?

What must be done at A?

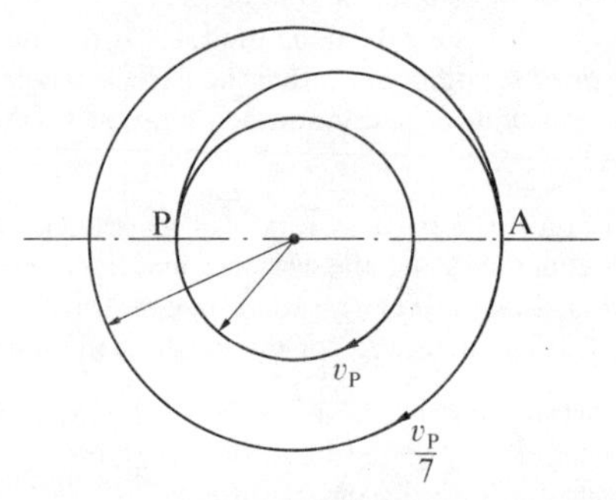

FIG. 6.21

21. A smooth, straight, rigid rod is fastened at right-angles to a rotating vertical axis so that it rotates in a horizontal plane. The axis is driven by whatever torque is necessary to maintain a constant angular velocity Ω.

Along the rod can slide freely a small 1 kg mass, and this mass is attached to the axis by a tension/compression spring of unstretched length 0·1 m and stiffness 10 N m^{-1}.

Show that the equation of motion of the mass in the radial direction in terms of its distance r from the axis is

$$\ddot{r} + (10 - \Omega^2)r = 1.$$

Find expressions for the equilibrium value of r and the natural frequency of radial oscillations in terms of Ω. What happens to these as Ω is slowly increased?

22. A particle of mass m is attached to a fixed point O by a light spring of stiffness k and unstretched length l. Initially it is moving in a circular path of radius r_O with O as centre. Show that the angular velocity of the centre-line of the spring is given by Ω where

$$\Omega^2 = \frac{k}{m}\left(1 - \frac{l}{r_O}\right).$$

The particle is displaced slightly from its steady orbit and allowed to move freely. By considering the length of the spring to be $r_O + x$, where x is small, and noting that (unlike the rod in question 21) the angular velocity of the centre-line of the spring does *not* remain constant, show that the frequency of small fluctuations of spring length is

$$\frac{1}{2\pi}\sqrt{\frac{k}{m}}\left(4-3\frac{l}{r_O}\right).$$

23. A particle P of mass m rests in a smooth straight groove in a circular disc. The groove is distant a from the centre of the disc and the position of the particle is given by the variable distance x from the mid-point A of the groove, as shown in Fig. 6.22. The disc rotates about a vertical axis perpendicular to its plane at a constant angular speed Ω. Express the position vector $\mathbf{r}$ of P with respect to O in terms of the unit vectors $\mathbf{e}_1$ (along OA) and $\mathbf{e}_2$ (along AP). Hence derive expressions for the components of acceleration of P.

The particle is attached to B by a spring of stiffness s and natural length AB. Show that it can oscillate relative to the disc at a frequency

$$\frac{1}{2\pi}\sqrt{\left(\frac{s}{m}-\Omega^2\right)}.$$

What happens if $\Omega > \sqrt{\frac{s}{m}}$?

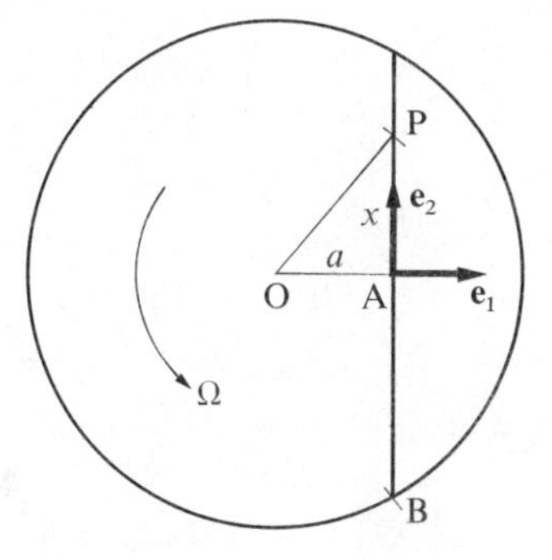

FIG. 6.22

24. In the motion of a particle system, under what circumstances is angular momentum conserved (a) about a fixed point, (b) about a fixed axis?

A small body is attached to a fixed point by a light string of length 1 m. The string is initially held taut at 60° to the downward vertical and the body is projected at 3 m s^{-1} horizontally and perpendicular to the string. Show that this speed is too slow to maintain a circular path. Fing the maximum speed during the subsequent motion.

25. A smooth, thin-walled bowl is in the form of a sphere of radius a and centre O with a hole of radius $-a\sqrt{3}$ cut out of it. The axis of symmetry is vertical and the hole is uppermost.

(i) A particle is projected along the inner surface of the bowl from its lowest point. Show that the particle will not fall back into the bowl if its initial velocity is greater than $\sqrt{(5\,ga)}$.

(ii) The particle is projected horizontally along the surface of the bowl from a point $a/2$ from the axis. What is its greatest possible initial velocity if it is not to rise above the level of O?

26. In the motion of a particle, under what circumstances is (a) total mechanical energy conserved, and (b) moment of momentum conserved (i) about a point, and (ii) about an axis?

A smooth conical vessel of height 2 m is fixed with its vertex downwards and its axis vertical. A particle is projected with horizontal velocity V on the inner surface at a height of 1 m measured vertically above the vertex. Show that if V is not greater than $5{\cdot}1\ \mathrm{m\,s^{-1}}$, the particle will remain inside the vessel.

Describe the path followed by the particle.

7. Multiparticle systems

SO FAR we have concentrated on the dynamics of a single particle. It was not felt necessary at any stage to give a formal definition of what is meant by a particle, for in every situation it was clearly implied that we were considering a body whose dimensions were insignificant in relation to the magnitude of its motion. The observer on the ground finds it hard to consider the motion of a spacecraft in terms other than those of a particle. An astronaut in the same spacecraft has a different perspective. The general question to be considered is whether it is valid to apply the laws of particle motion to a body of finite size.

7.1. The mass and mass centre of a set of particles

We define the mass M of a set of particles as the sum of the masses of the individual particles:

$$M = \sum m_i \tag{7.1}$$

the summation being extended over all the particles.

We define the mass centre G as the point whose position $\mathbf{r}_G$ relative to an arbitrary point O is given by the equation

$$M\mathbf{r}_G = \sum m_i\mathbf{r}_i \tag{7.2}$$

where $\mathbf{r}_i$ locates the particle of mass m_i relative to O.

Although it is necessary to choose a point of reference O to locate G, the absolute position of G does not depend on the choice of reference point. If in Fig. 7.1 O′ is chosen instead of O

$$\begin{aligned} M\mathbf{r}'_G &= \sum m_i\mathbf{r}'_i \\ &= \sum m_i(\mathbf{r}_i - \mathbf{d}). \end{aligned}$$

But in this equation $\mathbf{d}$ is a constant so

$$\sum m_i\mathbf{d} = (\sum m_i)\mathbf{d} = M\mathbf{d}$$

and

$$\sum m_i\mathbf{r}_i = M\mathbf{r}_G$$

so that

$$M\mathbf{r}'_G = M(\mathbf{r}_G - \mathbf{d}).$$

Hence $\mathbf{r}'_G = (\mathbf{r}_G - \mathbf{d})$ so that G and G′ are the same point.

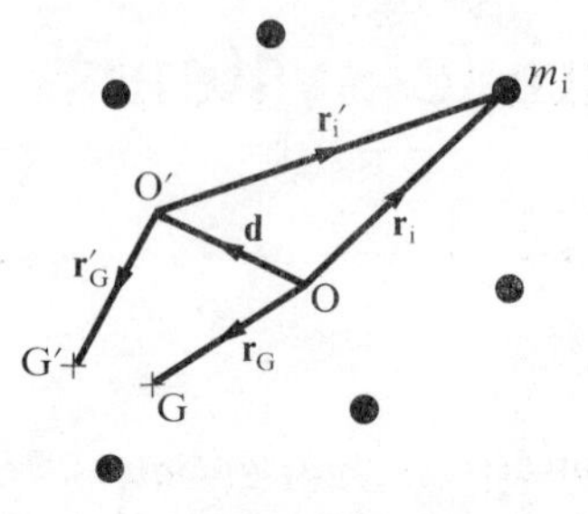

FIG. 7.1

7.1.1. Acceleration of the mass centre

Let the particle be subjected to external forces **P** and internal forces **p**, the internal forces being gravitational or magnetic attraction between particles, or due to light springs or rods between particles. For a typical particle

$$\mathbf{P}_i + \mathbf{p}_i = m_i \ddot{\mathbf{r}}_i.$$

Summing over all particles

$$\sum (\mathbf{P}_i + \mathbf{p}_i) = \sum m_i \ddot{\mathbf{r}}_i.$$

From Newton's Third Law the internal force **p** on any one particle must be associated with forces on other particles such that $\sum \mathbf{p}_i = 0$.

As all the masses m_i are constant

$$\sum m_i \ddot{\mathbf{r}}_i = \frac{d^2}{dt^2} \sum m_i \mathbf{r}_i = \frac{d^2}{dt^2} (M\mathbf{r}_G) = M\ddot{\mathbf{r}}_G.$$

So

$$\sum \mathbf{P}_i = \mathbf{P} = M\ddot{\mathbf{r}}_G = M\mathbf{a}_G \tag{7.3}$$

Hence the mass centre of a set of particles moves as if all the forces and all the mass were concentrated there. It follows that as far as the motion of the mass centre is concerned a set of particles or a rigidy body can be treated as a single particle whose mass is equal to the total mass.

Eqn (7.3) reveals nothing about the motion of the particles relative to G and, in particular, gives no indication of any rotation about G.

The mass centre of a set of particles is usually taken to be synonymous with its centre of gravity, the point through which the resultant gravitational force acts. Indeed the common practice of using the letter G to denote the centre of mass implies this. The two centres are coincident when the masses are in a constant gravitational field but are separate when the inverse-square law applies to the gravitational attraction. The distinction is important and can be exploited in maintaining the orientation of a satellite with respect to the earth, but is insignificant in most normal circumstances.

7.2. Linear momentum

The linear momentum of a set of particles is

$$\sum m_i\dot{\mathbf{r}}_i = \sum m_i(\dot{\mathbf{r}}_G + \dot{\boldsymbol{\rho}}_i)$$

$$= (\sum m_i)\dot{\mathbf{r}}_G + \frac{d}{dt}\sum m_i\boldsymbol{\rho}_i.$$

where $\boldsymbol{\rho}_i$ is the vector from G to the ith mass, Fig. 7.2. It follows from the definition of G that $\sum m_i\boldsymbol{\rho}_i = 0$ so

$$\sum m_i\dot{\mathbf{r}}_i = M\dot{\mathbf{r}}_G = M\mathbf{v}_G. \tag{7.4}$$

The linear momentum of a set of particles is therefore simply the velocity of its centre of mass multiplied by its total mass.

From eqn (7.3)

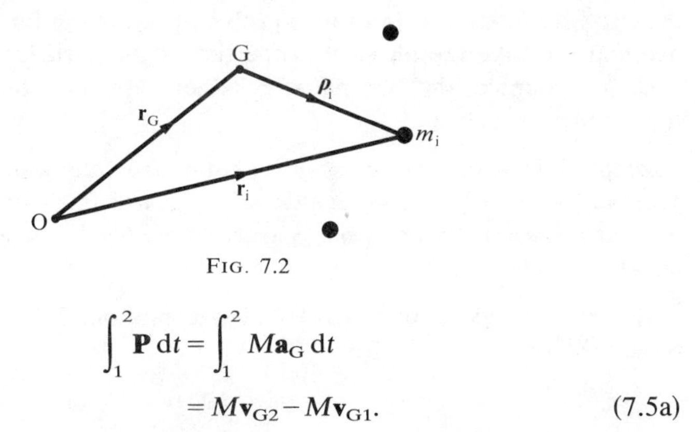

FIG. 7.2

$$\int_1^2 \mathbf{P}\,dt = \int_1^2 M\mathbf{a}_G\,dt$$

$$= M\mathbf{v}_{G2} - M\mathbf{v}_{G1}. \tag{7.5a}$$

Thus, forces acting on a set of particles cause a change in linear momentum equal to the impulse of the forces.

Eqn (7.5a) is most useful when the set of particles form a solid body. When the particles are truly separate it is usually more convenient to write eqn (7.5a) in the alternative form

$$\int_1^2 \mathbf{P}\,dt = (\sum m_i\mathbf{v}_i)_2 - (\sum m_i\mathbf{v}_i)_1 \tag{7.5b}$$

7.2.1. Conservation of linear momentum

If the resultant external force on a set of particles is zero, $\mathbf{P} = 0$, the total linear momentum is constant. In particular, there is no net loss of momentum solely as a result of collisions or other interactions between particles of the same set.

7.2.2. Collisions

The principle of conservation of linear momentum is not by itself sufficient to determine the motion of two particles after they have collided, except in special circumstances.

$\mathbf{v}_{A1}$ $\mathbf{v}_{B1}$ $\mathbf{v}_{A2}$ $\mathbf{v}_{B2}$

Before impact Impact After impact

FIG. 7.3

In Fig. 7.3 particle A is assumed to be moving in the same direction as particle B and, moving faster than B, crashes into it. After the collision the two particles have the velocities shown. In these circumstances linear momentum is conserved and

$$m_A\mathbf{v}_{A1}+m_B\mathbf{v}_{B1}=m_A\mathbf{v}_{A2}+m_B\mathbf{v}_{B2}.$$

The velocities on the left-hand side of this equation are known; those on the right are unknown. In order to solve the equation for $\mathbf{v}_{A2}$ and $\mathbf{v}_{B2}$ it is necessary to take the physical properties of the particles into account.

If, for example, the two particles adhere together then $\mathbf{v}_{B2}=\mathbf{v}_{A2}$ and the problem is solved.

Example 7.1: A bullet of mass m is fired horizontally with velocity $\mathbf{v}$ *into a stationary block of mass M supported so that it is free to move horizontally without restriction. What is the velocity of the block immediately after the bullet has lodged in it?*

As no external forces act to hinder motion, linear momentum is conserved and

$$m\mathbf{v}=(m+M)\mathbf{v}'$$

$$\Rightarrow \qquad \mathbf{v}'=\frac{m}{m+M}\mathbf{v}.$$

Although momentum is conserved, energy is not. The kinetic energy after impact is

$$\begin{aligned}&\tfrac{1}{2}(m+M)v'^2\\ &=\frac{1}{2}\frac{m^2}{m+M}v^2\\ &=\frac{1}{2}mv^2-\frac{1}{2}\frac{Mm}{M+m}v^2\end{aligned}$$

the loss of energy being $\frac{1}{2}\frac{Mm}{M+m}v^2$. In physical terms the energy dissipated by the bullet as it tears into the block provides an obvious explanation for this loss. On the other hand, both energy and momentum are derived by integration of the equation of motion so that it is not immediately obvious how the loss of energy is explained in mathematical terms. Consider first momentum, treating the two masses separately and letting the forces at any instant be $-\mathbf{p}$ on the bullet and $+\mathbf{p}$ on the block.

Then

$$-\int \mathbf{p}\,\mathrm{d}t = m\mathbf{v}' - m\mathbf{v}$$

and

$$+\int \mathbf{p}\,\mathrm{d}t = M\mathbf{v}' - 0.$$

On adding these two equations the two integrals cancel and the momentum equation follows:

$$m\mathbf{v} = (M+m)\mathbf{v}'.$$

The energy equations are

$$\tfrac{1}{2}m\mathbf{v}^2 - \int \mathbf{p}\,.\,\mathrm{d}\mathbf{r}_1 = \tfrac{1}{2}m\mathbf{v}'^2$$

and

$$0 + \int \mathbf{p}\,.\,\mathrm{d}\mathbf{r}_2 = \tfrac{1}{2}M\mathbf{v}'^2$$

for the bullet and the block respectively. At every instant the forces on the bullet and the block are equal and opposite, hence the mutual cancellation of the two integrals in the momentum equation, but the two particles do not move the same distance in the same period of time. The bullet moves faster than the block during the period of penetration so that $\mathrm{d}\mathbf{r}_1$ for the bullet does not equal $\mathrm{d}\mathbf{r}_2$ for the block. The two integrals in the energy equations are consequently unequal and do not balance each other out when the equations are added together. The discrepancy equals the loss of energy.

Example 7.2: A railway truck of mass m and velocity v runs into a stationary truck also of mass m. The buffers are restrained by simple springs of stiffness s, without precompression. The buffers and springs may be considered to be massless. Determine the maximum amount of energy stored in the buffer springs and the velocities of the trucks after they have separated.

Let the compression of a buffer spring at a typical instant be x. At maximum compression $\mathrm{d}x/\mathrm{d}t$ is zero and the two trucks have the same velocities, v' say. Linear momentum is conserved so that

$$mv = 2mv'$$

$$\Rightarrow \qquad v' = v/2.$$

As there is no impact between masses, the buffers being ‘light’, there is no loss of energy and

$$\tfrac{1}{2}mv^2 = 2\times\tfrac{1}{2}m(v/2)^2 + (\text{energy stored in buffers})$$

$$\Rightarrow \qquad \text{stored energy} = \tfrac{1}{4}mv^2.$$

After separation, let the velocity of the truck that was initially moving be v_1 and of the other v_2. Equating momentum before and after collision

$$mv = mv_1 + mv_2$$

$$v = v_1 + v_2.$$

Equating energy before and after collision,

$$\tfrac{1}{2}mv^2 = \tfrac{1}{2}mv_1^2 + \tfrac{1}{2}mv_2^2$$

$$\Rightarrow \qquad v^2 = v_1^2 + v_2^2.$$

On squaring and subtracting the two equations for velocities we find

$$0 = 2v_1v_2.$$

So, either v_1 or v_2 is zero. Unless the first truck has passed through the second $v_1 = 0$ and $v_2 = v$.

In an analogous situation two identical simple pendulums hang side by side with hardened steel bobs just touching. If the bob of one pendulum is pulled to one side and released so as to have normal impact with the other bob it will be found that the pendulums are alternately in motion and stationary between successive impacts. Starting with the assumption that there is no loss of energy in the collisions between the bobs the apparently strange behaviour is fully explained by the truck analysis above. The full reasons for there being virtually no loss of energy in this case are complex, though making the balls of material that is not readily damaged clearly makes a substantial contribution.

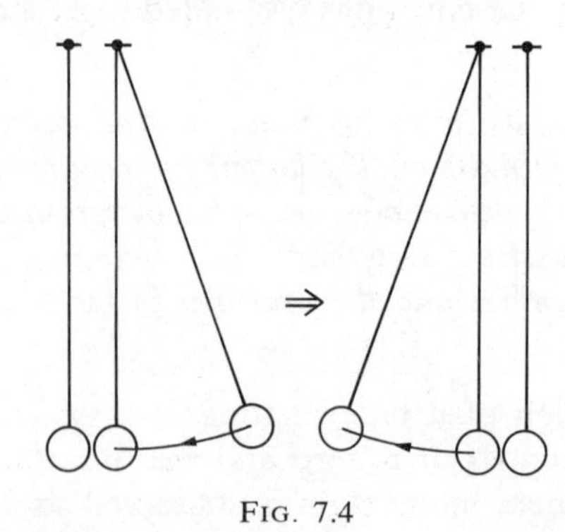

FIG. 7.4

In general, a collision between two massive bodies results in some permanent damage at the point of impact and after separation the two bodies may be vibrating. The energy associated with the vibration must come from the original kinetic energy of the system and, as there is no way of converting the vibrational energy back into kinetic energy of translation, it is 'lost'. The losses of energy due to damage and residual vibration are usually not readily calculable.

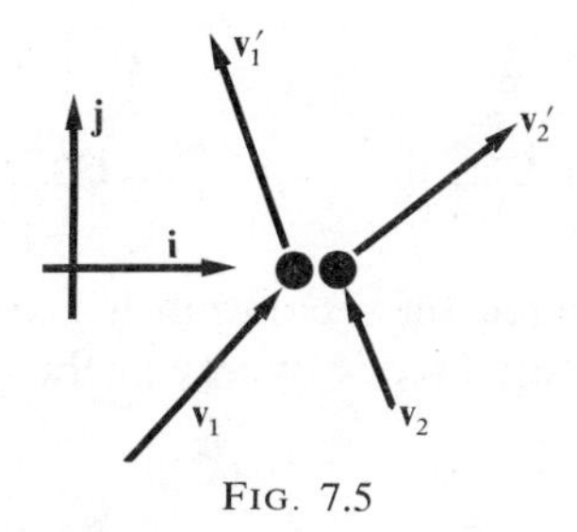

FIG. 7.5

In order to make the calculation of post-collision velocities practicable it is usual to introduce an experimental factor, the *coefficient of restitution.* In Fig. 7.5 two particles initially have velocities $\mathbf{v}_1$ and $\mathbf{v}_2$. They collide so that the direction for normal impact is $\mathbf{i}$. The coefficient of restitution e is defined by the equation

$$(\mathbf{v}_2' - \mathbf{v}_1') \cdot \mathbf{i} = e(\mathbf{v}_1 - \mathbf{v}_2) \cdot \mathbf{i}. \tag{7.6}$$

In the direction of glancing impact it is commonly assumed that the momentum of each particle is conserved, thus

$$\mathbf{v}_1' \cdot \mathbf{j} = \mathbf{v}_1 \cdot \mathbf{j} \quad \text{and} \quad \mathbf{v}_2' \cdot \mathbf{j} = \mathbf{v}_2 \cdot \mathbf{j}. \tag{7.7}$$

The coefficient of restitution depends on the materials, size and shape of the two bodies. It is related to the ratio of energy recovered as two bodies separate from a normal collision to that absorbed as they squash together in a normal collision.

Example 7.3: A mass m runs at velocity v into a stationary mass nm. Determine the maximum amount of energy absorbed in the impact and the proportion that is recovered on separation if the coefficient of restitution is e.

Maximum compression at the point of impact occurs when the two bodies have the same velocity v'. The momentum equation up to this instant is

$$mv + 0 = (m + nm)v'$$

$$\Rightarrow \qquad v' = v/(1+n).$$

The energy absorbed U_a is the difference between the kinetic energy at this instant and the initial kinetic energy:

$$\begin{aligned} U_a &= \tfrac{1}{2}mv^2 - \tfrac{1}{2}(m+nm)v^2/(1+n)^2 \\ &= \tfrac{1}{2}mv^2 n/(1+n). \end{aligned}$$

Let the velocities after separation be v_1 for the mass m and v_2 for the mass nm. Conservation of linear momentum gives

$$mv = mv_1 + nmv_2.$$

For a coefficient of restitution e,

$$v_2 - v_1 = ev.$$

Hence

$$v_2 = \left(\frac{1+e}{1+n}\right)v \quad \text{and} \quad v_1 = \left(\frac{1-ne}{1+n}\right)v.$$

The energy U_r recovered on separation is the difference between the final kinetic energy and the kinetic energy at the instant when the masses have a common velocity.

$$\begin{aligned} U_r &= \tfrac{1}{2}m\left(\frac{1-ne}{1+n}\right)^2 v^2 + \tfrac{1}{2}nm\left(\frac{1+e}{1+n}\right)^2 v^2 - \tfrac{1}{2}mv^2/(1+n) \\ &= \tfrac{1}{2}e^2mv^2n/(1+n) \\ &= e^2U_a. \end{aligned}$$

This analysis assumes that the bodies part without either having any residual vibration.

7.3. Kinetic energy of a set of particles

The concept of kinetic energy was introduced as a consequence of integrating the equation of motion

$$\mathbf{P} = m\mathbf{a}.$$

On taking the scalar product with an infinitesimal displacement $\delta\mathbf{r}$ and then integrating we have

$$\int_A^B \mathbf{P}\,.\,\mathrm{d}\mathbf{r} = \int_A^B m\mathbf{a}\,.\,\mathrm{d}\mathbf{r}.$$

In developing eqn (6.10) we noted that

$$\begin{aligned} \mathbf{a}\,.\,\mathrm{d}\mathbf{r} &= \frac{\mathrm{d}\mathbf{r}}{\mathrm{d}t}\cdot\frac{\mathrm{d}\mathbf{r}}{\mathrm{d}t}\,\mathrm{d}t. \\ &= \frac{1}{2}\frac{\mathrm{d}}{\mathrm{d}t}(\mathbf{v}\,.\,\mathbf{v})\,\mathrm{d}t. \end{aligned}$$

So the integration becomes

$$\begin{aligned} \int_A^B \mathbf{P}\,.\,\mathrm{d}\mathbf{r} &= \int_A^B \tfrac{1}{2}m\,\mathrm{d}(\mathbf{v}\,.\,\mathbf{v}) \\ &= \tfrac{1}{2}m\mathbf{v}_A\,.\,\mathbf{v}_A - \tfrac{1}{2}m\mathbf{v}_B\,.\,\mathbf{v}_B \end{aligned}$$

where the left-hand side of the equation is the work done by the resultant force that is applied to the particle and the right-hand side is the change of kinetic energy.

Recognition that the expression $\frac{1}{2}mv^2$ for the kinetic energy of a particle comes from $\frac{1}{2}m\mathbf{v}\cdot\mathbf{v}$ is helpful in determining an expression for the kinetic energy of a set of particles.

$$T=\sum \tfrac{1}{2}m_i v_i^2=\sum \tfrac{1}{2}m_i \mathbf{v}_i\cdot\mathbf{v}_i.$$

Putting $\mathbf{r}_i=\mathbf{r}_G+\boldsymbol{\rho}_i$, where $\mathbf{r}_G$ locates the centre of mass of the set and $\boldsymbol{\rho}_i$ locates the ith mass relative to the centre of mass,

$$\begin{aligned}\mathbf{v}_i=\dot{\mathbf{r}}_i&=(\dot{\mathbf{r}}_G+\dot{\boldsymbol{\rho}}_i)\\&=\mathbf{v}_G+\dot{\boldsymbol{\rho}}_i\end{aligned}$$

and

$$\begin{aligned}T&=\sum \tfrac{1}{2}m_i(\mathbf{v}_G+\dot{\boldsymbol{\rho}}_i)\cdot(\mathbf{v}_G+\dot{\boldsymbol{\rho}}_i)\\&=\sum \tfrac{1}{2}m_i\mathbf{v}_G^2+\sum m_i\mathbf{v}_G\cdot\dot{\boldsymbol{\rho}}_i+\sum \tfrac{1}{2}m_i\dot{\boldsymbol{\rho}}_i\cdot\dot{\boldsymbol{\rho}}_i.\end{aligned}$$

Now $\mathbf{v}_G$ does not vary from particle to particle and at any given instant can be regarded as a constant to be taken outside the summation signs, so that

$$T=(\sum \tfrac{1}{2}m_i)\mathrm{v}_G^2+\mathbf{v}_G\cdot(\sum m_i\dot{\boldsymbol{\rho}}_i)+\sum \tfrac{1}{2}m_i\dot{\rho}_i^2.$$

Now $m_i\dot{\boldsymbol{\rho}}_i=\dfrac{\mathrm{d}}{\mathrm{d}t}(m_i\boldsymbol{\rho}_i)$, and from the definition of the centre of mass, $\sum m_i\boldsymbol{\rho}_i=0$, so that the second term is zero. $\sum m_i$ is the total mass M so that

$$T=\tfrac{1}{2}Mv_G^2+\tfrac{1}{2}\sum m_i\dot{\rho}_i^2. \tag{7.8}$$

The kinetic energy of a set of particles is consequently the sum of two terms, one the kinetic energy which it would have if all the mass were concentrated at the centre of mass and had velocity $\mathbf{v}_G$, and another, the energy due to the motion that each particle has relative to the centre of mass. It should be noted that $\dot{\rho}_i^2=\dot{\boldsymbol{\rho}}_i\cdot\dot{\boldsymbol{\rho}}_i$ and that $\dot{\boldsymbol{\rho}}_i$ is not necessarily zero if ρ_i is constant.

It is rarely necessary to make direct use of eqn (7.8). Most important applications arise when the set of particles form a rigid body. The further development of eqn (7.8) to meet this important special case will be considered in Chapter 8.

7.4. Moment of momentum of a set of particles

The linear momentum of a set of particles has been shown to be $M\mathbf{v}_G$. It does not follow that the mass of the set can be regarded as if it were concentrated at the centre of mass when the moment of momentum of the set is calculated. The moment of momentum $\mathbf{h}$ of a single particle about O is

$$\mathbf{h}_O=\mathbf{r}\times(m\mathbf{v}).$$

For a set of particles

$$\mathbf{h}_O=\sum \mathbf{r}_i\times(m_i\mathbf{v}_i).$$

As in the previous sections we replace $\mathbf{r}_i$ by $\mathbf{r}_G+\boldsymbol{\rho}_i$, where $\mathbf{r}_G$ locates the centre of mass of the set of particles and $\boldsymbol{\rho}_i$ locates the ith particle relative to the centre of mass, so that

$$\mathbf{h}_O=\sum(\mathbf{r}_G+\boldsymbol{\rho}_i)\times\{m_i(\dot{\mathbf{r}}_G+\dot{\boldsymbol{\rho}}_i)\}.$$

It is helpful to write moment of momentum in the form $\mathbf{r}\times(m\mathbf{v})$ to remind us that we really are taking the moment of a momentum vector, but it would be equally valid to write $m\mathbf{r}\times\mathbf{v}$. On rearranging in this way, and taking advantage of the fact that $\mathbf{r}_G$ does not vary from particle to particle we find on expanding the above expression

$$\mathbf{h}_O=\mathbf{r}_G\times\dot{\mathbf{r}}_G\sum m_i+\mathbf{r}_G\times\sum m_i\dot{\boldsymbol{\rho}}_i+\left(\sum\boldsymbol{\rho}_i m_i\right)\times\dot{\mathbf{r}}_G+\sum\boldsymbol{\rho}_i\times(m_i\dot{\boldsymbol{\rho}}_i).$$

From the definition of the centre of mass $\sum m_i\boldsymbol{\rho}_i=0$, and $\sum m_i\dot{\boldsymbol{\rho}}_i=0$. Also $\sum m_i=M$, so that

$$\mathbf{h}_O=\mathbf{r}_G\times(M\mathbf{v}_G)+\sum\boldsymbol{\rho}_i\times(m_i\dot{\boldsymbol{\rho}}_i). \tag{7.9}$$

It would be permissible to regard the mass of a set of particles as being concentrated at the centre of mass for the purposes of calculating its moment of momentum only if the second term on the right-hand side of eqn (7.9) were zero. This happens only in special circumstances. A simple example illustrates the point. In Fig. 7.6 the two particles spin about their combined centre of mass with angular velocity $\boldsymbol{\omega}$. The centre of mass itself is stationary so that $\mathbf{r}_G\times(m\mathbf{v}_G)=0$ whilst $\mathbf{h}_O$ is clearly $2ma^2\boldsymbol{\omega}$.

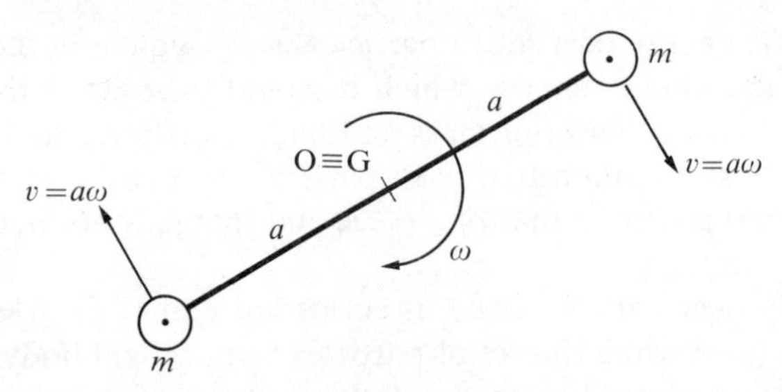

FIG. 7.6

As in the case of a single particle $\mathbf{Q}_O=\mathrm{d}\mathbf{h}_O/\mathrm{d}t$, where $\mathbf{h}_O$ is given by eqn (7.9), and the most important special case is when $\mathbf{Q}_O$ is zero so that $\mathbf{h}_O$ is constant. Even so, direct application of eqn (7.9) is not often required and the main purpose in introducing it here is as a forerunner to the study of rigid body motion.

7.5. Problems

1. A light rod, of length $2l$, has small masses attached at each end which slide on a rough horizontal plane of friction coefficient μ. At a particular time, the centre of the rod G is moving with velocity v and the rod is oriented as shown in Fig. 7.7, at 30° to the direction of motion of G. If the angular velocity of the rod is then v/l, find the acceleration of G and the force in the rod.

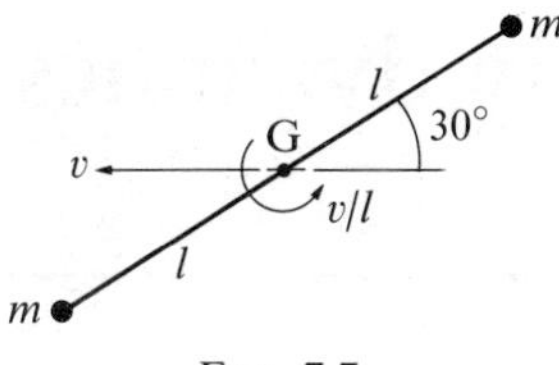

FIG. 7.7

2. Masses of 3 and 5 kg are connected by a taut, light, inextensible string passing over a light pulley. The masses are released, and 2 s latter the smaller picks up a rider of mass 4 kg. Neglecting all friction, find how high the 3 kg mass has risen above its initial position when the system first comes instantaneously to rest.

3. Two particles each of mass m are connected by a light, inextensible string. The string passes through a small hole in a smooth horizontal plane; initially one particle is held in contact with the plane at a distance a from the hole, and the other is suspended below the plane.

The particle on the plane is given an initial horizontal velocity v in a direction perpendicular to the string. Show that the distance r of this particle from the hole is determined by the equation

$$\frac{d^2r}{dt^2} - \frac{a^2v^2}{2r^3} + \frac{g}{2} = 0.$$

Show that if $v = \sqrt{(8ag/3)}$ the maximum value of r is $2a$.

4. A railway truck of mass m is connected to a second truck of mass $2m$ by means of a light spring of stiffness s. The two trucks are travelling along a level track with uniform velocity v when the leading truck, which has mass $2m$, runs into a stationary truck of mass m. The two colliding trucks couple together on impact and relative motion between them ceases.

Show that during the subsequent motion the maximum compression in the spring is $v\sqrt{(m/12s)}$, and determine the maximum and minimum velocities of the trucks.

Assume that the track is smooth.

5. A truck of mass 8 Mg moving at 2 m s^{-1} collides with a stationary truck of mass 12 Mg. A pair of light buffers operates on each truck and the force required to compress one buffer a distance x is $P_0 + kx$, where P_0 is 5000 N. If each buffer reaches a maximum compression of 0·2 m, find the value of the constant k.

If the velocity of the heavier truck on separation of the buffers is 1·25 m s^{-1}, find the velocity of the other. How much energy has been absorbed in the buffers?

6. A truck of mass 8 Mg moving at 2 m s^{-1} collides with a truck of 10 Mg initially at rest. Each truck has two buffers at each end. Each buffer spring is initially uncompressed and has a stiffness of 0.6 kN mm^{-1}. There is a constant friction force of 10 kN which resists motion of each buffer in either direction.

Determine the maximum compressive force developed in the buffer springs, and the velocities of the trucks after they separate.

8. Rigid-body dynamics

A FULL study of the dynamics of a rigid body in three-dimensional motion would be inappropriate here. Consideration will be given to a number of classes of problem, some of substantial practical importance. The problems are intended to encourage appeal to the basic principles of mechanics rather than to be regarded as a set of standard examples.

8.1. Simple rotors

A rotor is taken to mean any solid body that is constrained to rotate about a particular axis. Often the axis may be treated as having a fixed direction in space, and this will be assumed in our initial examples. The term rotor usually implies a high degree of symmetry about the axis of rotation, as with a flywheel or the rotor of a turbine, but this is not invevitably so. The crank shaft of a motor-car engine is a rotor having a certain regularity in its geometry, but it is not axi-symmetrical.

The rotor in Fig. 8.1 consists of a rigid shaft with a number of concentrated masses attached rigidly to it. The shaft is supported in bearings in which it turns with constant angular velocity $\boldsymbol{\omega}$. The acceleration of a typical particle m_i is $-\mathbf{r}_i\omega^2$, where $\mathbf{r}_i$ is the normal from the axis of the shaft to the mass. In accordance with Newton's Second Law the force on the shaft is $m_i\mathbf{r}_i\omega^2$. The shaft is in equilibrium under a set of such forces and the reactions $\mathbf{R}_1$ and $\mathbf{R}_2$ at the bearings so that

$$\sum m_i\mathbf{r}_i\omega^2+\mathbf{R}_1+\mathbf{R}_2=0$$

$$\Rightarrow \qquad M\mathbf{r}_G\omega^2+\mathbf{R}_1+\mathbf{R}_2=0$$

where $\mathbf{r}_G$ is the normal from the axis of the shaft to the centre of mass of the attached set of masses. The resultant force on the bearings $(\mathbf{R}_1+\mathbf{R}_2)$ is zero if $\mathbf{r}_G$ is zero. No account has been taken here of the dead weight of the shaft and masses. This does not imply that the weight is unimportant, but being always in one direction it is readily allowed for. The resultant force due to $\mathbf{r}_G\neq 0$ rotates with the shaft.

If the shaft is horizontal and can turn freely in the bearings it will take up the position in which $\mathbf{r}_G$ is vertically downwards. If the centre of mass is on the axis of the shaft there is no such preferred orientation and the shaft is said to be *statically balanced.* This state can always be achieved by the addition of a single concentrated mass m_b at radius $\mathbf{r}_b$ such that

$$m_b\mathbf{r}_b=-M\mathbf{r}_G.$$

The attainment of statical balance is not sufficient to ensure that $\mathbf{R}_1$ and

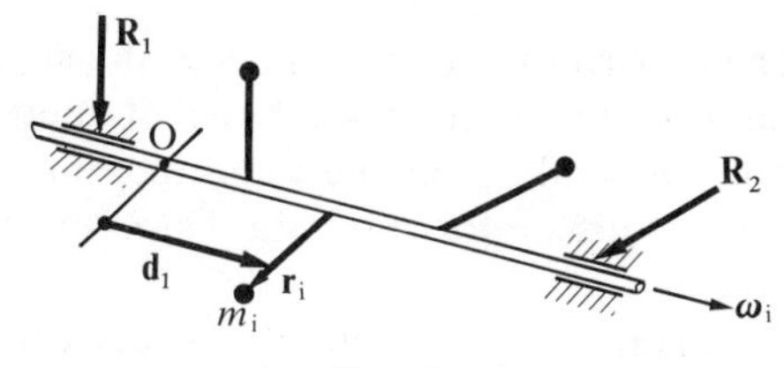

FIG. 8.1

$\mathbf{R}_2$ are separately zero. On taking moments about a point O on the axis of the shaft. Fig. 8.1, we have

$$\sum \mathbf{d}_i \times m_i \mathbf{r}_i \omega^2 + (\mathbf{d}_1 \times \mathbf{R}_1 + \mathbf{d}_2 \times \mathbf{R}_2) = 0$$

where $\mathbf{d}_1$ and $\mathbf{d}_2$ are vectors along the axis from O to bearings 1 and 2.

It is possible to make the term $(\mathbf{d}_1 \times \mathbf{R}_1 + \mathbf{d}_2 \times \mathbf{R}_2)$ zero by the addition of a single mass m_b' of the right amount at a convenient radius $\mathbf{r}_b'$ and at an arbitrary section located by $\mathbf{d}_b'$ along the shaft, so that

$$\sum \mathbf{d}_i \times m_i \mathbf{r}_i + \mathbf{d}_b' \times m_b' \mathbf{r}_b' = 0.$$

A second mass m_b'' at $\mathbf{r}_b''$ in the tranverse plane through O can now be sited to achieve statical balance without upsetting the balance of moments about O. Now with $\mathbf{R}_1 + \mathbf{R}_2 = 0$ and $(\mathbf{d}_1 \times \mathbf{R}_1 + \mathbf{d}_2 \times \mathbf{R}_2) = 0$ it follows that $\mathbf{R}_1$ and $\mathbf{R}_2$ are separately zero.

If $\mathbf{R}_1$ and $\mathbf{R}_2$ are both zero the rotor is said to be in a state of *dynamic balance.*

It will be noted that a state of balance can always be achieved by the addition of the appropriate $m_b \mathbf{r}_b$ in a pair of arbitrarily chosen transverse planes. This is quite independant of the number and distribution of original masses and therefore applies also when the mass is distributed continuously as with a solid rotor. Standard balancing machines are used to measure the amount of unbalance and to show what needs to be done to correct it. Sometimes balance is achieved, as has been implied here, by the addition of balance weights. The standard balancing procedure for car wheels is an example of this. Sometimes balancing is done by the removal of material by drilling: the balancing of small electric motors is an example of this.

Items such as motor-car wheels are designed to be axi-symmetrical so that balancing is a matter of correcting for manufacturing tolerances. With more irregular objects, such as engine crank shafts it is usual to seek local as well as overall balance by providing counterweights for individual eccentric masses. This is to avoid the creation of the bending moments to which the shaft would otherwise be subjected. Such counterweights are manufactured as part of the rotor. Testing for balance of the finished article is still necessary.

If it is left uncorrected, lack of balance can result in very heavy loads on the rotor itself and on the supporting bearings. The result is inevitably a high rate of wear in bearings and vibrations in the supporting structure. The vibrations can lead to failure by fatigue of metal, and can cause fastenings to loosen. Depending on the amplitude and frequency vibrations can be a source of discomfort or even distress to the human beings that suffer them.

Even if perfect balance were attainable, a perfectly balanced rotor would not remain balanced for long because the slightest change of temperature distribution or stress would cause distortions to upset the ideal state of balance. The standard of balance that is sought is related to the use. For example, there is little point in trying to balance the wheels of a motor car beyond the point where the effects of unbalance are small compared with the other sources of noise and vibration. To do so would increase the cost pointlessly.

8.2. Kinetic energy, momentum, and moment of momentum of a rigid body

There is no reason in principle why rigid bodies should not be regarded as connected sets of particles to which the results that we have already deduced can be applied, as in the discussion of the balancing of rotors. In practice it is usually better to modify the equations to take account of the fact that the set of particles that make up a rigid body are all fixed in the body, and so in relation to each other.

The equation of motion for the acceleration of the centre of mass

$$\mathbf{P} = M\mathbf{a}_{\mathrm{G}}$$

and the linear momentum

$$\int \mathbf{P}\,\mathrm{d}t = \Delta(Mv_{\mathrm{G}})$$

are independent of the relative motion of the particles and so apply equally to freely moving sets of particles and to rigid bodies.

The kinetic energy of a set of particles was shown, eqn (7.8) to be

$$T = \tfrac{1}{2}Mv_{\mathrm{G}}^2 + \tfrac{1}{2}\sum m_{\mathrm{i}}\dot{\rho}_{\mathrm{i}}^2.$$

Now $\boldsymbol{\rho}_{\mathrm{i}}$ is the vector drawn from the centre of mass to the ith particle. In a rigid body $\boldsymbol{\rho}_{\mathrm{i}}$ is of fixed magnitude and can change only if the body has an angular velocity so that $\dot{\boldsymbol{\rho}}_{\mathrm{i}} = \boldsymbol{\omega} \times \boldsymbol{\rho}_{\mathrm{i}}$. Remembering that the term $\frac{1}{2}\sum m_{\mathrm{i}}\dot{\rho}_{\mathrm{i}}^2$ is derived from $\frac{1}{2}\sum m_{\mathrm{i}}\dot{\boldsymbol{\rho}}_{\mathrm{i}} \cdot \dot{\boldsymbol{\rho}}_{\mathrm{i}}$ we have

$$\begin{aligned}\tfrac{1}{2}\sum m_{\mathrm{i}}\dot{\rho}_{\mathrm{i}}^2 &= \tfrac{1}{2}\sum m_{\mathrm{i}}(\boldsymbol{\omega} \times \boldsymbol{\rho}_{\mathrm{i}}) \cdot (\boldsymbol{\omega} \times \boldsymbol{\rho}_{\mathrm{i}}) \\ &= \tfrac{1}{2}\sum m_{\mathrm{i}}\{\omega^2\rho_{\mathrm{i}}^2 - (\boldsymbol{\omega} \cdot \boldsymbol{\rho}_{\mathrm{i}})^2\} \\ &= \tfrac{1}{2}\omega^2 \sum m_{\mathrm{i}}\rho_{\mathrm{i}}^{*2}\end{aligned}$$

where $\boldsymbol{\rho}_{\mathrm{i}}^{*}$ is the component of $\boldsymbol{\rho}_{\mathrm{i}}$ perpendicular to $\boldsymbol{\omega}$. Hence

$$T = \tfrac{1}{2}Mv_{\mathrm{G}}^2 + \tfrac{1}{2}I_{\omega}\omega^2 \tag{8.1}$$

where $I_\omega = \sum m_i \rho_i^{*2}$ is the *moment of inertia* of the body about the axis through G that is parallel to ω.

In a similar manner, the expression for the moment of momentum of a set of particles

$$\mathbf{h}_0 = \mathbf{r}_G \times (M\mathbf{v}_G) + \sum \boldsymbol{\rho}_i \times (m_i \dot{\boldsymbol{\rho}}_i)$$

also depends on the relative motion of the particles, this time in the term $\sum \boldsymbol{\rho}_i \times (m_i \dot{\boldsymbol{\rho}}_i)$. On substituting $\boldsymbol{\omega} \times \boldsymbol{\rho}_i$ for $\dot{\boldsymbol{\rho}}_i$ we have

$$\mathbf{h}_0 = \mathbf{r}_G \times (M\mathbf{v}_G) + \sum m_i \boldsymbol{\rho}_i \times (\boldsymbol{\omega} \times \boldsymbol{\rho}_i). \tag{8.2}$$

The term $\sum m_i \boldsymbol{\rho}_i \times (\boldsymbol{\omega} \times \boldsymbol{\rho}_i)$ is referred to as the *angular momentum* of the body. It does not simplify quite as readily as does the corresponding term in the expression for the kinetic energy.

8.3. Moment of momentum of a rigid lamina moving in its own plane

On expanding the triple vector product in $m_i \boldsymbol{\rho}_i \times (\boldsymbol{\omega} \times \boldsymbol{\rho}_i)$, eqn (8.2) becomes

$$\mathbf{h}_0 = \mathbf{r}_G \times (M\mathbf{v}_G) + \sum m_i \{\boldsymbol{\omega} \rho_i^2 - \boldsymbol{\rho}_i (\boldsymbol{\omega} \cdot \boldsymbol{\rho}_i)\}.$$

In planar motion $\boldsymbol{\omega} \cdot \boldsymbol{\rho}_i = 0$ so that

$$\mathbf{h}_0 = \mathbf{r}_G \times (M\mathbf{v}_G) + \boldsymbol{\omega} \sum m_i \rho_i^2$$

or

$$\mathbf{h}_0 = \mathbf{r}_G \times (M\mathbf{v}_G) + I\boldsymbol{\omega} \tag{8.3}$$

where

$$I = \sum m_i \rho_i^2 \tag{8.4}$$

and is the moment of inertia of the body about the axis that is through G, and is perpendicular to the plane of motion. The moment of inertia involves only the mass distribution and dimensions of a body and is independent of its motion so that it can be calculated without reference to any particular motion. It will be helpful for future reference to determine the moments of inertia for a few regular shapes.

Example 8.1: Determine I for a thin circular hoop of mass M and radius a about the central axis perpendicular to its plane, Fig. 8.2.

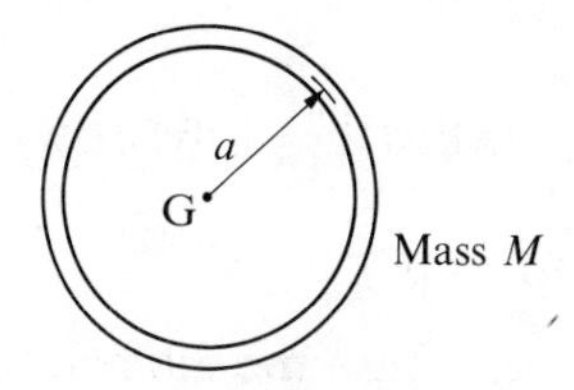

FIG. 8.2

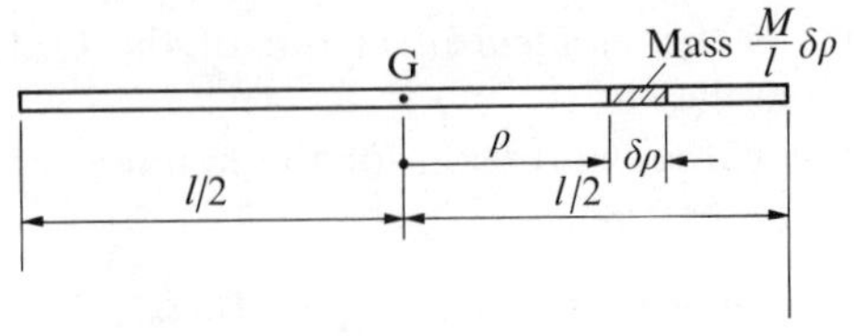

FIG. 8.3

As all the mass is at the same radius a

$$I = \sum m_i \rho_i^2 = Ma^2$$

Example 8.2: Determine I for a uniform bar of mass M and length l about an axis through G and normal to itself, Fig. 8.3.

$$I = \sum m_i \rho_i^2 = \sum \rho^2 (M/l)\delta\rho$$

$$= \int_{-l/2}^{l/2} \rho^2 (M/l)\, d\rho$$

$$= \tfrac{1}{12} Ml^2.$$

Example 8.3: Determine I for a uniform circular disc of mass M and radius a about the axis through its centre and perpendicular to its plane, Fig. 8.4.

$$I = \sum m_i \rho_i^2 = \sum \rho^2 (M/\pi a^2)\rho\delta\theta\delta\rho.$$

After summing for all the elements round the hoop of radius ρ

$$I = \int_0^a (2M/a^2)\rho^3\, d\rho$$

$$= \tfrac{1}{2} Ma^2.$$

Example 8.4: Determine I for a uniform rectangular plate of mass M and sides b×d about the axis through its centre and normal to its plane, Fig. 8.5.

$$I = \sum m_i \rho_i^2 = \sum \rho^2 (M/bd)\delta x \delta y$$

$$= \int_{-b/2}^{b/2} \int_{-d/2}^{d/2} (M/bd)(x^2 + y^2)\, dx\, dy$$

$$= \int_{-b/2}^{b/2} (M/bd)[\tfrac{1}{3}x^3 + xy^2]_{-d/2}^{d/2}\, dy$$

$$= \int_{-b/2}^{b/2} (M/bd)\left(\frac{d^3}{12} + dy^2\right) dy$$

$$= \tfrac{1}{12} M(b^2 + d^2).$$

It can be seen that in every case the moment of inertia is equal to the total mass multiplied by a factor whose dimensions are (length) squared.

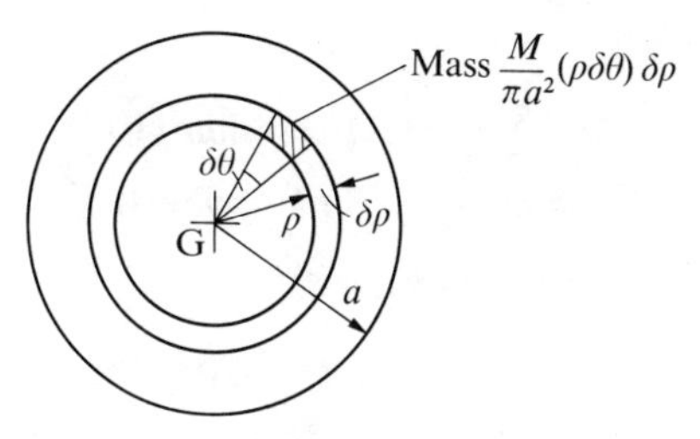

FIG. 8.4

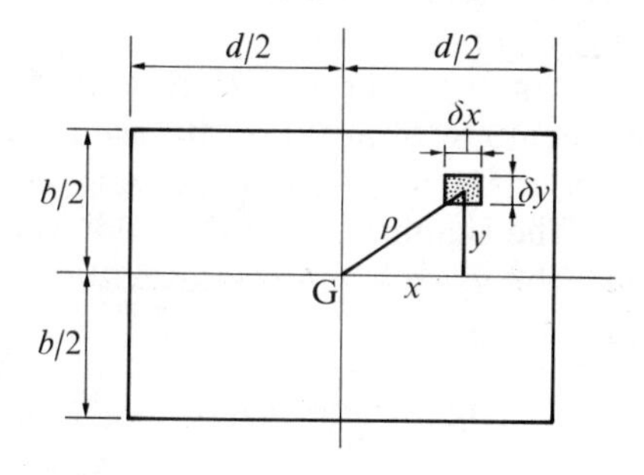

FIG. 8.5

It is often convenient to express the moment of inertia in the form

$$I = Mk^2$$

where k is thus defined as the *radius of gyration* of the body.

8.4. Moment of momentum of a lamina rotating about an axis in its own plane and through its centre of mass

A lamina, Fig. 8.6, rotates with angular velocity $\omega_x\mathbf{i}$. By definition, the moment of momentum about G is

$$\mathbf{h} = \sum \mathbf{r} \times (\delta m\,\mathbf{v})$$

where $\mathbf{v}$ is the velocity of the element δm. In this case $v = \omega_x y\mathbf{k}$. On

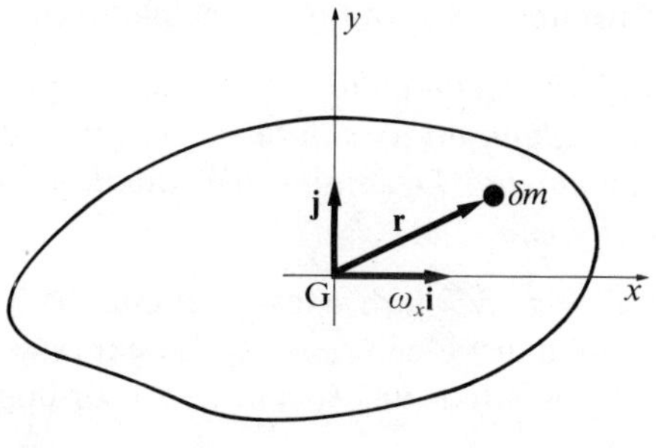

FIG. 8.6

substituting $x\mathbf{i}+y\mathbf{j}$ for $\mathbf{r}$

$$\mathbf{h}=\sum(x\mathbf{i}+y\mathbf{j})\times(\delta m\omega_x y\mathbf{k})$$
$$=\omega_x\sum(y^2\delta m\mathbf{i}-xy\delta m\mathbf{j}).$$

Letting $\sum y^2\delta m=I_{xx}$ and $\sum xy\delta m=I_{xy}$

$$\mathbf{h}=\omega_x(I_{xx}\mathbf{i}-I_{xy}\mathbf{j}). \tag{8.5}$$

I_{xx} is the moment of inertia of the lamina about Gx, and I_{xy} is called the product of inertia. Although it is not immediately relevant, we can similarly define $I_{yy}=\sum x^2\delta m$.

The conclusion to be drawn from eqn (8.5) is that rotation about Gx induces angular momentum about Gy as well as Gx. Whilst the angular momentum about Gx would be expected, the component about Gy may be surprising. The significance of this additional term is realized when it is recalled that the axis system rotates with the lamina at angular velocity $\omega_x\mathbf{i}$. So, from eqn (6.9),

$$\mathbf{Q}_G=\frac{d\mathbf{h}}{dt}=\omega_x\mathbf{i}\times\mathbf{h}$$

$$\Rightarrow \qquad \mathbf{Q}_G=-\omega_x^2 I_{xy}\mathbf{k}. \tag{8.6}$$

This moment is applied through the bearing reactions and with a change of sign is identical to the out of balance couple calculated in Section 8.1. Dynamic balance is obtained when $I_{xy}=0$.

The set of axes for which I_{xy} is zero are called *principal axes*. If the lamina has an axis of symmetry Gx, say, it is a principal axis. This is because every element δm on one side of Gx is balanced by a similar element on the other side. In one case y is positive, and in the other it is negative so that for that pair of elements $\sum \delta mxy$ is zero. If I_{xy} is zero in eqn (8.5) it is also zero in the corresponding equation

$$\mathbf{h}=\omega_y(I_{yy}\mathbf{j}-I_{xy}\mathbf{i})$$

so that Gy is then also a principal axis.

If the principal axes for a lamina are known it is not necessary to calculate the product of inertia to calculate the unbalanced couple when the lamina is rotated about a non-principal axis.

Example 8.5: A uniform rectangular lamina of length d and breadth b, and of mass M rotates about an axis through G that is at an angle θ to the lengthwise axis of symmetry. Determine the unbalanced couple due to a steady angular velocity ω.

In this case GXY, Fig. 8.7, are axes of symmetry and are therefore principal axes. The angular velocity $\omega\mathbf{i}$ can be expressed in terms of the unit vectors $\mathbf{e}_1$ and $\mathbf{e}_2$ as $\omega(\cos\theta\mathbf{e}_1-\sin\theta\mathbf{e}_2)$. Accordingly

$$\mathbf{h}=I_{XX}\omega\cos\theta\mathbf{e}_1-I_{YY}\omega\sin\theta\mathbf{e}_2.$$

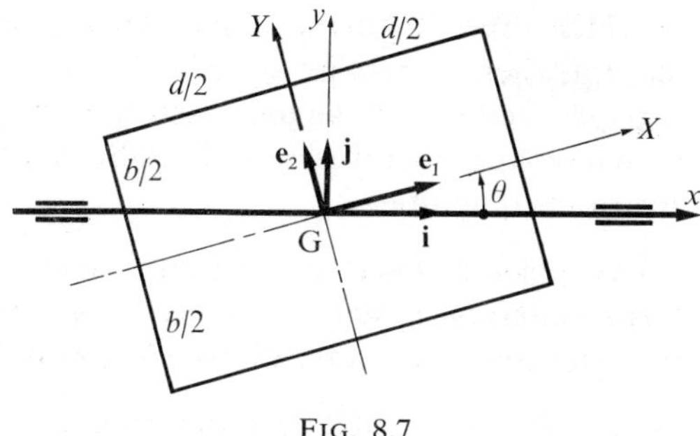

FIG. 8.7

Now

$$\mathbf{e}_1 = \mathbf{i}\cos\theta + \mathbf{j}\sin\theta$$

and

$$\mathbf{e}_2 = -\mathbf{i}\sin\theta + \mathbf{j}\cos\theta$$

so that

$$\mathbf{h} = (I_{XX}\cos^2\theta + I_{YY}\sin^2\theta)\omega\mathbf{i} + \tfrac{1}{2}(I_{XX} - I_{YY})\sin 2\theta\omega\mathbf{j}.$$

The torque applied to the lamina through the bearings is

$$\mathbf{Q} = \frac{\mathrm{d}\mathbf{h}}{\mathrm{d}t} = \omega\mathbf{i} \times \mathbf{h}$$

$$= \tfrac{1}{2}\omega^2(I_{XX})\sin 2\theta\mathbf{k}.$$

It remains to calculate I_{XX} and I_{YY}. If we regard the lamina as a number of uniform rods side by side we can use the result from page 184 for a uniform bar of mass M. From this result it follows that

$$I_{XX} = \tfrac{1}{12}Mb^2 \quad \text{and} \quad I_{YY} = \tfrac{1}{12}Md^2$$

so the unbalanced couple is

$$-\mathbf{Q} = -M(d^2 - b^2)\sin 2\theta\mathbf{k}.$$

It may be noted that $\mathbf{Q}$ is zero if $b = d$. It follows that all in-plane axes through the centre of a square lamina are principal axes.

The relationships between moments of inertia about different sets of axes are treated in detail in Appendix 2.

8.5. Various applications of $\mathbf{Q} = \mathrm{d}\mathbf{h}/\mathrm{d}t$ and conservation of moment of momentum to planar dynamics

For any set of particles or rigid body the moment of momentum $\mathbf{h}_O$ about a fixed point O is constant if the net moment about O of the forces on the body is zero. If $\mathbf{Q}_0$ is not zero then $\mathbf{Q}_0 = \mathrm{d}\mathbf{h}/\mathrm{d}t$. The purpose of this section is to draw attention to circumstances in which the concept of moment of momentum can lead to useful results. We will proceed by way of examples each of which seeks to make a particular point. It is to be understood that, except where otherwise stated, all velocities are parallel to the plane of the paper, and that all angular velocities are perpendicular to the plane of the paper. It is therefore usually sufficient to write h_0 for $\mathbf{h}_0$ and similarly for other quantities.

Example 8.6: A body pivoted about a horizontal axis executing small oscillations about its equilibrium position is called a compound pendulum, Fig. 8.8. Derive an expression for the natural frequency of the oscillations.

The only forces that act on the body are the weight, which is known, and the unknown reaction at the pivot O.

$$\mathbf{h}_0 = \mathbf{r}_G \times (M\mathbf{v}_G) + I\boldsymbol{\omega}.$$

Taking the anticlockwise direction as positive

$$h_0 = Ma^2\dot{\theta} + Mk^2\dot{\theta}$$
$$= M(a^2 + k^2)\dot{\theta}.$$

$$\frac{dh_0}{dt} = M(a^2 + k^2)\ddot{\theta} = Q_0$$
$$= -Mga \sin\theta.$$

Hence the equation of motion is

$$M(a^2 + k^2)\ddot{\theta} + Mga \sin\theta = 0.$$

If θ is small, say <15°, this equation approximates to

$$M(a^2 + k^2)\ddot{\theta} + Mga\theta = 0.$$

This can be recognized as the equation for simple harmonic motion with circular frequency

$$\omega_n = ga/(a^2 + k^2).$$

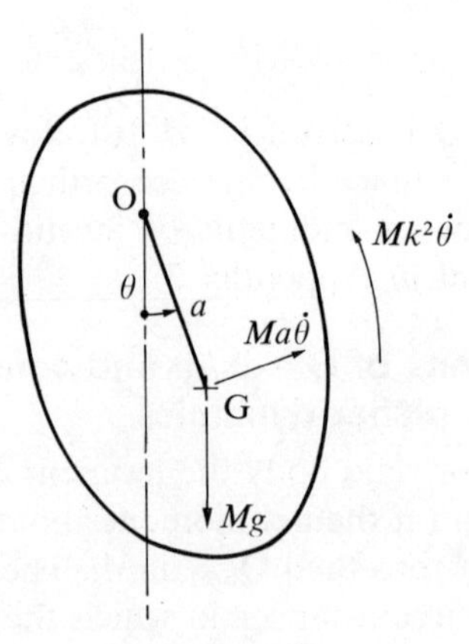

FIG. 8.8

It is always the case that the momentum of a body of mass M rotating about a fixed axis O that is distance a from the centre of mass is $M(a^2+k^2)$. It is consequently natural to refer to $M(a^2+k^2)$ as the moment of inertia about O, or strictly, about the axis through O. Then $(a^2+k^2)^{\frac{1}{2}}$ is the radius of gyration about O. Although this is quite proper, and indeed is the embodiment of a theorem called the parallel axis theorem, the practice is not recommended because it is too easy to misapply it.

Example 8.7: A mass m is supported by a light string wrapped round the rim of a flywheel. Ignoring friction, determine an expression for the angular acceleration of the flywheel, Fig. 8.9.

$$h_0 = Mk^2\omega + ma^2\omega$$

$$\frac{\mathrm{d}h_0}{\mathrm{d}t} = (Mk^2 + ma^2)\dot{\omega} = Q_0$$

$$= mga$$

$$\Rightarrow \qquad \dot{\omega} = mga/(Mk^2 + ma^2).$$

Again, by taking moments about the axis of the flywheel we are able to leave out of account the unknown reaction at the bearings. We have also left out of account friction in the bearings and windage. These give rise to a moment about O so that the expression for $\dot{\omega}$ must be in error to some extent. The tensile force T in the string is an internal force that acts equally but in opposite direction on the mass and the flywheel and is rightly ignored as an internal force.

Example 8.8: A child projects a hoop with a forward velocity V but with a backward spin Ω, Fig. 8.10. Assuming that the plane of the hoop stays

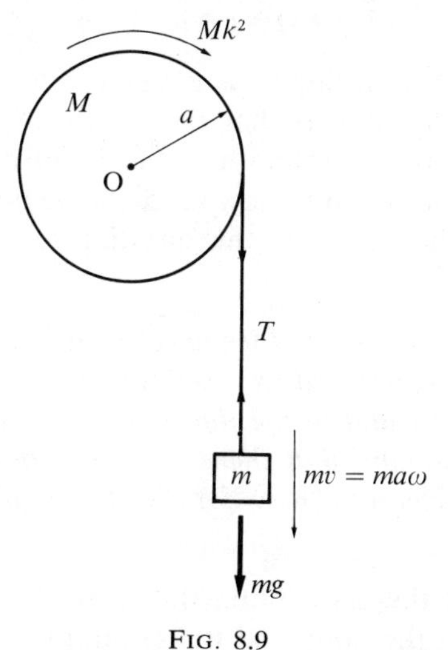

FIG. 8.9

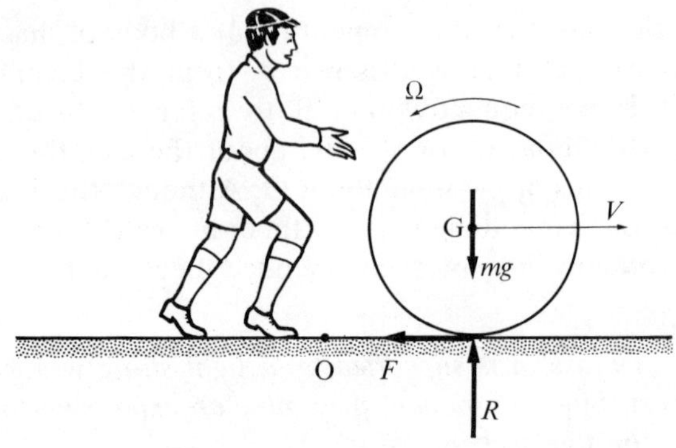

Fig. 8.10

vertical, and that the ground is level, what is the condition that the hoop should return to the child?

The external forces on the hoop are its weight mg, the normal reaction from the ground R and a friction force F. As there is no vertical motion mg and R balance.

Let O be an arbitrary point on the ground in the plane of the hoop. As F has then no moment about O, $Q_O = 0$, and h_0 is constant at its initial value. So for all time

$$h_O = amV - I\Omega.$$

Eventually the hoop rolls without slipping with a forward velocity v and an angular velocity $\omega = v/a$, so that

$$amV - I\Omega = amv + I\omega.$$

For a hoop $I = ma^2$ so that

$$V - a\Omega = v + a\omega = 2v.$$

The hoop returns if v is negative, as happens when $\Omega > V/a$.

The interesting thing here is that the nature and the amount of the friction force between the ground and the hoop does not matter. It is different if the question posed is, how far does the hoop go before starting its return journey? The answer to this question requires a full solution of the equation of motion.

Example 8.9: A hoop of radius R is projected so as to roll without slip. The rough ground can be represented as a series of sharp ridges, all at the same height, distance a apart and in the direction perpendicular to the plane of the hoop, $a \ll R$. Assuming that there is no slip or rebound as the hoop strikes each ridge, determine how far the hoop rolls before its forward velocity is halved.

The significance of the assumption that there is no slip or rebound at each contact between the hoop and the ground is that at every stage the

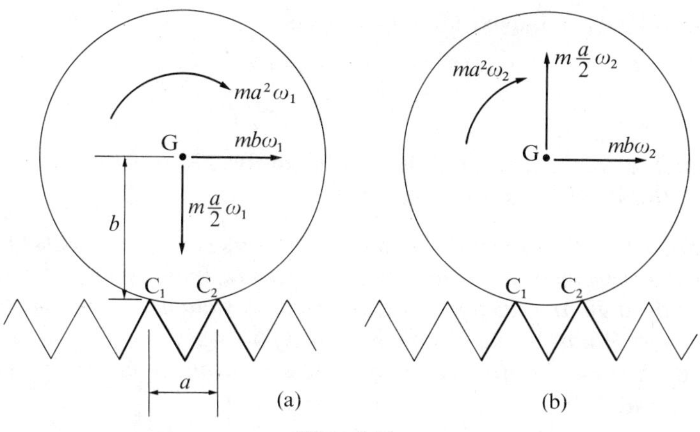

FIG. 8.11

momentum of the hoop can be expressed in terms of the single variable ω, its angular velocity. Fig. 8.11(a) shows the hoop turning about point C_1 just before contact is made at C_2. Fig. 8.11(b) shows the hoop turning about C_2 immediately after contact has been made.

The moments of momenta about C_2 before and after contact are related by the first integral of

$$\mathbf{Q} = \mathrm{d}\mathbf{h}/\mathrm{d}t$$

namely

$$\int_0^{\Delta t} \mathbf{Q}\,\mathrm{d}t = \mathbf{h}_2 - \mathbf{h}_1$$

where Δt is the time interval between the two instants. If the impulsive moment $\int_0^{\Delta t} \mathbf{Q}\,\mathrm{d}t$ tends to zero when Δt is infinitesimal the moment of momentum is conserved and $\mathbf{h}_2 = \mathbf{h}_1$.

The weight mg acts through G and has a moment of magnitude $mga/2$ about C_2 so that its contribution to the impulsive moment is $\int_0^{\Delta t} -mga\,\mathrm{d}t = -mga\Delta t$ and tends to zero as Δt tends to zero.

The reaction P_1 at C_1 is finite in the first instance and, provided that ω_2 is positive, drops to zero as the hoop starts to turn about C_2. So again, as $\Delta t \to 0$ there is no contribution to an impulsive moment about C_2. The reaction at C_2 has no moment about C_2. Hence, the net impulsive moment about C_2 is zero so that the moment of momentum is conserved and

$$h_2 = h_1$$

$$\Rightarrow \quad ma^2\omega_1 + mb^2\omega_1 - \tfrac{1}{2}m(a/2)^2\omega_1 = ma^2\omega_2 + mb^2\omega_2 + \tfrac{1}{2}m(a/2)^2\omega_2$$

$$\Rightarrow \quad \frac{\omega_2}{\omega_1} = \frac{a^2 + b^2 - a^2/8}{a^2 + b^2 + a^2/8}$$

$$\approx 1 - a^2/4R^2.$$

If the velocity is halved after n impacts

$$(1-a^2/4R^2)^n = \tfrac{1}{2}$$

$$\Rightarrow \qquad n = 2R^2/a^2.$$

The total distance travelled is consequently $2R^2/a$: the rougher the ground, the shorter the distance.

Example 8.10: Two coaxial shafts that carry rotors with moments of inertia I_1 and I_2 respectively are connected by a clutch. Initially the rotor I_1 turns freely with angular velocity. The other rotor is stationary, the clutch being disengaged. What is the common velocity ω after the clutch has been engaged. Assume that friction is negligible. How much energy is dissipated in the clutch?

As no external moment is applied to the rotors the angular momentum about the central axis is conserved and

$$I_1\Omega = (I_1+I_2)\omega$$

$$\Rightarrow \qquad \omega = I_1\Omega/(I_1+I_2).$$

Initially the kinetic energy is $\tfrac{1}{2}I_1\Omega^2$. Finally it is $\tfrac{1}{2}(I_1+I_2)\omega^2$ so that the energy U that is dissipated is

$$\begin{aligned} U &= \tfrac{1}{2}I_1\Omega^2 - \tfrac{1}{2}(I_1+I_2)\omega^2 \\ &= \tfrac{1}{2}\Omega^2\{I_1 - I_1^2/(I_1+I_2)\} \\ &= \tfrac{1}{2}I_1I_2\Omega^2/(I_1+I_2). \end{aligned}$$

It is to be noted that the energy loss does not depend on the characteristics of the clutch, nor is there any implication that the common velocity is attained instantaneously on engagement.

Example 8.11. A disc of mass M and radius R turns freely on an axle through its centre. A second disc has mass $M/4$ and radius $R/2$ and is supported in bearings that are parallel to those of the first rotor but which can move bodily sideways to press the edges of the two discs together. Initially the axes are pulled apart so that there is no contact between the two discs. The larger disc spins with angular velocity Ω and the smaller disc is stationary. The edges of the two discs are then pressed together. Determine the angular velocities of the discs after all slipping has ceased.

The temptation to say that angular momentum is conserved must be resisted. If the two discs had the same radius the final angular momentum would be zero because the two discs would be rotating with the same speed, but in opposite directions. In this special case the failure to conserve angular momentum is obvious.

The reason why angular momentum is not conserved is that the bearing reactions apply a couple to the discs during the period of slip. During slip

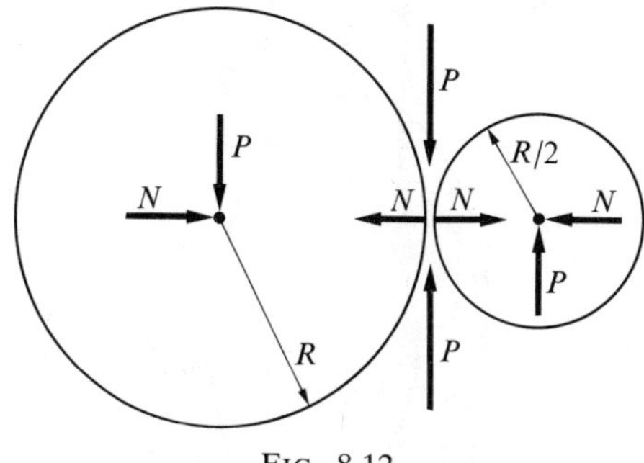

FIG. 8.12

there are equal and opposite friction forces P tangential to the discs at the point of contact. As there is no motion of the centre of mass of either disc forces P must be induced at the bearings to preserve equilibrium for each disc as shown in Fig. 8.12. The bearing reactions apply an overall couple to the discs which causes the change in total angular momentum.

Let the final angular velocity for the larger disc be ω. Then for that disc, whose moment of inertia is $\frac{1}{2}MR^2$,

$$\tfrac{1}{2}MR^2\Omega - \int RP \, \mathrm{d}t = \tfrac{1}{2}MR^2\omega.$$

For the smaller disc

$$\int (R/2)P \, \mathrm{d}t = -(M/4)(R/2)^2 \times 2\omega.$$

On eliminating $\int P \, \mathrm{d}t$ between the two equations we find

$$\omega = 0.8\Omega.$$

It is always worth verifying that there has indeed been a loss of kinetic energy. In this case

$$\text{Initial k.e.} = \tfrac{1}{2}(\tfrac{1}{2}MR^2)\Omega^2$$

$$\text{Final k.e.} = \tfrac{1}{2}(\tfrac{1}{2}MR^2)(0.8\Omega)^2 + \tfrac{1}{2}(M/4)(R/2)^2(1.6\Omega)^2$$
$$= \tfrac{1}{4}MR^2\Omega^2 \times 0.96.$$

$\Rightarrow$ Energy lost = 4 per cent of initial kinetic energy.

Example 8.12. A uniform disc is supported in horizontal bearings in which it spins with angular velocity $\boldsymbol{\omega}$. *The bearings are carried in a frame that rotates with constant angular velocity* $\boldsymbol{\Omega}$ *about a vertical axis. Determine the forces that must be applied to the system to maintain equilibrium.*

At any instant the angular velocity of the rotor has two components $\boldsymbol{\Omega}$ and $\boldsymbol{\omega}$ which are multiplied by the appropriate moments of inertia to give the angular momentum as

$$\mathbf{h} = I'\boldsymbol{\Omega} + I\boldsymbol{\omega}.$$

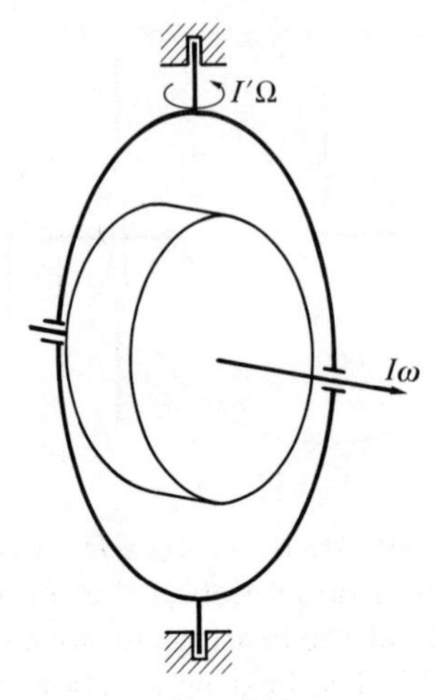

FIG. 8.13

An instant later the only change that is perceived is a change in the direction of $I\boldsymbol{\omega}$ due to the rotation $\boldsymbol{\Omega}$. Hence

$$Q = \frac{d\mathbf{h}}{dt} = I\boldsymbol{\Omega} \times \boldsymbol{\omega}.$$

This is the couple that must be supplied by transverse forces at the vertical bearings, and is usually referred to as the *gyroscopic couple*. It has magnitude $I\Omega\omega$, and acts about the axis that is normal to both Ω and ω.

Example 8.13: The two-bladed propeller of an aircraft has a moment of inertia I about its spin axis. The aircraft turns in a horizontal plane with angular velocity $\boldsymbol{\Omega}$ *about the vertical axis. Determine the resulting couple on the aircraft when the propeller is at an angle* θ *to the horizontal.*

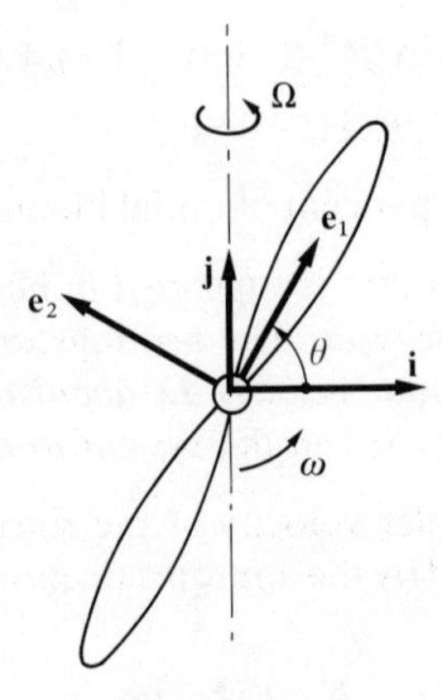

FIG. 8.14

If the propeller is thin enough its moment of inertia about its longitudinal axis will be negligible. Expressing $\mathbf{\Omega}$ in terms of the unit vectors $\mathbf{e}_1$ and $\mathbf{e}_2$ which rotate with the propeller, thus

$$\mathbf{\Omega} = \Omega \sin\theta \mathbf{e}_1 + \Omega \cos\theta \mathbf{e}_2,$$

gives

$$\mathbf{h} = I(\boldsymbol{\omega} + \Omega \cos\theta \mathbf{e}_2).$$

The first component of $\mathbf{h}$ is due to the spin of the propeller and the second is due to the turning of the aircraft. The couple that must be applied to the propeller to sustain the motion is

$$\mathbf{Q} = \frac{\mathrm{d}\mathbf{h}}{\mathrm{d}t} = I\left\{\mathbf{\Omega} \times \boldsymbol{\omega} - \Omega\dot{\theta} \sin\theta \mathbf{e}_2 + \Omega \cos\theta \frac{\mathrm{d}\mathbf{e}_2}{\mathrm{d}t}\right\}.$$

$\mathbf{e}_2$ changes direction due to $\mathbf{\Omega}$ and $\boldsymbol{\omega}$ so that

$$\begin{aligned}\frac{\mathrm{d}\mathbf{e}_2}{\mathrm{d}t} &= (\mathbf{\Omega} + \boldsymbol{\omega}) \times \mathbf{e}_2 \\ &= (\Omega \sin\theta \mathbf{e}_1 + \Omega \cos\theta \mathbf{e}_2) \times \mathbf{e}_2 + \boldsymbol{\omega} \times \mathbf{e}_2.\end{aligned}$$

if $\mathbf{k}$ is the unit vector pointing along the axis of the aircraft and such as to make a right-handed set with the other unit vectors $\mathbf{i}$ and $\mathbf{j}$ shown in Fig. 8.14 we find, on noting $\boldsymbol{\omega} \times \mathbf{e}_2 = -\dot{\theta}\mathbf{e}_1 = -\omega \mathbf{e}_1$, that

$$\begin{aligned}\mathbf{Q} &= I\mathbf{\Omega} \times \boldsymbol{\omega} - I\Omega\omega(\cos\theta \mathbf{e}_1 + \sin\theta \mathbf{e}_2) + I\Omega^2 \sin\theta \cos\theta \mathbf{k} \\ &= (I\Omega\omega + \tfrac{1}{2} I\Omega^2 \sin 2\theta)\mathbf{k} - I\Omega\omega(\cos\theta \mathbf{e}_1 + \sin\theta \mathbf{e}_2).\end{aligned}$$

The term $I\Omega\omega k$ is the simple gyroscopic torque revealed in the previous example. The additional component in the $\mathbf{k}$-direction, $\frac{1}{2}I\Omega^2 \sin 2\theta\mathbf{k}$, fluctuates with frequency 2ω. The double frequency arises because when the blade is, for example, horizontal there is no distinction between the two ends, and the blade is horizontal twice a revolution. Bearing in mind that normally $\Omega \ll \omega$ this term is normally negligible. The effect of the second bracketed term is seen by expressing $\mathbf{e}_1$ and $\mathbf{e}_2$ in terms of $\mathbf{i}$ and $\mathbf{j}$, which are fixed to the aircraft.

$$-I\Omega\omega(\cos\theta \mathbf{e}_1 + \sin\theta \mathbf{e}_2) = -I\Omega\omega(\mathbf{i} \cos 2\theta + \mathbf{j} \sin 2\theta).$$

This is a transverse couple whose direction rotates at twice the speed of the propellor; it is horizontal when the propeller is horizontal. We have determined the couple that must be applied to the propeller. The couples on the aircraft are equal and opposite to those that we have determined, which are the couples on the propeller.

When a propeller has three or more blades at equal angular spacing the couples that act at twice the blade frequency are all balanced and only the main gyroscopic torque $I\Omega\omega$ remains.

8.6. Moving frames of reference

Our mechanics is based on the idea that all motion is referred ultimately to an absolutely stationary frame of reference. The simple statements that a particle has velocity $\mathbf{v}$ or that a body has angular velocity $\boldsymbol{\omega}$ have always carried the the tacit qualification 'with respect to an absolute, or Newtonian, frame of reference'. This has not prevented us from using moving frames of reference, as in the last example where unit vectors attached to the propeller have motion relative to a frame fixed in the aircraft and moves with the aircraft. But we have been careful to take account of the changes in these vectors due to their motion, recognizing that a failure to do so would lead to incorrrect conclusions.

The general matter to be considered now is the extent to which failure to take account of the movement of a frame of reference leads to wrong answers to problems in dynamics. The commonest example is our normal assumption that the earth provides an absolute frame of reference.

Let $\mathbf{R}_i$ locate a typical particle of mass m_i of a set of moving particles relative to a fixed origin. If $\mathbf{P}_i$ and $\mathbf{p}_i$ are the forces applied respectively to the particle by external agencies and the system itself, the equation of motion is

$$\mathbf{P}_i + \mathbf{p}_i = m_i \ddot{\mathbf{R}}_i.$$

If O is the origin of a moving frame of reference then from Fig. 8.15

$$\mathbf{R}_i = \mathbf{r}_O + \mathbf{r}_i$$

and

$$\dot{\mathbf{R}}_i = \dot{\mathbf{r}}_O + \boldsymbol{\omega} \times \mathbf{r}i + \partial \mathbf{r}_i / \partial t$$

where $\boldsymbol{\omega}$ is the angular velocity of the moving frame of reference and $\partial \mathbf{r}_i/\partial t$ is the velocity of m_i relative to the moving frame. On differentiating

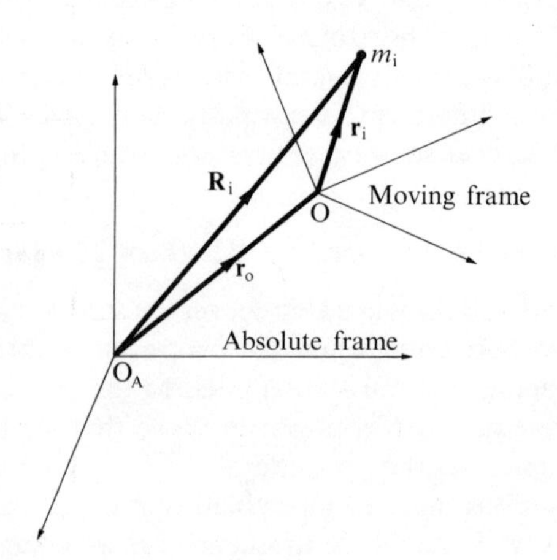

FIG. 8.15

again

$$\ddot{\mathbf{R}}_i = \ddot{\mathbf{r}}_0 + \dot{\boldsymbol{\omega}} \times \mathbf{r}_i + \boldsymbol{\omega} \times (\boldsymbol{\omega} \times \mathbf{r}_i) + 2\boldsymbol{\omega} \times \frac{\partial \mathbf{r}_i}{\partial t} + \frac{\partial^2 \mathbf{r}_i}{\partial t^2}.$$

$\partial^2\mathbf{r}_i/\partial t^2$ is the apparent acceleration of the particle as seen by an observer who is situated within the moving frame of reference. Such an observer will properly infer that this is the absolute acceleration only if (i) the moving axes do not rotate, so that $\dot{\omega} = 0 = \omega$, and (ii) the velocity of translation of the moving system is constant, $\ddot{\mathbf{r}}_O = 0$. Given these conditions, summation of forces on the whole set of particles yields

$$\mathbf{P} = M\ddot{\mathbf{r}}_G = \frac{d}{dt}(M\mathbf{v}_G)$$

as before, except that $\mathbf{v}_G$ is now measured relative to the moving frame of reference.

If we assume that the earth provides an absolute frame of reference we ignore all the terms in the expression for $\ddot{\mathbf{R}}_i$ except the last. There is little error for most purposes in ignoring the first two terms. Assuming that the earth moves round the sun, assumed itself to be fixed, in a circular path of radius $r_0 \approx 150 \times 10^6$ km and that the period is 365·25 days $\ddot{r}_0 \approx 10^{-3}$ m s^{-2}. The angular acceleration $\dot{\boldsymbol{\omega}}$ of the earth due to the changing direction of its N–S axis of rotation is infinitesimal. The significance of the remaining two terms $\boldsymbol{\omega} \times (\boldsymbol{\omega} \times \mathbf{r})$ and the Coriolis term $2\boldsymbol{\omega} \times \partial\mathbf{r}_i/\partial t$ have already been discussed on page 119. The point was made there that except where large scale phenomena are concerned, for example tidal motion, or very refined measurements are involved, as with inertial navigation, it is acceptable to treat the earth as being fixed in space for the purposes of earth-bound mechanics.

There will be occasions when it is useful to utilize a moving frame of reference whose motion is significant. The problem of the aircraft propeller above is an example. In such cases it can be important to know the circumstances in which it is still permissible to write $\mathbf{Q}_0 = \partial\mathbf{h}_0/\partial t$, where $\mathbf{h}_0$ is measured in the moving frame of reference and $\partial\mathbf{h}_0/\partial t$ is the rate of change of $\mathbf{h}_0$ as seen within the moving frame.

In Fig. 8.16 the slider of mass m moves along a guide that is fixed in the frame $(\mathbf{i}, \mathbf{j})$ which rotates with constant angular velocity $\boldsymbol{\omega}$ about the

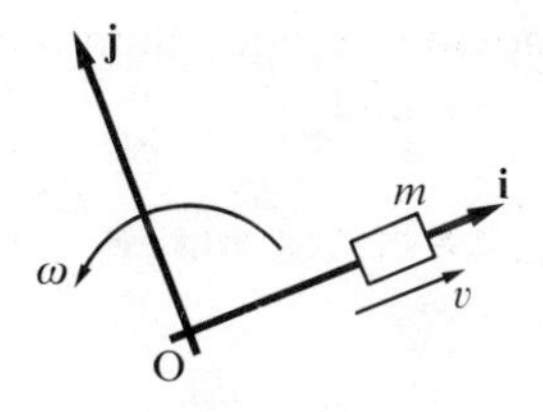

FIG. 8.16

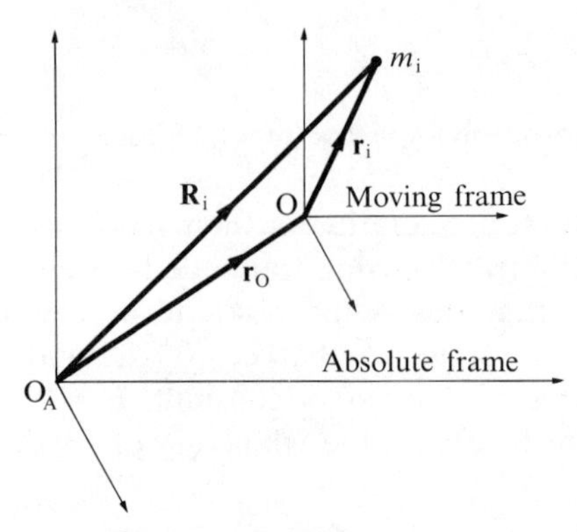

FIG. 8.17

fixed point O. In the moving frame $\mathbf{h}_O = 0$ and $\partial \mathbf{h}_O/\partial t = 0$. But we know from previous work that this motion involves a sideways reaction between the mass and the guide so that $\mathbf{Q}_O \neq 0$. Evidently we cannot write $\mathbf{Q}_0 = \partial \mathbf{h}_0/\partial t$ when the frame of reference rotates and we will exclude this possibility from further consideration.

In Fig. 8.17 the frame of reference moves without rotation. For a typical particle m_i

$$\mathbf{P}_i + \mathbf{p}_i = m_i \ddot{\mathbf{R}}_i$$
$$= m_i \frac{\mathrm{d}}{\mathrm{d}t}(\dot{\mathbf{r}}_O + \dot{\mathbf{r}}_i)$$

where we understand $\dot{\mathbf{r}}_i = \partial \mathbf{r}_i/\partial t$.

Taking moments about O

$$\mathbf{r}_i \times (\mathbf{P}_i + \mathbf{p}_i) = m_i \mathbf{r}_i \times \left(\frac{\mathrm{d}\dot{\mathbf{r}}_O}{\mathrm{d}t} + \frac{\partial \dot{\mathbf{r}}_i}{\partial t}\right).$$

Now, $\mathbf{r}_i \times \dfrac{\partial \dot{\mathbf{r}}_i}{\partial t} = \dfrac{\partial}{\partial t}(\mathbf{r}_i \times \dot{\mathbf{r}}_i) - \dot{\mathbf{r}}_i \times \dot{\mathbf{r}}_i$, so that, as $\mathbf{r}_i \times \mathbf{r}_i = 0$,

$$\mathbf{r}_i \times (\mathbf{P}_i + \mathbf{p}_i) = m_i \mathbf{r}_i \times \frac{\mathrm{d}\dot{\mathbf{r}}_O}{\mathrm{d}t} + \frac{\partial}{\partial t}\{\mathbf{r}_i \times (m_i \dot{\mathbf{r}}_i)\}$$
$$= m_i \mathbf{r}_i \times \frac{\mathrm{d}\dot{\mathbf{r}}_O}{\mathrm{d}t} + \frac{\partial \mathbf{h}_i}{\partial t}$$

where $\mathbf{h}_i$ is the moment of momentum about O of the ith particle as seen by an observer in the moving system. Summing over all particles, and after allowing $\sum \mathbf{r}_i \times \mathbf{p}_i = 0$, we find

$$\mathbf{Q}_O = \sum \mathbf{r}_i \times \mathbf{P}_i = (\sum m_i \mathbf{r}_i) \times \frac{\mathrm{d}\dot{\mathbf{r}}_O}{\mathrm{d}t} + \frac{\partial \mathbf{h}_O}{\partial t}.$$

This reduces to

$$\mathbf{Q}_0 = \frac{\partial \mathbf{h}_0}{\partial t}$$

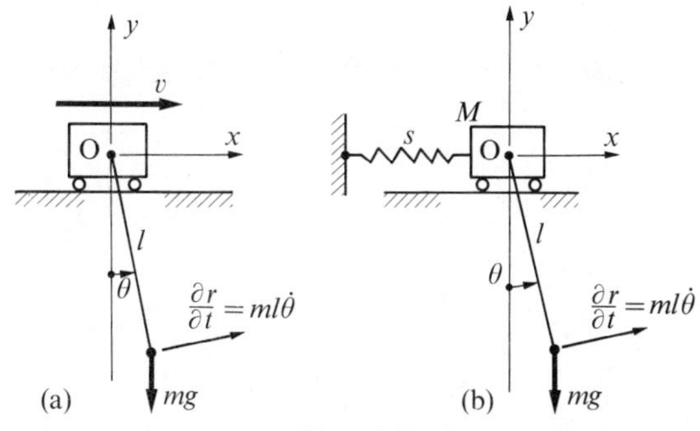

FIG. 8.18

when either (i) $\sum m_i \mathbf{r}_i = 0$, that is O coincides with G, or (ii) the moving origin O has constant velocity, or (iii) $\sum m_i \mathbf{r}_i$ is parallel to $\mathrm{d}\dot{\mathbf{r}}_0/\mathrm{d}t$.

The third possibility can be ignored as being too special. The second possibility, namely that $\mathbf{Q}_0 = \partial \mathbf{h}_0/\partial t$ when O has constant velocity, provided that the frame of reference is not rotating, is a valuable conclusion that is easy to apply.

If the frame of reference is accelerating, but not rotating, we can still use the simple expression provided that the origin is located at the centre of mass of the system, that is $\mathbf{Q}_G = \partial \mathbf{h}_G/\partial t$. If $\mathbf{Q}_G = 0$ then $\mathbf{h}_G$ is constant, and the angular momentum is conserved irrespective of the motion of G itself.

A few examples may help erroneous use of $\mathbf{Q}_0 = \partial \mathbf{h}_0/\partial t$ to be avoided. For simplicity the examples are chosen from the field of particle mechanics.

In Fig. 8.18(a) a simple pendulum is suspended from a body that is moving with constant velcoity v. In this case it is correct to refer the motion to the moving system Oxy and to write

$$-mgl \sin \theta = \frac{\mathrm{d}}{\mathrm{d}t}(ml^2\dot{\theta})$$

$$\Rightarrow \qquad \ddot{\theta} + (g/l)\sin \theta = 0.$$

The result is precisely the same as when the point of support is stationary.

It would be incorrect to write the same equation for the system in Fig. 8.18(b) because here the origin O does not have constant velocity. This embargo is often loosely expressed by stating that it is not permissible to take moments of momentum about a moving point. This is not quite correct, for whoever is doing the calculation is free to take moments about any point he chooses. The error lies in using an inadmissible statement of the momentum.

Referring to Fig. 8.18(b), the velocity of the pendulum has two components $\dot{x}$ due to the motion of the pivot and $l\dot{\theta}$. It is in order to write

$$\mathbf{Q}_O = -mgl \sin \theta = \frac{\mathrm{d}}{\mathrm{d}t}(ml^2\dot{\theta} + m\dot{x}l \cos \theta)$$

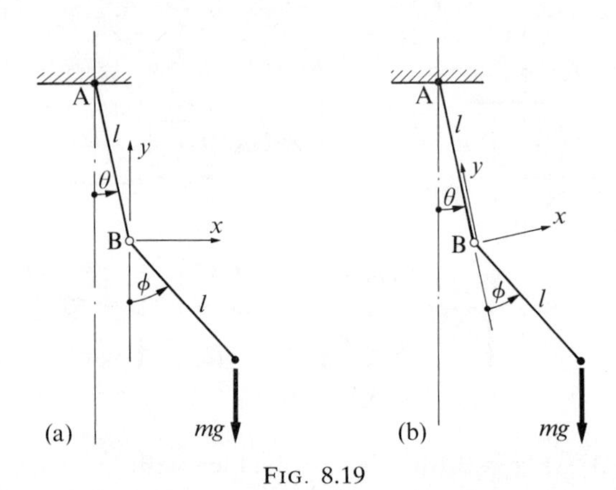

FIG. 8.19

because we are no longer measuring the momentum relative to the moving Oxy frame of reference. If θ is small the equation reduces to

$$\ddot{\theta}+\ddot{x}/l+(g/l)\theta=0.$$

A second equation is needed to determine the motion. This is provided by equating the spring force to the rate of change of horizontal momentum of the whole system:

$$-sx=(M+m)\ddot{x}+ml\ddot{\theta}.$$

In similar vein, it would be incorrect to express the momentum of the bob of the jointed pendulum in Fig. 8.19(a) as $ml\dot{\phi}$. This is the momentum relative to the Bxy set of axes, but the origin B is accelerating. It would be doubly wrong to represent the displacement as in Fig. 8.19(b) and express the momentum of the bob as $ml\dot{\phi}$. In this case not only is the origin B accelerating, but the axis system Bxy is rotating with angular velocity $\dot{\theta}$.

8.7. Inertia forces in planar mechanisms

Appeal to the principle of conservation of moment of momentum is very effective in the solution of problems that are concerned mainly with motion about a particular axis, and which involve only one or two bodies. When there are several interconnected bodies, as in linkage mechanisms, it is usually advantageous to take a different approach.

For any one body, there are two basic equations of motion:

$$\mathbf{P}=M\mathbf{a}_{\mathrm{G}} \quad \text{and} \quad \mathbf{Q}_{\mathrm{G}}=\frac{\partial \mathbf{h}_{\mathrm{G}}}{\partial t}.$$

We have seen above that the second of these equations holds irrespective of the motion of G. For planar motion it can be expressed alternatively as

$$\mathbf{Q}_{\mathrm{G}}=I_{\mathrm{G}}\dot{\boldsymbol{\omega}}.$$

As quantities on either side of an equation must be expressed in the same physical units, we can regard $M\mathbf{a}_G$ as a force and $I_G\dot{\boldsymbol{\omega}}$ as a couple. We can then go a stage further and write the equations in the form

$$\mathbf{P}+M(-\mathbf{a}_G)=0 \quad \text{and} \quad \mathbf{Q}_G+I_G(-\dot{\boldsymbol{\omega}})=0$$

so that our equations of motion have now become effectively equilibrium equations which state that the applied force $\mathbf{P}$ is in equilibrium with the inertia force $M(-\mathbf{a}_G)$ and the applied moment $\mathbf{Q}_G$ is in equilibrium with the inertia couple $I_G(-\dot{\boldsymbol{\omega}})$. The inertia force/couple is also referred to as the d'Alembert force/couple.

8.7.1. D'Alembert's principle

D'Alembert's principle states that a mechanical system is in equilibrium under the action of its applied and d'Alembert forces. On first reading, this statement of principle might seem to be nothing more than a play with words. Nevertheless the idea proves to be very useful in practice. The general effect is to transform a problem in dynamics to one in statics. This means that any procedure that is valid for statics is also valid for dynamics. Some examples will illustrate the use of d'Alembert's principle.

Example 8.14: Circular motion of a particle centrifugal force.

The acceleration of a particle moving with constant angular velocity ω round a circular path of radius a is $-a\omega^2$. For a mass m, the inertia force associated with this motion is $ma\omega^2$, and is commonly referred to as the *centrifugal* force. As there seems to be little advantage in singling out this particular example of inertia force we shall not use this term.

Example 8.15: Derive the equation of motion for the compound pendulum in Fig. 8.20.

It is advisable to draw two diagrams in all but the simplest of examples, one to show the accelerations, the other to show the applied and inertia

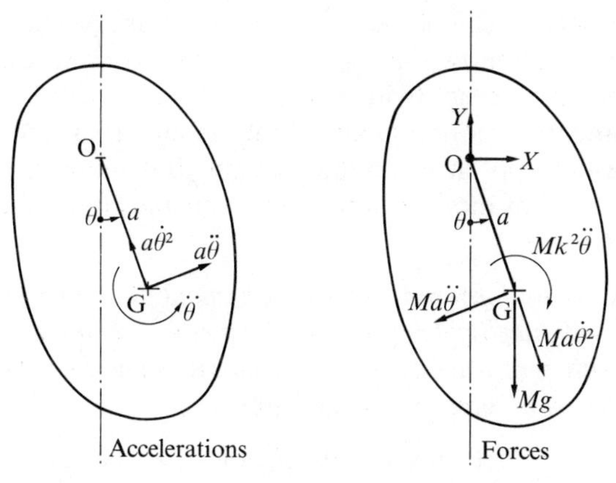

FIG. 8.20

forces. It is also advisable to show all the forces that act even those that are believed to be irrelevant. This has been done for the compound pendulum in Fig. 8.20. The forces in the right-hand figure are in equilibrium. By taking moments about O we eliminate the unknown reactions X and Y at the support, and so obtain as the equation of motion

$$M(a^2+k^2)\ddot{\theta}+Mga \sin \theta = 0.$$

If θ is small enough for the approximation $\sin \theta = \theta$ to be made, the equation becomes the equation for simple harmonic motion, as on p. 188.

For large values of θ we must be content with obtaining an integral only as far as $\dot{\theta}$. After making the substitution

$$\ddot{\theta} = \dot{\theta}\, \mathrm{d}\dot{\theta}/\mathrm{d}\theta = \frac{\mathrm{d}}{\mathrm{d}\theta}\,(\tfrac{1}{2}\dot{\theta}^2)$$

we find

$$\tfrac{1}{2}M(a^2+k^2)\dot{\theta}^2 - Mga \cos \theta = C.$$

If $\theta = \theta_0$ when $\dot{\theta} = 0$, $C = -mga \cos \theta_0$

$$\Rightarrow \qquad (a^2+k^2)\dot{\theta}^2 = 2ga(\cos \theta - \cos \theta_0),$$

for which there is no solution for θ in terms of simple functions.

The component reactions at the pivot are obtained by resolution of the forces:

$$X = Ma\ddot{\theta} \cos \theta - ma\dot{\theta}^2 \sin \theta$$

$$Y = Mg + Ma\ddot{\theta} \sin \theta + Ma\dot{\theta}^2 \cos \theta.$$

On substituting for $\ddot{\theta}$ and $\dot{\theta}^2$ from the equation of motion and its first integral respectively, X and Y are obtained as functions of θ.

A comparison of this analysis of the compound pendulum with our earlier one by way of $\mathbf{Q} = \mathrm{d}\mathbf{h}_0/\mathrm{d}t$ evokes some comment. Firstly, by comparison with Fig. 8.8, Fig. 8.20 is more complicated. We have felt it necessary to draw two diagrams where one sufficed before, and more vectors are involved. This is because for every one vector in a velocity diagram there are often two in the acceleration diagram. The increase in complexity does pay some dividends. Some readers will feel that inertia forces are somehow more tangible than momentum vectors and will prefer the present approach for that reason. But more importantly, the present analysis yields more information; it enables the force at the pivot to be calculated.

Example 8.16: A uniform rod slides in a vertical plane, on the ground and against a wall with negligible friction at the points of contact, Fig. 8.21. The motion starts with zero velocity when θ is small. Show that the rod will lose contact with the wall before it hits the ground.

For convenience, let the length of the rod be $2a$, so that $k^2 = (2a)^2/12 = a^2/3$. As long as the rod is in contact with the wall $x_G = a \sin \theta$

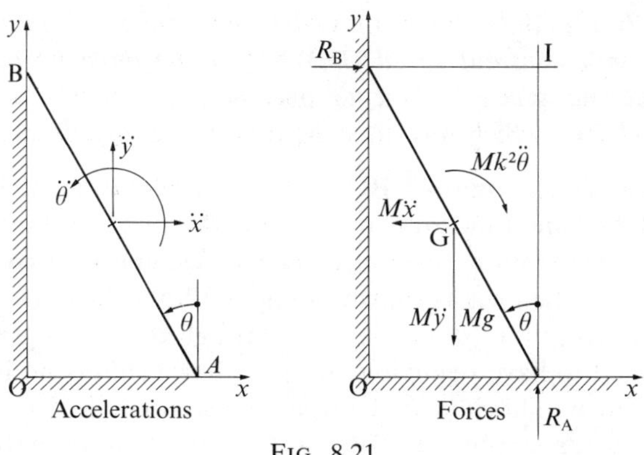

FIG. 8.21

and $y_G = a\cos\theta$, where Ox and Oy are along the ground and wall respectively. On differentiating we find

$$\ddot{x} = -a\dot{\theta}^2 \sin\theta + a\ddot{\theta}\cos\theta$$

and

$$\ddot{y} = -a\theta^2\cos\theta - a\ddot{\theta}\sin\theta.$$

By taking moments about I we eliminate the reactions R_A and R_B from the equation of motion:

$$Mk^2\ddot{\theta} + yM\ddot{x} - x(M\ddot{y} + Mg) = 0$$

which reduces to

$$(k^2 + a^2)\ddot{\theta} - ga\sin\theta = 0,$$

the terms in $\dot{\theta}^2$ having conveniently cancelled each other out. On solving this equation we find

$$\tfrac{2}{3}a^2\dot{\theta}^2 + ga\cos\theta = \mathrm{C}.$$

$\dot{\theta} = 0$ when $\theta = 0$, so $\mathrm{C} = ga$ and

$$\dot{\theta}^2 = 3g(1 - \cos\theta)/2a.$$

On equating forces in the direction Ox we find

$$R_B = M\ddot{x} = Ma(-\dot{\theta}^2\sin\theta + a\ddot{\theta}\cos\theta).$$

On substituting for $\dot{\theta}^2$ and $\ddot{\theta}$ this reduces to

$$R_B = \tfrac{3}{2}Mg\sin\theta(3\cos\theta - 1).$$

Separation occurs when $R_B = 0$, that is when $\theta = \cos^{-1}(2/3) = 70{\cdot}5°$.

In this problem, momentum analysis or the energy equation could have been used to find the equation of motion, but an analysis of the inertia forces would still have been necessary to determine the condition for separation at the wall.

Example 8.17: Fig. 8.22 shows a four-bar mechanism with the input crank AB rotating with constant speed ω. What torque must be applied to the crank AB at the given instant to overcome the inertia effects of the connecting rod BC, which has mass M and radius of gyration k?

The angular acceleration of BC and the acceleration of its centre of gravity are determined in terms of ω_{AB} for the given instant by drawing velocity and acceleration diagrams. The d'Alembert force $-Ma_G$ and couple $-Mk^2\dot{\omega}_{BC}$ then act as shown in Fig. 8.22(a). These are balanced by the forces applied to the connecting rod through the bearings at B and C.

CD is deemed to have negligible mass so that it suffers no inertia forces and must be in equilibrium under the forces applied at C and D, Fig. 8.22(b). The force applied to BC at C must consequently be in the direction CD.

In a similar way a force is applied at B along AB. But in this case there is also a transverse force $\mathbf{F}_B$ due to the torque $\mathbf{Q}$ applied at A, and such that $Q = F_B \, . \, AB$.

The set of forces on CB is therefore made up of the inertia force and couple, the forces along AB and CD respectively, and $-\mathbf{F}_B$ transversely to

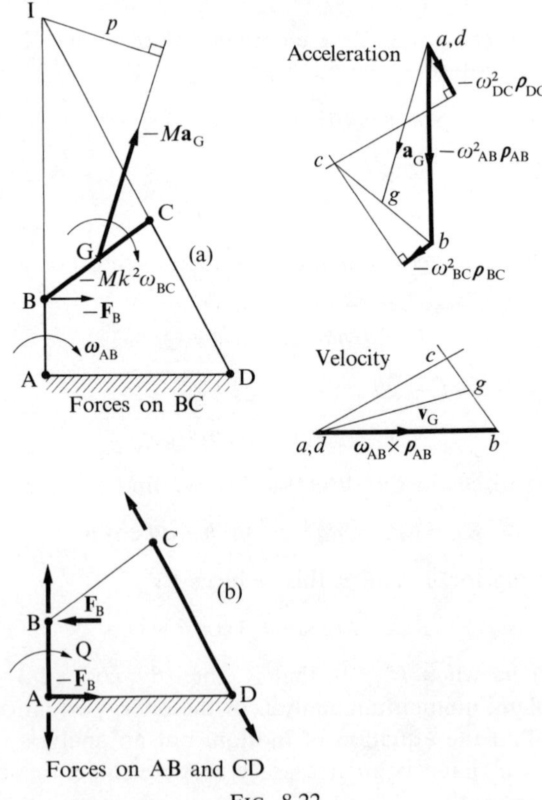

FIG. 8.22

AB at B. The sum of the moments of these forces about any point is zero. By taking moments about I we dispose of the unknown forces along AB and CD and have as the equilibrium equation

$$-\mathrm{IB}\times F_{\mathrm{B}}-p\times Ma_{\mathrm{G}}+Mk^{2}\dot{\omega}_{\mathrm{BC}}=0.$$

Hence
$$Q=\mathrm{AB}\times F_{\mathrm{B}}$$
$$=\mathrm{AB}\times(Mk^{2}\dot{\omega}_{\mathrm{BC}}-pMa_{\mathrm{G}})\div\mathrm{IB}.$$

If, as can happen, Q turns out to be negative it means that it is exerting a restraining influence on the motion of the mechanism at the given instant rather than driving it.

Links AB and CD cannot, of course, really be massless as assumed here. Provided that AB rotates at constant speed, its mass does not affect the driving torque. The mass of CD has significant effect and must be included in practice. It has been omitted here to simplify the presentation but can be included by a treatment similar to that used for BC.

When allowance has to be made for the masses of several links, the calculations are simplified by using the principle of virtual work.

8.7.2. Combined use of the principles of d'Alembert and virtual work

D'Alembert's principle states that a mechanical system is in equilibrium under its applied and inertia forces. The principle of virtual work says that the net work done by a set of forces in equilibrium in any virtual displacement is zero.

With the four-bar mechanism that we have just considered there is a very convenient set of virtual displacements at hand. These are the displacements which occur due to the actual velocities in a small time interval δt.

Taking only the mass of BC into consideration, the elements of the virtual work equation are: $\mathbf{Q}\cdot\boldsymbol{\omega}_{\mathrm{AB}}\delta t$ from the input torque, $(-M\mathbf{a}_{\mathrm{G}})\cdot\mathbf{v}_{\mathrm{G}}\delta t$ due to the motion of the centre of gravity of the connecting rod, and $(-Mk^{2}\dot{\boldsymbol{\omega}}_{\mathrm{BC}})\cdot\boldsymbol{\omega}_{\mathrm{BC}}\,\delta t$ due to its rotation. Hence the virtual work equation is

$$\mathbf{Q}\cdot\boldsymbol{\omega}_{\mathrm{AB}}\delta t-M\mathbf{a}_{\mathrm{G}}\cdot\mathbf{v}_{\mathrm{G}}\delta t-Mk^{2}\dot{\boldsymbol{\omega}}_{\mathrm{BC}}\cdot\boldsymbol{\omega}_{\mathrm{BC}}\delta t=0$$

$$\Rightarrow\qquad Q\omega_{\mathrm{AB}}=M\mathbf{a}_{\mathrm{G}}\cdot\mathbf{v}_{\mathrm{G}}+Mk^{2}\dot{\boldsymbol{\omega}}_{\mathrm{BC}}\cdot\boldsymbol{\omega}_{\mathrm{BC}}.$$

As $\dot{\boldsymbol{\omega}}_{\mathrm{BC}}$ and $\boldsymbol{\omega}_{\mathrm{BC}}$ are in the same direction, their scalar product is a simple multiplication. The product $\mathbf{a}_{\mathrm{G}}\cdot\mathbf{v}_{\mathrm{G}}$ must take into account the different directions of $\mathbf{v}_{\mathrm{G}}$ and $\mathbf{a}_{\mathrm{G}}$ as determined by the velocity and acceleration diagrams.

If several of the links of a mechanism have mass, all that needs to be done is to add the corresponding terms to the right-hand side of the virtual work equation so that,

$$Q\boldsymbol{\omega}_{\mathrm{AB}}=\sum M_{\mathrm{i}}\mathbf{a}_{\mathrm{Gi}}\cdot\mathbf{v}_{\mathrm{Gi}}+\sum M_{\mathrm{i}}k_{\mathrm{i}}^{2}\dot{\boldsymbol{\omega}}_{\mathrm{i}}\cdot\boldsymbol{\omega}_{\mathrm{i}}.$$

Although for the purposes of calculation the equation is used in the form just derived, it is of interest to consider its implications. Noting that

$$\mathbf{a}_G \cdot \mathbf{v}_G = \frac{1}{2}\frac{d}{dt}(\mathbf{v}_G \cdot \mathbf{v}_G) = \tfrac{1}{2}v_G^2$$

and that

$$\dot{\boldsymbol{\omega}} \cdot \boldsymbol{\omega} = \frac{1}{2}\frac{d}{dt}(\boldsymbol{\omega} \cdot \boldsymbol{\omega}) \quad = \tfrac{1}{2}\omega^2$$

we have

$$Q\omega_{AB} = \frac{d}{dt}\left(\sum \tfrac{1}{2}M_i \mathbf{v}_{Gi}^2 + \sum \tfrac{1}{2}M_i k_i^2 \boldsymbol{\omega}_i^2\right)$$

so that the rate of work input $Q\omega_{AB}$ equals the rate of change of kinetic energy of the mechanism.

8.8. Problems

1. The crankshaft ABCD of a four-cylinder two-stroke engine may be regarded as having four concentrated masses, each m, at the ends of cranks of radius r. The cranks are attached to the shaft at A, B, C, and D, with AB = BC = CD = a. In an end view of the shaft the cranks are at 90° intervals. Show that the resultant force on the crank-shaft bearings due to an angular velocity ω of the shaft is zero. Investigate the effect of different orders of crank angles (e.g. crank B at 90° or 180° to crank A etc.) on the resultant couple imposed on the bearings. What arrangement gives the least couple, and what is its magnitude and direction.

2. Determine the moments of inertia for the following bodies about the stated axis through their centres of mass. In every case the mass distribution is uniform throughout the body and the total mass is M.
   (a) A sphere of radius r about a diameter.
   (b) A solid rectangular prism with side lengths a, b, and c about the axis parallel to the sides of length c.
   (c) A solid right circular cone of base radius r about the central axis.
   (d) A solid right circular cone of height h and base radius r about an axis through G perpendicular to the central axis.
   Note that the perpendicular axis theorem (Appendix 2) applies only to the laminae.
   (e) A lamina in the form of an equilateral triangle of side length a about an axis through G parallel to one of the sides. Confirm that the moment of inertia is the same about an axis through G perpendicular to one of the sides.
   What is the diameter and location of Mohr's circle (Appendix 2)?
   (f) A lamina in the form of a 3, 4, 5 triangle about axes through G parallel to the sides of length 3 and 4. Determine also the product of inertia for these two axes and hence find the principal axes.

3. A flywheel requires a torque of 1·8 N m to overcome the total resistance to rotation at 500 rev min^{-1}. Friction at the bearings contributes a constant torque of 0·6 N m, and the remainder is due to air resistance which may be assumed to be proportional to the square of the speed.

If the flywheel is spinning freely at 500 rev min^{-1}, find the time taken for it to come to rest. Its moment of inertia is 4 kg m^{-2}.

4. Two horizontal shafts A and B are geared together so that the speed of B is four times that of A. Their moments of inertia are 20 kg m^{-2} for shaft A and 1·2 kg m^{-2} for shaft B. A

drum of 1 m diameter is keyed to shaft A, and a rope wound round the drum carries a mass of 120 kg at its free end.

Initially the system is held at rest, and is then released. Calculate the velocity of the mass after it has fallen 3 m. Friction at the bearing exerts a torque of 20 N m on each shaft.

5. A solid uniform sphere of radius a is at rest on a horizontal surface, the coefficient of friction between them being 0·2. The surface is given a constant horizontal acceleration α. Find the maximum value of α if the sphere is to roll without slipping on the surface, and show that if α has this magnitude the sphere will make its first complete revolution in time $(8\pi a/g)^{\frac{1}{2}}$.

6. Fig. 8.23 shows a connecting rod free to swing about a horizontal knife-edge at A with a periodic time of 2·3 s. When it is inverted and supported at B the time for a complete oscillation is 2·5 s. If the distance between A and B is 1·2 m, find the position of the centre of gravity G of the rod, and the radius of gyration about the axis through G parallel to the knife-edge.

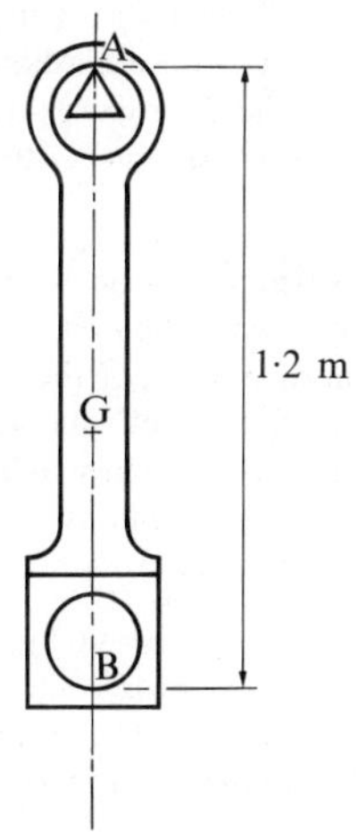

FIG. 8.23

7. A straight uniform rigid bar APB of length l and mass m per unit length is free to turn about a transverse horizontal axis at P, where $AP = l/4$ and $PB = 3l/4$. The bar is released from rest in the horizontal position.

Show that the angular acceleration immediately after release is $12g/7l$.

What is the angular velocity at the bottom of the swing?

8. A uniform disc of radius r is supported in a horizontal plane by three equal strings attached to equally spaced points on its perimeter. The upper ends of the strings are attached to fixed points height z above the disc so that when the disc is in equilibrium the strings are vertical. Find the period of small torsional oscillations of the disc.

What is the period when the points of suspension are equally spaced round a circle of radius R if z is still the height of these points above the disc?

9. A tractor climbs up a slope which makes an angle of $\sin^{-1}0{\cdot}1$ to the horizontal. The combined mass of the rear driving wheels is 275 kg, their radius of gyration is 0·75 m, and their radius is 1 m. The mass of the remainder of the tractor is 900 kg.

Assuming no slip occurs between the rear wheels and the surface, calculate the acceleration of the tractor up the slope when the torque transmitted to the rear axle is 2700 N m.

10. A sun and planet gear-train consists of the wheels P and R connected by the arm Q as

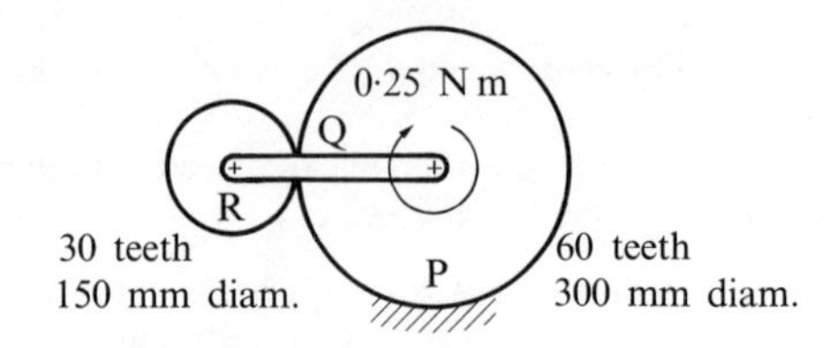

FIG. 8.24

shown in Fig. 8.24. P is fixed with its axis vertical and a steady torque of 0·25 N m is applied to Q.

Show that, when Q has turned through an angle θ, R has turned through 3θ. Find the angular velocity of Q when it has turned through two revolutions from rest.

The mass of R is 1 kg and its radius of gyration about its central axis is 50 mm. The moment of inertia of Q about the axis of P is $0{\cdot}06\ \mathrm{kg\,m^2}$.

11. The door of a railway carriage, which has hinges (supposed smooth) towards the engine, is at right angles to the train which starts with a acceleration f.

Find the angular velocity with which the door closes if its width is $2a$ and relevant radius of gyration about an axis through G is k. Assume G to be at the geometric centre of the door.

12. A rough belt is moving in a horizontal plane with constant velocity u when a solid cylinder of radius a is placed in contact with the belt. The axis of the cylinder is horizontal and is perpendicular to the direction of motion of the belt.

If the coefficient of friction between the belt and the cylinder is μ, show that when slipping ceases the belt will have moved a distance $u^2/3\mu g$ and that the cylinder will be rolling backwards relative to the belt with a velocity $2u/3$.

What will be the final velocity of the cylinder if the belt stops?

13. Fig. 8.25 shows two wheel pairs, each of mass 90 kg, radius 0·3 m and radius of gyration 0·25 m, which are connected by two equal rods each of mass 9 kg freely pivoted to the wheels at similar points 0·2 m from their axes. The system is placed on level ground rough enough to prevent slipping. Find the frequency of oscillations about the position of stable equilibrium.

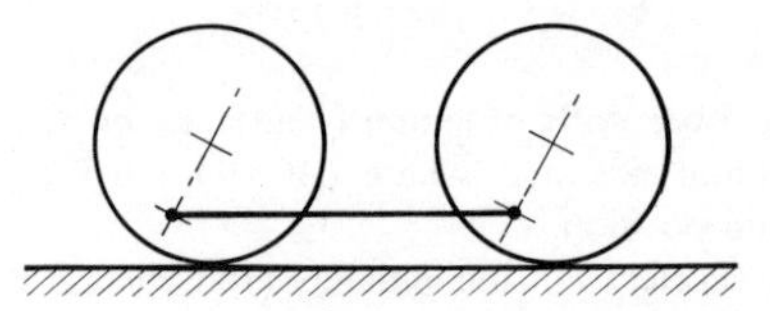

FIG. 8.25

14. A light horizontal arm AB of length l rotates about a vertical axis through A. A uniform arm BC of mass M and length l is pinned to AB at B, and is free to rotate relative to it in a horizontal plane. AB rotates with constant angular velocity ω about A and initially C coincides with A and is then slightly displaced.

Show that the angular velocity of BC relative to AB is

$$\omega\sqrt{6}\sin(\theta/2)$$

where θ is the angle ABC.

Determine also the kinetic energy of the system when $\theta = 90°$.

15. A fan rotor has three blades at 120° to each other and spins with constant angular speed $\boldsymbol{\omega}$. At the same time the spin axis is made to turn with constant angular velcoity $\boldsymbol{\Omega}$ about a fixed axis in the plane of the blades. The two axes intersect at the centre of the rotor.

Taking each blade to be equivalent to a concentrated mass at radius r, show that the angular momentum of the blades is $3mr^2\boldsymbol{\omega} + \frac{3}{2}mr^2\boldsymbol{\Omega}$.

What couple is imposed on the rotor bearings?

16. A tea chest is slid onto a floor from a platform down a ramp set at $\tan^{-1}1/4$ to the floor. The chest is a 1 m cube packed uniformly and has a radius of gyration of 0·4 m about an axis through the centre of mass.

Calculate the minimum velocity which the chest must attain if it overturns without slipping about the edge that first hits the floor.

17. A bat is supported by a clamp on a knife-edge as shown in Fig. 8.26. Its centre of gravity is at G, and it oscillates with a natural frequency of n c s^{-1}. The bat is to be set in oscillation by a horizontal blow; where must this be applied if no impulsive reactions are to occur at the points of support?

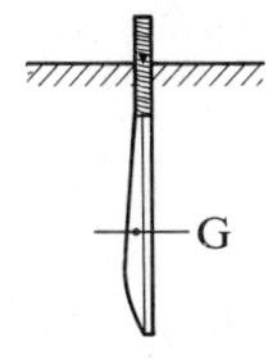

FIG. 8.26

18. A solid right circular cylinder of radius r has on its curved surface a small sharp ridge of height h ($\ll r$) and negligible mass running parallel with the longitudinal axis of the cylinder.

The cylinder rolls on a horizontal surface. Show that for each complete revolution of the cylinder the fractional loss of kinetic energy is approximately $8h/3r$. Assume that there is no slip or rebound between impacting surfaces and that friction losses are negligible.

19. A rigid uniform plank of length l and mass m is placed vertically with one end on the ground and allowed to overturn. When it has reached the horizontal position it comes into contact with a rigid horizontal bar placed transversely to the plank.

Show that the impulse of the bar will be greatest if the bar is at a distance $l/\sqrt{3}$ from the end of the plank in contact with the ground. What will then be its magnitude?

It may be assumed that the plank does not slip on the ground or rebound from the bar.

20. Two friction wheels A and B are mounted on parallel shafts. A has a moment of inertia of 0·2 kg m^{-2} and a diameter of 0·3 m, B has a moment of inertia of 0·1 kg m^{-2} and a diameter of 0·2 m.

Wheel A is rotating at 120 rev min^{-1} with B at rest when the rims are pressed together with a constant normal force of 40 N. If the coefficient of friction is 1/3, estimate the time of slip.

21. Two equal uniform rods AB and BC, each of length $2a$, are freely hinged together at B and lie on a smooth horizontal plane in a straight line ABC. An impulse perpendicular to ABC is applied to AB at a point D. Find the distance of D from A if rod AB has no angular velocity immediately after impact.

22. The bell-crank ABC shown in Fig. 8.27 moves in a horizontal plane. Its motion about its bearing B is controlled by the crank OD and connecting rod DC. Crank OD rotates clockwise at a constant speed of 40 rad s^{-1}. For the position shown, find the velocity of A and the angular acceleration of ABC.

There is a force of 450 N acting as shown at A, and the moment of inertia of ABC about B is 0·085 kg m^{-2}. The inertia of DC is negligible. Find the torque required about O on the crank.

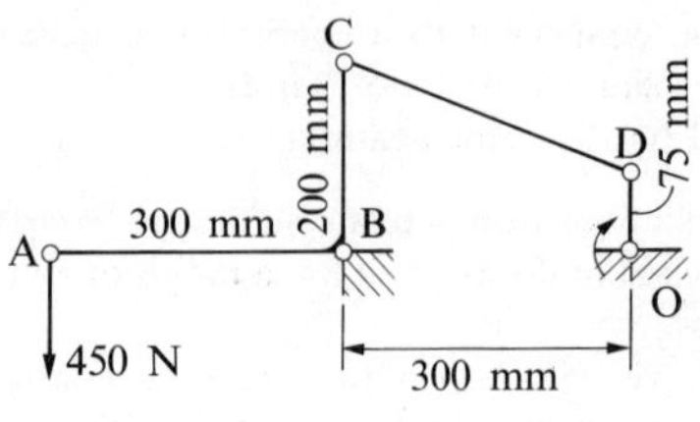

FIG. 8.27

23. In a light plane four-bar mechanism ABCD, the crank AB is of length a, link BC and crank CD are both of length $4a$, and the fixed link AD is of length $5a$. At a given instant angle DAB is zero. A concentrated mass m is carried by BC at E distant $4a$ from both B and C on the side of BC that is remote from D.

If crank AB is driven with constant angular velocity ω, determine:

(a) The velocity of E,
(b) The acceleration of E,
(c) The driving torque at A,
(d) The reactions at A and D.

24. The mechanism shown in Fig. 8.28 consists of two light links AB and CD and a uniform heavy link BC of mass m; the joints are frictionless. For the position shown in a horizontal plane, find the value of the torque necessary to maintain link AB rotating at a constant speed ω.

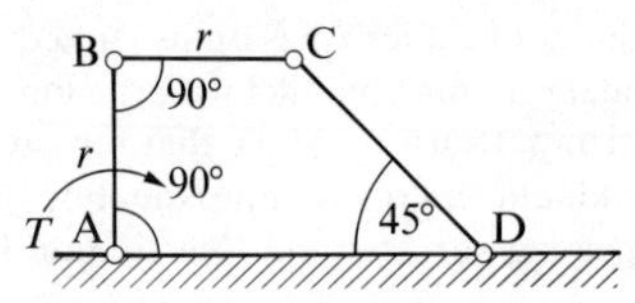

FIG. 8.28

25. A uniform rod of mass m and length l is suspended by two strings AB and AC attached to the same end A of the rod and to fixed points B and C. Each string is inclined at 30° to the vertical. If one string is cut, find the tension in the other string immediately afterwards.

9. Particle streams

So far we have studied problems that have been concerned with the dynamics of single particles, a particular set of particles, or a rigid body. In every case all the particles, even when they form a solid body, have in principle been identifiable. There might appear to be a problem of identification when the particles of the system are in a continuous stream. The solution lies in studying what happens to a selected set of particles in a definite time interval, usually small, but not necessarily so. The procedure follows a standard pattern and is best explained by giving a number of examples.

9.1. Steady change of linear momentum

All the problems in this chapter will be studied in relation to changes in linear and angular momentum. As the first example we will evaluate the force on a plate due to a jet of water hitting it. In Fig. 9.1 a jet of water issues from a nozzle and impinges with velocity $\mathbf{v}$ on a flat plate. There is no splashing and the liquid runs away laterally as shown. It is quite proper to argue that for a mass-flow rate m, the amount of liquid hitting the plate in a time δt is $m\,\delta t$ and that in the same interval momentum $m\mathbf{v}\,\delta t$ is destroyed. The rate of change of momentum in the direction of flow is therefore $m\mathbf{v}=-\mathbf{F}$ where $\mathbf{F}$ is the force that the plate applies to the liquid. As it is not always easy to apply this line of reasoning to more complicated situations we will adopt a slightly different argument.

First we identify a control surface, Fig. 9.1. This must be large enough to contain the whole region in which the change of momentum is taking place so that we can see exactly what is happening as the fluid enters and leaves the control volume. The particular group of particles on which we will concentrate are all the particles within the control volume at the given instant together with the group that will enter the control volume during the time interval τ that follows. The latter group indicated by the shaded block in the incoming jet in Fig. 9.1 has mass $m\tau$. At the end of the time interval τ all the block is just inside the control surface and an equivalent mass has left having been deflected by the plate, Fig. 9.1(b). The momentum of the set of particles in the direction normal to the plate has two components at the start of the time interval τ; there is $\mathbf{p}$, say, due to the motion of the particles within the control volume and $m\tau\mathbf{v}$ attributable to the block of particles in the incoming jet. The picture within the control surface is precisely the same at the end of the time interval τ as it was at the beginning and the momentum there is still $\mathbf{p}$.

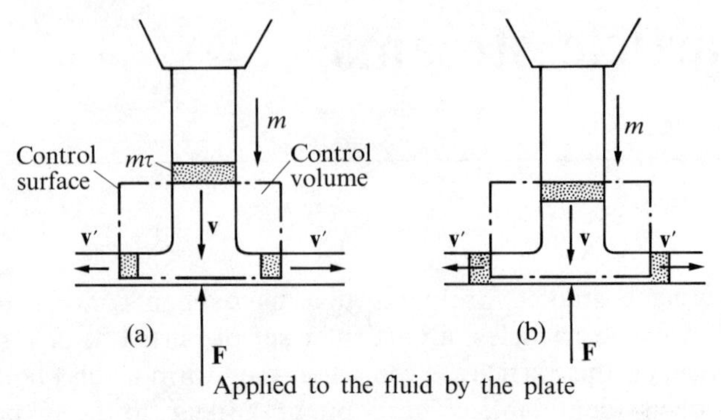

FIG. 9.1

Some of the particles that were originally within the control volume have now left it but with no momentum in the direction normal to the plane. The momentum equation for the whole set of particles is consequently

$$\int_0^{\tau} \mathbf{F}\, dt = \mathbf{p} - (\mathbf{p} + m\mathbf{v}\tau)$$

$$\Rightarrow \qquad \mathbf{F}\tau = -m\mathbf{v}\tau$$

$$\Rightarrow \qquad \mathbf{F} = -m\mathbf{v}.$$

It is worth noting that the net momentum of the water leaving the control volume must be zero in the direction parallel to the plane, for there is no force to change it.

A second example on the same theme is given in Fig. 9.2. Here material discharges from a hopper at a steady rate on to a conveyor belt. The problem is to determine the power needed to drive the conveyor belt with velocity **V**. Figs. 9.2(a) and (b) show the control surface and the set

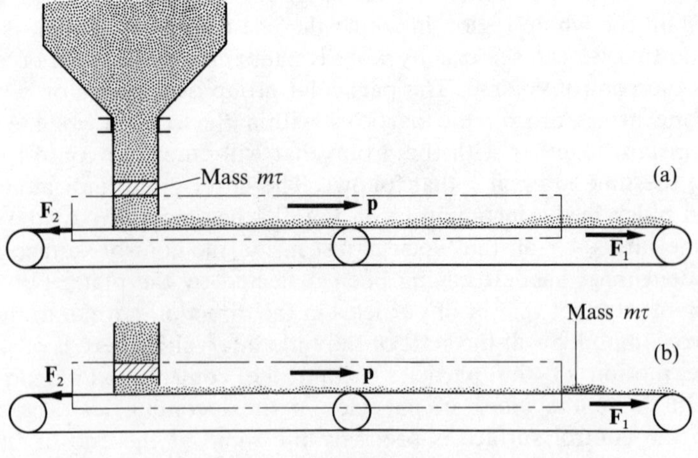

FIG. 9.2

of particles being considered at the beginning and end of a time interval τ. The right-hand boundary must be sufficiently far down the conveyor belt for there to be no slip between the belt and the material on it, otherwise the velocity of the particles at exit would be unknown. The momentum equation for the set of particles is

$$(\mathbf{F}_1 - \mathbf{F}_2)\tau = (m\mathbf{V}\tau + \mathbf{p}) - \mathbf{p}$$

$$\Rightarrow \qquad \mathbf{F}_1 - \mathbf{F}_2 = m\mathbf{V}$$

where $\mathbf{F}_1$ and $\mathbf{F}_2$ are the tensions at either end of the belt. The power needed to drive the belt is $(F_1 - F_2)V = mV^2$, that is twice the rate at which kinetic energy is acquired by the material on the belt.

Except for requiring conditions within the control surface to remain constant in time, the details of what is happening there are irrelevant. The particles are dumped on to the conveyor with zero horizontal velocity so that initially there is slip between them and the belt. They are brought up to speed by friction between them and the belt. The amount of slip depends on how rough the belt is but it has no effect on the power needed to drive the belt.

9.2. Steady change of moment of momentum

Fig. 9.3 shows a simplified irrigation device. Water is pumped along a horizontal pipe of length a which rotates about a vertical axis through the supply end with angular velocity Ω. The calculation of the torque Q needed to maintain the motion is very similar to the calculation of the driving force for the conveyor belt above. Because of the changing direction of the water leaving the end of the pipe and because there is an unknown sideways reaction at the supply end we consider moment of momentum balance over an infinitesimal time interval δt. During this time interval a mass of water $m\,\delta t$ leaves the pipe with velocity $a\Omega$ and has

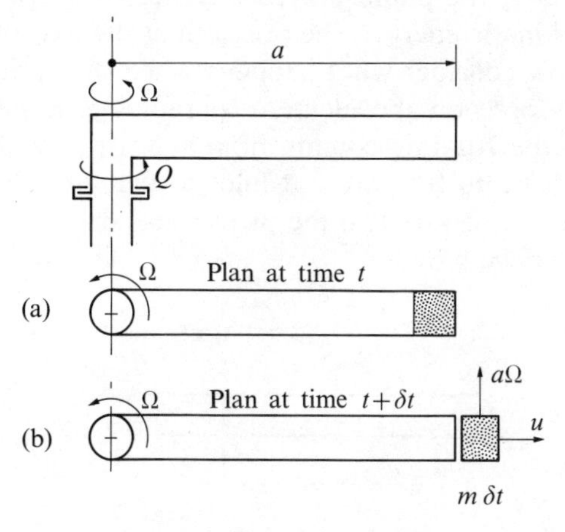

FIG. 9.3

moment of momentum about the axis of rotation of an amount $ma^2\Omega t$. In the meantime a similar mass of water has entered the pipe on the axis of rotation and consequently with zero moment of momentum. The picture of the water in the pipe, and its moment of momentum h, are unchanged so that the equation for the moment of momentum is

$$Q\,\delta t = (h + ma^2\Omega\,\delta t) - h$$

$$\Rightarrow \qquad Q = ma^2\Omega.$$

The power needed to drive the system is $ma^2\Omega^2$.

There may be two queries concerning this result. Firstly, what is the mechanism whereby the pipe supplies power to the water? Secondly, as there is no obvious dissipation of energy why is the power required to drive the sprinkler twice the rate at which the water acquires kinetic energy?

To answer the first question we must consider the forces on an element of water in the pipe. Let the element fill the cross-sectional area A and be of length δx so that its mass is $\rho A\,\delta x$. Its Coriolis acceleration is $2u\Omega$, where u is the velocity of flow. The force necessary to give this acceleration is $2u\Omega\rho A\,\delta x = 2m\Omega\,\delta x$. The total torque is therefore

$$\int_0^a 2m\Omega x\,\mathrm{d}x = ma^2\Omega.$$

The explanation of the apparent loss of power is less obvious. Our previous calculation involves no assumptions about friction between the water and the pipe. We are therefore free to assume that there is no friction loss within the rotating piece of pipe, and it simplifies the reasoning to do so. There must be some means of supplying the water and we will envisage that it is pumped through a supply pipe. Let us assume that the pressure drop along the pipe, due to friction, is Δp. With the sprinkler stationary, the pump power is ΔpuA to overcome friction and $\frac{1}{2}mu^2$ to supply kinetic energy. The pressure at the axis of the sprinkler is atmospheric. Now consider what happens when the sprinkler is rotating. The centrifugal force on the element at radius x is $\rho Ax\Omega^2$. The total radial force on the rotating column of water is $\int_0^a \rho Ax\Omega^2\,\mathrm{d}x = \frac{1}{2}\rho Aa^2\Omega^2$, so that the pressure in the water at inlet to the sprinkler is now $\frac{1}{2}\rho a^2\Omega^2$ below atmospheric pressure and the pressure against which the pump has to operate is reduced by the same amount. The reduction in power

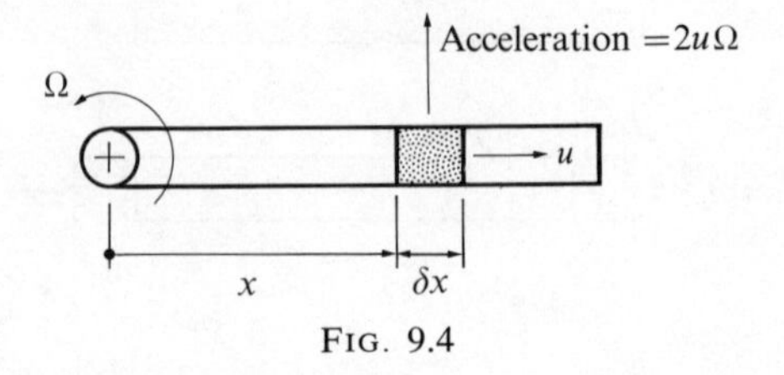

FIG. 9.4

required to drive the pump is therefore $\frac{1}{2}\rho a^2\Omega^2 uA = \frac{1}{2}ma^2\Omega^2$. This accounts for the difference between the power $ma^2\Omega^2$ needed to drive the sprinkler and the $\frac{1}{2}ma^2\Omega^2$ kinetic energy that the water acquires by virtue of its tangential motion at exit from the sprinkler. The pump continues to supply the $\frac{1}{2}mu^2$ due to the flow along the pipe.

9.3. 'Variable-mass' systems

In a so-called variable-mass system there is variation in the mass of the body whose motion is being considered. A good example is a rocket whose total mass diminishes as its fuel is burnt and ejected as exhaust gas whose subsequent motion is of no interest. The equation of motion of such a body is best derived by applying momentum considerations to the total mass over an infinitesimal time interval.

In Fig. 9.5 a rocket is shown travelling vertically against gravity at times t and $(t+\delta t)$. As it makes for confusion of the mathematics to presuppose that $\mathrm{d}m/\mathrm{d}t$ is negative, as in reality it is, we assume that the mass m at time t becomes $m+\delta m$ at time $t+\delta t$. This means that the mass of the exhaust gases ejected in the time interval δt is $-\delta m$. The velocity with which the products of combustion are ejected relative to the rocket depends on the chemical reaction in the combustion chamber, and is normally assumed to be constant at u, say. If there is no air resistance the only force acting is that due to gravity so that the momentum equation for the total mass is

$$mg\,\delta t = mv - \{(m+\delta m)(v+\delta v) + (-\delta m)(v+\delta v-u)\}.$$

On neglecting products of two infinitesimal quantities this equation reduces to

$$\delta v = -g\,\delta t - (u/m)\,\delta m$$

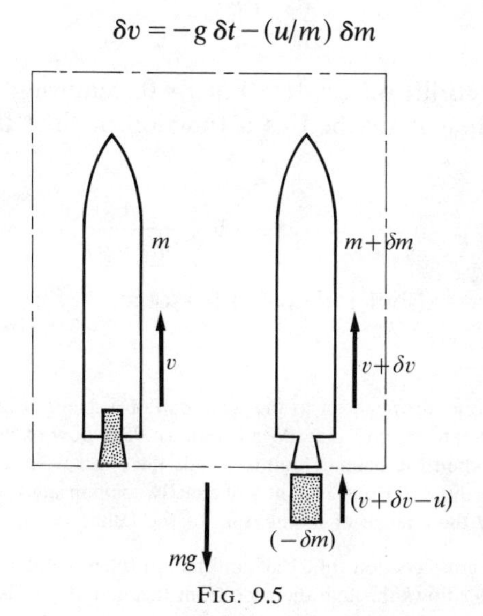

FIG. 9.5

and on proceeding to the limit as $t \to 0$

$$dv = -g\,dt - (u/m)\,dm.$$

On integrating, we have

$$v = -gt - u \log_e m + C.$$

If $m = m_0$ and $v = 0$ at $t = 0$ this becomes

$$v = -gt + u \log_e(m_0/m).$$

This analysis makes the tacit assumption that the rocket has axial symmetry, so that there is no tendency to rotate about a transverse axis. It is also assumed that the fuel burns in such a way as to keep the centre of mass at a constant position in the body of the rocket.

It is worth emphasizing that no assumption has been made concerning the rate at which the fuel is burnt, or even that the rate of burning is constant. The velocity at time t is the same whatever the manner of burning.

The rate of burning does, however, affect the maximum height reached by the rocket and determines whether the rocket can take off.

The differential equation of motion is found above to be

$$dv = -g\,dt - (u/m)\,dm$$

or

$$\frac{dv}{dt} = -g - \frac{u}{m}\frac{dm}{dt}.$$

If the rate of burning μ is constant so that $m = (m_0 - \mu t)$ and $dm/dt = -\mu$, then

$$\frac{dv}{dt} = \frac{\mu u}{m} - g.$$

If the rocket is to lift off $dv/dt > 0$ at $t = 0$, requiring $\mu > gm_0/u$.

To obtain the height reached as a function of time it is necessary to integrate

$$v = \frac{dh}{dt} = -gt + u \log_e \frac{m_0}{m_0 - \mu t}.$$

This is straightforward but rather cumbersome.

9.4. Problems

1. Coffee beans are being dropped on to the scale pan of a spring balance from a constant height h at a constant rate m, and they do not bounce. The flow of beans is cut off at the source at the instant when the balance pointer reads the exact weight required. Show that the weight of beans in the air at that instant will exactly compensate for the false reading of the balance caused by the change of momentum of the falling beans.

2. A pipe of uniform cross-section $0{\cdot}001\ \mathrm{m}^2$ and length $0{\cdot}6$ m rotates about a vertical axis through one end. Water flows through the pipe from this end at a constant rate of $5\ \mathrm{kg\,s^{-1}}$.

Find the torque required to make the pipe rotate at 600 rev min^{-1} assuming (a) that the pipe is straight, and (b) that there is a small 90° bend at the outlet so that the water issues in the direction of motion of that end.

What is the ratio of the rate at which the water gains kinetic energy to the power input?

3. A ping-pong ball of mass 1 g is supported by a jet of water which issues vertically at 4 m s^{-1} from a nozzle of cross-sectional area 1 mm^2. Estimate the height at which the ball remains stationary.

4. A sprinkler is shown in the plan in Fig. 9.6. It is free to rotate about O. Water is supplied through a pipe at the centre at a rate of 10^5 mm^3 s^{-1}. A horizontal nozzle on the end of each arm produces a jet of diameter 2·5 mm.

(a) Calculate the torque necessary to prevent the arms from rotating.

(b) Show that, if the pivot is frictionless, the arms will rotate at approximately 778 rev min^{-1}.

(c) If the arms are observed to rotate at 500 rev min^{-1}, calculate the friction torque.

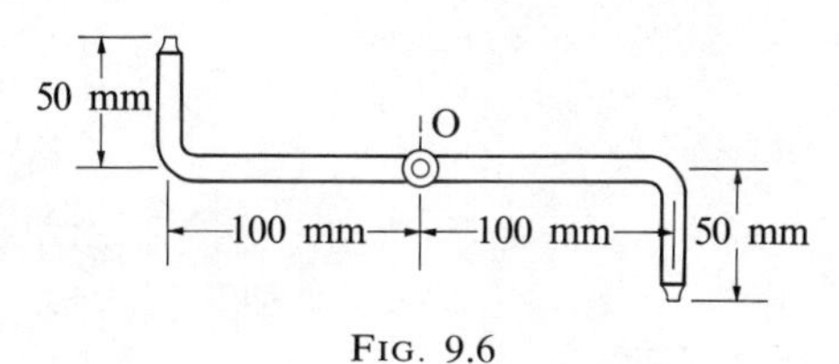

FIG. 9.6

5. A straight horizontal pipe AB of length l and of uniform cross-section can rotate freely about a vertical axis AZ. Water enters the pipe at A and flows outward through the pipe with a constant velocity v. The pipe is given an initial angular velocity ω_0 about AZ. Show that after a time t the angular velocity is

$$\omega = \omega_0 e^{-kt}$$

where $k = 3Mv/(M+m)l$ in which m is the mass of the pipe and M is the total mass of water that it contains at any instant.

6. During a test with the rotor held stationary, a jet of water of area 5000 mm^2 strikes the fixed 'bucket' of a pelton wheel at 50 m s^{-1} and is slowed down by friction to 44 m s^{-1} and turned through 160° in two symmetrical streams as shown in Fig. 9.7. Find the force exerted by the jet on the bucket.

If the bucket then moves in the direction of the jet at 24 m s^{-1} and the relative velocity undergoes the same percentage reduction in magnitude and the same change of direction, what is the force then exerted on the bucket, and what power is transmitted to it?

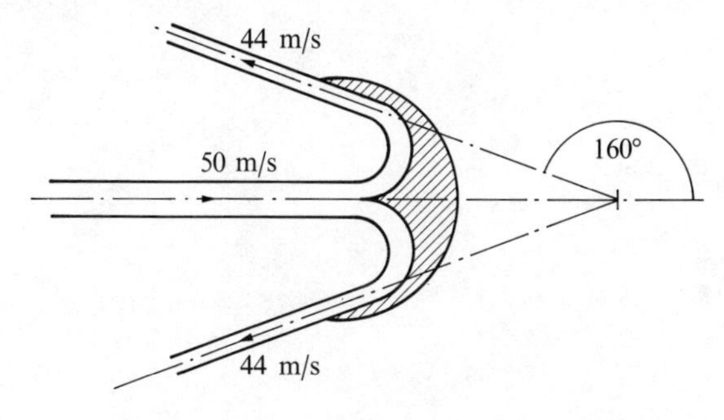

FIG. 9.7

If the wheel consists of a series of buckets moving at 24 m s^{-1} and so arranged that all the water emerging from the nozzle strikes one bucket or another, what mean power is transmitted to the wheel?

7. A rocket, the total initial mass (fuel and shell) of which is M_0, ejects combustion products at a constant rate CM_0 at a speed u backwards relative to the shell. Show that the lowest rate of fuel consumption which will permit the rocket to rise at once is given by $C = g/u$. Assuming that C has this value, and taking the final mass to be M_1, find the greatest speed and the greatest height attached by the rocket. Neglect variations in g.

8. A jet-propelled aircraft of weight 7 Mg travels in straight horizontal flight in still air. The engine takes in 45 kg s^{-1} of air which discharges with a velocity of 615 m s^{-1} relative to the aircraft. Fuel is burnt at the steady rate of 0·7 kg s^{-1}, the products of combustion being passed out with the exhausted air. Assuming that the pressure difference between inlet and exhaust is negligible, and that the resistance to the motion of the aircraft due to drag forces acting on its outside surface is 13 kN, calculate the acceleration when the speed is 960 km h^{-1}.

10. Power transmission

THE principles that have been introduced and developed in the previous chapters find application in the design and analysis of machinery for the transmission of motion and power. We will conclude this book with a consideration of some of the applications. The particular basic problem to be considered is that of transmitting power between two rotating shafts. The decision as to the method of connecting the two shafts depends on several factors such as: are the shafts coaxial or parallel? If not, do they intersect or are they skew?

If the shafts are coaxial and run at the same speed it may be simply a matter of making a rigid joint between the ends of the shafts, but the practicability of this depends on there being no misalignment to be accommodated. Also we must consider whether the connection is required to be permanent or whether it may be necessary to disconnect the shafts during the course of normal running. If such disconnection is necessary a clutch is required. We have also to ask whether the speeds of the two shafts are required to be the same at all times. If so, a rigid joint would be the obvious choice, but if there is possibility of stalling the output, and thereby causing damage or other trouble upstream of the driving shaft, a fluid drive may be appropriate as in some automobile and marine applications.

It will be evident that a single chapter can hardly suffice to deal with all such matters in detail, and that we will have to restrict our attention. We will concentrate on two aspects: firstly, friction drives and secondly, gear and other mechanism drives between parallel shafts, taking account of friction in such drives.

10.1. Clutches

A clutch is a device which allows two coaxial shafts to be connected together, or not, as required. The shafts carry discs at their ends, and the connection is made by pressing the discs together. Slip between the two discs is usually prevented in one of two ways. In the dog clutch there are projections, or teeth, on the surfaces of the two plates which engage when the plates are pressed together. In such an arrangement there is an absolute assurance that there will be no slip, but there are problems in making the teeth engage when there is relative motion between the shafts. The second type, which can be engaged without difficulty when the shafts are already in motion, relies only on friction between the two surfaces that are pressed together, Fig. 10.1.

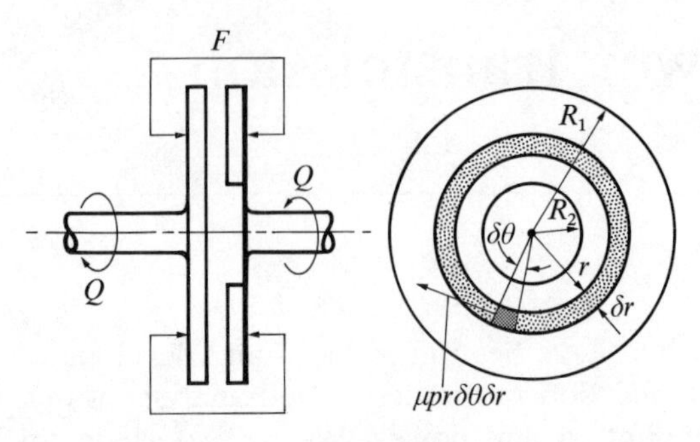

FIG. 10.1

If the total force pressing the two surfaces together is evenly distributed over the area of contact, thus causing a uniform contact pressure p, the maximum friction that can be generated over an element of the contact area at radius r is $\mu p(r\,\delta\theta)\,\delta r$. The contribution to the transmitted torque for the shaded annulus, Fig. 10.1, is $r\mu p(2\pi r)\,\delta r$. If the limiting radii for contact are R_1 and R_2, as shown, the maximum torque that can be transmitted is

$$Q=\int_{R_2}^{R_1} 2\pi\mu p r^2\,\mathrm{d}r$$

$$=\frac{2\pi}{3}\mu p(R_1^3-R_2^3).$$

But

$$p=F\div\pi(R_1^2-R_2^2),$$

hence

$$Q=\frac{2\mu(R_1^3-R_2^3)}{3(R_1^2-R_2^2)}F. \tag{10.1}$$

This result depends on the assumption that the pressure between the two plates is even. If this is not the case, then slip may occur at a greater or lesser value of Q. The actual pressure distribution depends on the conformity of the two surfaces and their flexibilities. Detailed calculations to allow for the flexibility of the surfaces would be inappropriate here, but it is worth pausing to consider the effect of non-conformity on the transmitted torque.

If, for example, one disc is slightly conical so that contact with the other disc occurs only at the outer edge, where $r=R_1$, the torque would be $\mu R_1 F$. Similarly, if contact occurs only at the inner edge the torque is $\mu R_2 F$. These are respectively the maximum and minimum possible values for the torque. The value given by eqn (10.1) lies between these two extremes.

It can be visualized that if contact is made only at the outer edge, say, there will be rapid wear as the clutch is engaged and disengaged so that the zone of contact will spread inwards with a consequential reduction in the torque that can be transmitted. If the initial cone angle is very flat it may be that in due course there will be contact over the whole of the surfaces and that eqn (10.1) will again hold. Another possibility is that the situation will tend to one where the rate of wear is uniform over the surfaces. The rate of wear at a given point can be assumed to be proportional to the rate at which energy is dissipated there and so proportional to the product of the local pressure and the sliding velocity. Hence

$$\text{rate of wear} = k \times r\omega \times p = \text{constant},$$

where ω is the relative angular velocity of the two discs. Putting $rp = \lambda$, where λ is a constant, we have

$$F = \int_{R_2}^{R_1} 2\pi rp \, \mathrm{d}r$$

$$= \int_{R_2}^{R_1} 2\pi\lambda \, \mathrm{d}r$$

$$= 2\pi\lambda(R_1 - R_2)$$

$$\Rightarrow \qquad \lambda = F/\{2\pi(R_1 - R_2)\}.$$

$$Q = \int_{R_2}^{R_1} 2\pi\mu pr^2 \, \mathrm{d}r$$

$$= \pi\mu\lambda(R_1^2 - R_2^2)$$

$$= \tfrac{1}{2}\mu F(R_1 + R_2).$$

The torque calculated on the assumption of constant pressure always exceeds that calculated for constant wear for all values of R_2/R_1 less than 1, so that the latter calculation will be safest for assessing the size of clutch needed for a given torque. The difference is not very great except when R_2/R_1 is very small. For R_2/R_1 greater than 0·3 the difference between the two values is less than 10 per cent.

The torque can be increased many-fold for the same value of F by using more than one friction surface. Fig. 10.2 shows a conventional

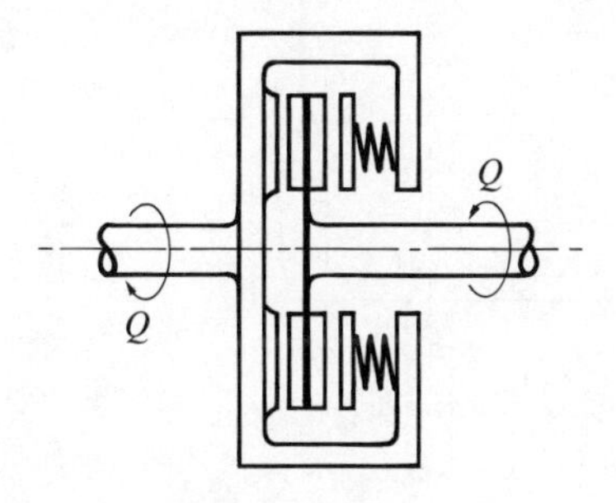

FIG. 10.2

arrangement whereby the torque is doubled by using both sides of a friction disc. The torque can also be increased by using a conical friction surface rather than a plane one, see Problem 2, page 243.

When it is operating correctly there is no slip between the operating surfaces of a clutch except during periods of engagement and disengagement. Ideally, the surfaces would always be engaged when their relative velocity is zero. The careful driver of a motor car, for example tries to achieve this to give a smooth drive and to lessen wear in the clutch.

Friction brakes operate in a very similar way to friction clutches by pressing one rough surface against another. A brake serves either to prevent motion occurring, as with the hand-brake of a motor car, or to arrest motion by converting kinetic energy into heat, as with the foot-brake. A substantial amount of heat may be generated when a brake is used to arrest motion and provision must be made for its disposal by suitable cooling arrangements. In a clutch heat is generated only incidentally during engagement and disengagement. Most friction devices involve a certain amount of energy dissipation.

10.2. Wheel and disc drive

In Fig. 10.3 a short cylinder of radius r, length a, is pressed by a force F against a circular disc. The cylinder (wheel) turns with angular velocity ω and as a result the disc turns with angular velocity Ω. The speed ratio Ω/ω is varied by altering the distance R from the centre of the disc to the mid-plane of the cylinder. This device is a satisfactory friction drive for many purposes. Some problems attend its use: the velocity ratio is not fixed for a given value of R but varies with the power transmitted; there is a continual dissipation of energy, and even if the output torque Q is zero, there still has to be a non-zero driving torque q.

To explain these points it is necessary to consider what happens along the line of contact between the cylinder and the disc. At the end of the cylinder closest to the centre of the disc the speed of the disc is $(R - a/2)$ and the speed of the surface of the cylinder is $r\omega$. For no slip these two speeds must be equal. At the outer end of the cylinder the speed of the

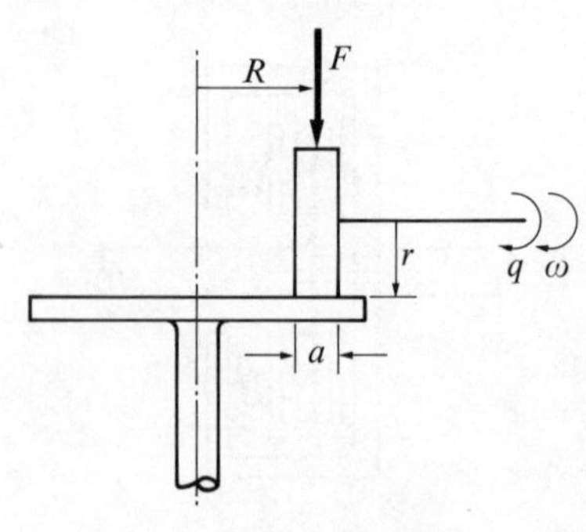

FIG. 10.3

disc is $(R+a/2)$ while the speed of the surface of the cylinder is again $r\omega$. Clearly, the no-slip condition cannot be satisfied simultaneously at both ends of the cylinder, and slip is unavoidable.

Assume that the cylinder and disc surface speeds are matched at a distance X from the centre of the disc, so that $X = r\omega/\Omega$. At a distance x from the centre of the disc where $x < X$ the cylinder overtakes the disc. When $X < x$ the disc overtakes the cylinder. The friction force per unit length of cylinder is everywhere $\mu F/a$, driving the disc and retarding the cylinder for $x < X$ and driving the cylinder while retarding the disc for $X < x$.

For the cylinder

$$\begin{aligned} q &= (\mu rF/a)\{[X-(R-a/2)]-[(R+a/2)-X]\} \\ &= 2(\mu rF/a)(X-R). \end{aligned}$$

For the disc

$$\begin{aligned} Q &= (\mu F/a)\int_{(R-a/2)}^{X} x\,\mathrm{d}x - (\mu F/a)\int_{X}^{(R+a/2)} x\,\mathrm{d}x \\ &= (\mu F/a)\{X^2-(R^2+a^2/4)\}. \end{aligned}$$

X cannot be greater than $(R+a/2)$ so that the maximum output torque is

$$\begin{aligned} Q_{\max} &= (\mu F/a)\{(R+a/2)^2-(R^2+a^2/4)\} \\ &= \mu FR. \end{aligned}$$

If the output torque $Q=0$, $X=(R^2+a^2/4)^{\frac{1}{2}}$. As Q is varied from zero up to the maximum X also varies, and with it Ω/ω. It will be noted that $q>0$ for all $Q>0$.

10.3. Efficiency

The efficiency of a machine is defined by

$$\eta = (\text{Output power}) \div (\text{Input power}). \tag{10.2}$$

It may be defined for a particular instant, or configuration of the machine, or as the average for a cycle of operation. A high efficiency results when the rate at which energy is dissipated in friction is small compared with the power being transmitted. It must be recognized, however, that the difference between input and output powers at a particular instant is not necessarily due to friction but may be attributable to storage or release of energy within the machine, kinetic energy being the main example of this in machines. No problem is caused when the power is averaged over a complete cycle or a number of complete cycles, for there can be no net storage or release of energy if the machine is operating steadily.

There is no capacity for energy storage in the wheel and disc drive

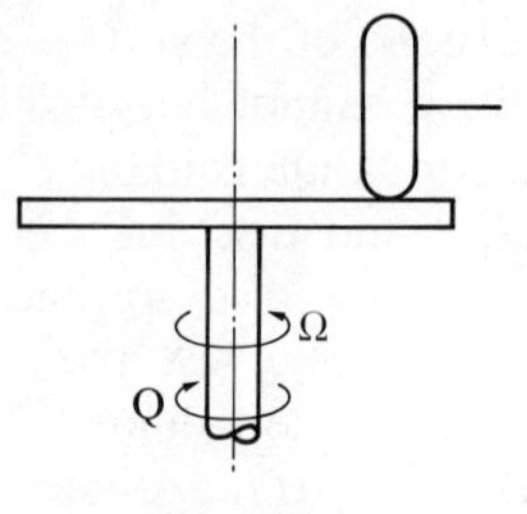

FIG. 10.4

analysed in the last section and for it

$$\eta = Q \div q\omega.$$

$\eta = 0$ when $Q = 0$, and it is not difficult to show that $\eta < 1$ for $0 < Q < Q_{max}$.

The efficiency is increased and approaches unity if $a \ll R$, and in a practical drive the cylinder might be replaced by a wheel with a rounded edge, Fig. 10.4, that makes a effectively zero. If the contact force F is increased to allow the transmission of a significant amount of power there will be deformation at the point of contact. This causes the load to spread over a finite area and energy is dissipated in the same general manner as has already been described for the disc and cylinder.

10.4. Belt drives

The belt and pulley drive is a very common and extremely important form of power transmission between parallel shafts. It consists simply of an endless belt wrapped under tension round two pulley wheels, Fig. 10.5. The belt may be made of leather, rubber, cotton, man-made fibres, steel, and so on; in fact any material that is flexible but reasonably inextensible. Friction between the belt and the pulley provides the means whereby the power is transmitted. The advantages of this method over others are that it is simple, relatively cheap, efficient, and can function with relatively large distances between the centres of the shafts. Its disadvantages are that it is not suitable for very large powers, it can be

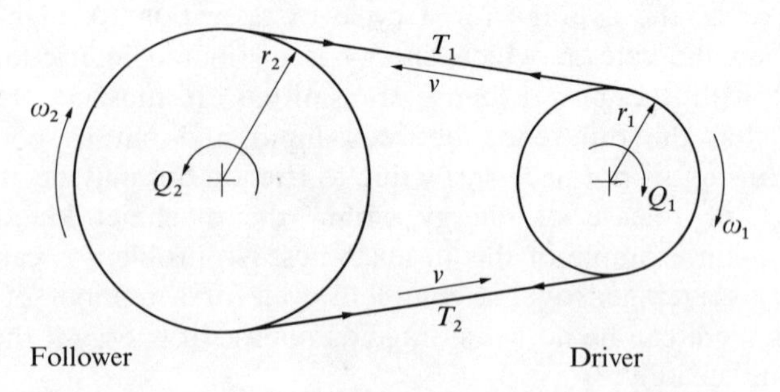

FIG. 10.5

noisy, takes up a lot of space, and does not give a positive relationship between the angular positions of the two shafts. Apart from the relative angular displacement that is allowed by extension of the belt there is a small amount of slip between the belt and each pulley. The mechanism of this slip is of interest and will be described below. It means that $r_1\omega_1$ is not quite equal to $r_2\omega_2$, and also that $\eta = Q_2\omega_2/Q_1\omega_1$ is slightly less than unity.

As a prelude to calculating the maximum power that can be transmitted by a belt drive let us consider first the relationship between the tensions on either side of a fixed pulley of a belt that is about to slip, Fig. 10.6. Imagine that initially it is held in contact with the curved surface by a tension T_2 at either end. The force is then held constant at T_2 at one end whilst at the other end it is gradually increased until slip occurs with T_2 being overpowered. Suppose that this occurs when the larger force is T_1.

A typical element AB of the belt, shown enlarged in Fig. 10.6, is subjected to forces T and $T+\delta T$ at either end, a resultant normal reaction R between the belt and the pulley, and a friction force μR. The angle between the normals to the surfaces at the ends A and B, and hence between the tensile forces T and $T+\delta T$, is $\delta\psi$. From the polygon of forces on the element

$$R = T\,\delta\psi$$

and

$$\delta T = \mu R.$$

In the limit when $\delta\psi \to 0$

$$\mathrm{d}T/T = \mu\ \mathrm{d}\psi.$$

On integrating

$$\log_e T = \mu\psi + \text{constant}.$$

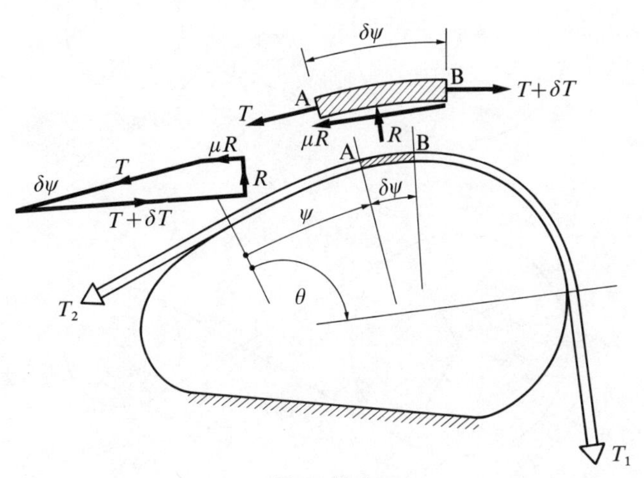

FIG. 10.6

When $\psi = 0$, $T = T_2$ so that

$$\log_e T/T_2 = \mu\psi.$$

When $\psi = \theta$, $T = T_1$ so that

$$T_1/T_2 = e^{\mu\theta}. \quad (10.3)$$

It should be noted that there is no need for the profile round which the belt is wrapped to be circular. The angles ψ and θ are simply the angles between normals and are not the angles subtended at particular points, or centres of curvature. Eqn (10.3) gives the limiting value for T_1 in terms of T_2, μ, and θ. If $T_1 < T_2 e^{\mu\theta}$ there will be no general slip of the belt round the fixed surface, although as we shall see below there must be some localized slip.

Eqn (10.3) is the basic equation for the design of flat-belt drives such as the one in Fig. 10.5. The power that is transmitted when the belt is on the point of slipping is

$$\begin{aligned} P &= (T_1 - T_2)r_1\omega_1 = (T_1 - T_2)r_2\omega_2 \\ &= (T_1 - T_2)v \\ &= T_1 v(1 - e^{\mu\theta}) \end{aligned}$$

where v is the speed of the belt. Maximum power is transmitted when T_1 has the largest value that considerations of strength will allow.

It is clear from Fig. 10.5 that the angle of embrace θ is less on the smaller pulley than it is on the larger one, so that if slip occurs it does so first at the smaller pulley. The angle of embrace at the smaller pulley can be increased by the use of an idler pulley, Fig. 10.7(a). The introduction

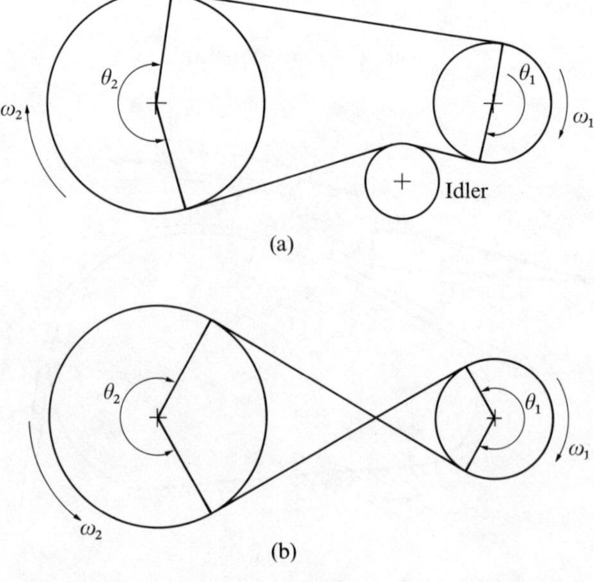

FIG. 10.7

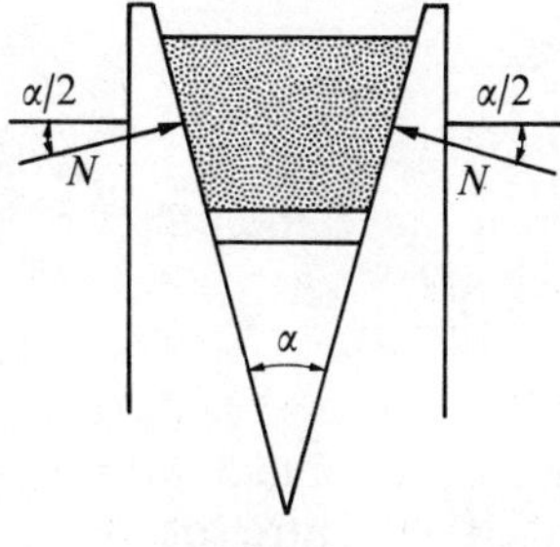

FIG. 10.8

of the idler pulley puts up the cost but it also provides an easy means of taking up slack due to stretch of the belt and of maintaining the mean tension at a prescribed value.

Crossing the belt as in Fig. 10.7(b) also increases the angle of embrace so as to give equal values on both pulleys. The direction of rotation of the follower is now opposite to that of the driver, so that the desire to achieve this effect is a more likely reason for using the crossed belt with the improved angle of embrace being a bonus.

10.4.1. Vee belts

The coefficient of friction is used more effectively by a vee-belt drive. In this the belt bears against the sides of a vee-shaped groove as shown in cross-section in Fig. 10.8.

If the normal force between an element of belt and the pulley is N, the radial and circumferential components of the force on the element are as shown in Fig. 10.9, so that

$$2N \sin \alpha/2 = T\,\delta\mu$$

and

$$\delta T = 2\mu N.$$

Hence

$$\mathrm{d}T/T = \mu'\,\mathrm{d}\psi$$

where

$$\mu' = \mu \operatorname{cosec} \alpha/2.$$

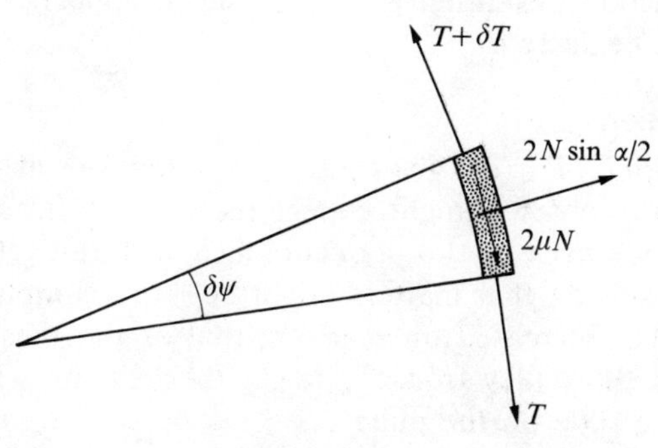

FIG. 10.9

On integrating we have

$$T_1/T_2 = e^{\mu'\theta}.$$

The vee belt is thus seen to have the effect of raising the coefficient of friction from μ to μ cosec $\alpha/2$.

10.4.2. Centrifugal effect

Each element of belt that is in contact with a pulley of radius r has an acceleration v^2/r towards the centre of the pulley. The belt tension provides the force necessary for this acceleration in addition to that needed to balance the radial reaction between the belt and the pulley. The additional element of force that is needed is $m(v^2/r)r\,\delta\psi$, where m is the mass per unit length of the belt. Taking this into account the equations of equilibrium for an element of vee belt are

$$2N \sin \alpha/2 + m(v^2/r)r\,\delta\psi = T\,\delta\psi$$

and

$$\delta T = 2\mu N.$$

Hence

$$\int \frac{\mathrm{d}T}{T - mv^2} = \mu'\,\mathrm{d}\psi.$$

On integration we find

$$\frac{T_1 - mv^2}{T_2 - mv^2} = e^{\mu'\theta}.$$

The power transmitted is

$$P = v(T_1 - mv^2)(1 - e^{-'\theta}).$$

This equation shows that there is a limit to the power that can be transmitted for a given amount of belt tension. Taking T_1 to be constant and differentiating P with respect to v we have

$$\mathrm{d}P/\mathrm{d}v = (T_1 - 3mv^2)(1 - e^{\mu'\theta}).$$

$\mathrm{d}P/\mathrm{d}v$ is zero, and P is a maximum, when $v = (T_1/3m)$. At this speed the power that can be transmitted is only $\frac{2}{3}$ what it appears to be when the inertia effect is neglected.

10.4.3. Belt creep

The relationship $T_1/T_2 = e^{\mu\theta}$ gives the ratio of tensions at which total slip occurs. If $T_1/T_2 < e^{\mu\theta}$ we might expect there to be no slip and consequently that a belt drive is 100 per cent efficient. Intuitively this seems to be unlikely. Suspicion that matters might be more complicated than has been suggested is increased on realizing that as the tension in the belt changes round the pulley from T_1 to T_2 the belt must contract elastically, and hence slide on the pulley.

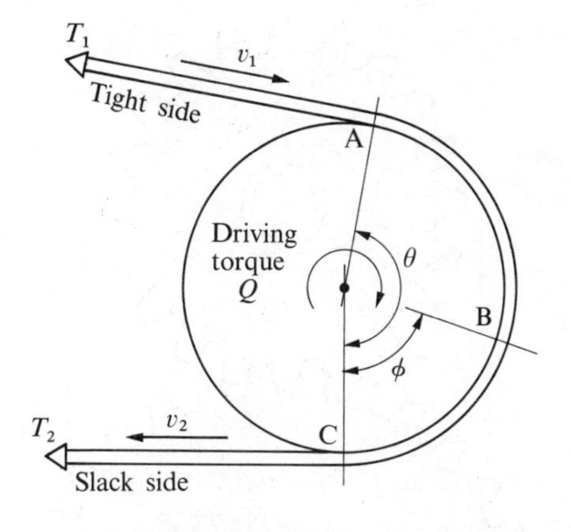

FIG. 10.10

Let us study in detail the way in which the tension changes as the belt passes round the driving pulley, Fig. 10.10. Contact starts at A and finishes at C. At A the belt is stretched relative to its state at C, but the mass flow-rate past the two points must be the same, so that $v_2 < v_1$. Now as there can be no overall slip if $T_1/T_2 < e^{\mu\theta}$ we may assume that there is no slip either at A or at C. We would expect the peripheral speed of the driving pulley to be slightly faster than the average belt speed so let us assume that there is no slip at A. If we assume further that there is no slip between A and B then there is no change in belt strain between these two points and the belt tension must stay constant at T_1. On this hypothesis the belt gradually contracts as it moves from B to C and so sliding back relative to the surface of the pulley. This means that the limiting friction-force opposes the motion of the pulley, so balancing the driving torque Q, and causes the exponential build-up of tension from C back to B with $T_1/T_2 = e^{\mu\phi}$.

If the power is increased ϕ increases. The limiting power is reached when $\phi = \theta$ and total slip then occurs. The alternative assumption of no slip at C is found to lead to the friction force being applied in the opposite direction, that is assisting Q, and is therefore unacceptable.

If this continuous slip, which varies in amount with the transmitted power, is unacceptable, but very high precision is not required, a chain drive may be a suitable alternative to a belt drive. The main disadvantage of chain drives is that they are relatively noisy in operation due to the impacts that occur as the links come into contact with the sprockets of the wheels. For high precision where the shafts are close together gearing will usually be preferred.

10.5. Involute gearing

Fig. 10.11 shows a pair of gears. The drive is provided by the meshing of teeth round the edges of circular discs. Assuming that there is no

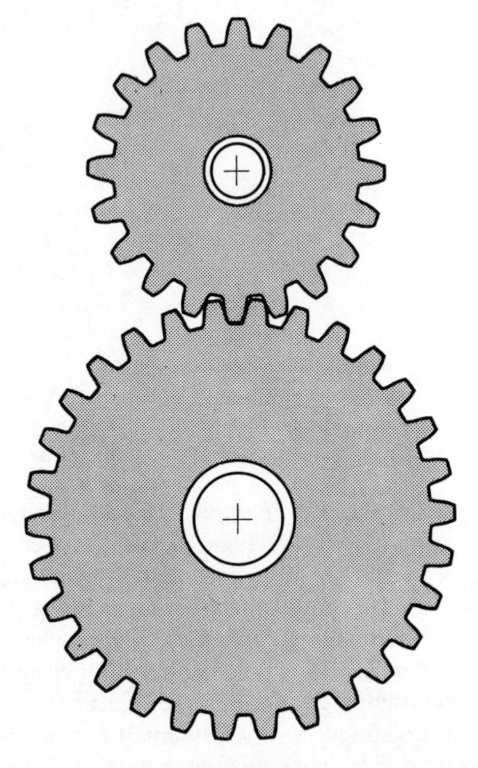

FIG. 10.11

backlash, a gear drive provides an exact relationship between the rotations of two shafts apart from errors due to distortion under load. If the overall motion is considered in terms of several turns of both shafts the drive appears to provide a constant-velocity ratio. But, unless the two sets of teeth are properly shaped there will be fluctuations in speed ratio superimposed on the gross motion.

A basic requirement for a gear drive is that there should be no variation in speed ratio during the period of contact between one pair of teeth. To derive a shape of tooth that will ensure this we will think first in terms of two pulleys connected by a crossed belt, Fig. 10.12. Only one length of belt is shown, others being omitted to clarify the diagram. It is assumed that the belt is inextensible and that there is no slip between it and either pulley.

As the pulleys turn keeping the belt taut a point C on the belt moves from A to B. If a piece of paper is stuck to one pulley, a pencil attached to the belt at C traces the path of C relative to that pulley. C also moves along a similar path relative to the other pulley. The two curves touch at the current position of C, as shown in the figure. If these curves are used as the profiles of gear teeth a constant speed ratio is automatically attained. A succession of similar teeth permits continuous motion, and if

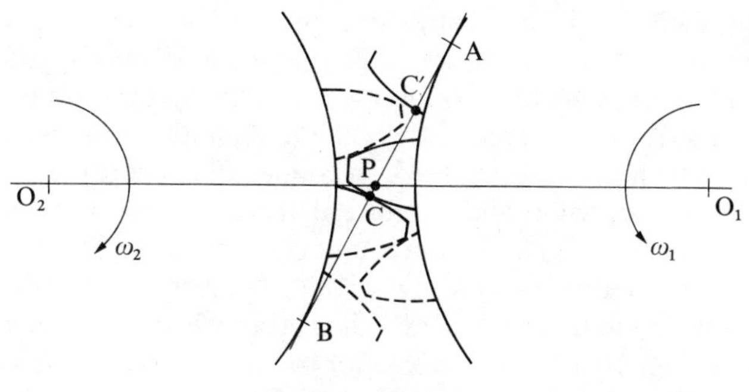

FIG. 10.12

the back surfaces of the teeth have similar profiles torques can be transmitted in either direction. There are further geometric requirements for a usable drive: the teeth must be sufficiently long for contact between one pair of teeth to be maintained until the next pair makes contact, and the tooth thicknesses must be matched to avoid backlash, but it would be inappropriate to examine these points in detail here.

The curve traced in the manner described is an *involute*, and gears shaped on this principle are called involute gears. Various geometrical properties are important. It can be seen that depending on which profile is considered C can at every stage be regarded as swinging about A or B. It follows that AC and BC are respectively the radii of curvature of the two curves at their point of contact, and that AB is the common normal to the curves at C. As AB is a fixed line it cuts the line of centres at a fixed point P. The involute gear tooth provides, therefore, an illustration of the proposition presented in Section 1.6.3, that, for a constant velocity ratio, the common normal at the point of contact between a cam and follower must pass through a fixed point on the line of centres. Here the driving tooth acts as a cam and the driven tooth as a follower. The point P is referred to as the *pitch point*. It is the point at which a friction drive between two cylinders would touch to give the same velocity ratio:

$$O_1P\omega_1 = O_2P\omega_2.$$

In the absence of friction the force between two teeth acts along the common normal. The total force along this line, obtained by adding forces if more than one pair of teeth are in contact, is constant when the driving force is constant. The reactions at the shaft bearings are equal in magnitude to the total force along the normal and so they too are constant when the driving torque is constant. This feature is unique to involute gear teeth, and is one of their virtues. The picture is somewhat upset by the presence of friction, but given the very high efficiency of gear drives it is not a point that we need examine in detail. Suffice it to note that the bearing reactions are sensibly constant, for constant torque, and that this means an absence of a possible source of vibration.

It can be seen that if the centres of a pair of involute gears are moved slightly apart, the basic geometry of Fig. 10.12 is in no way affected and the constant velocity feature is unimpaired. This feature too is unique to involute gears. It permits some relaxation in manufacturing requirements, but not too much because gaps are introduced between the teeth giving rise to a loss of precision and of noise if the direction is reversed.

The final feature to be noted here, and perhaps the one which above all others has led to the involute gear being almost universally adopted, relates to the involute rack. A rack is a gear wheel with infinite pitch-circle radius, Fig. 10.13. Point B cannot be shown because it is at infinity which means that the radius of curvature BC of a rack tooth is infinite and so its profile is a straight line.

10.5.1. Definitions

Two gears in contact are said to form a *pair.* The smaller, usually the driver, is referred to as a *pinion* and the larger is called a *wheel.* A succession of gears is called a *gear train.*

Referring to Fig. 10.12, the circles O_1B and O_2A are *base circles.* The distance CC′ between succeeding profiles as measured along the common normal at the point of contact is constant, the points C and C′ being firmly attached to the hypothetical string AB. As this distance is equal to the distance between profiles measured round the *base circle* it is called the *base-circle pitch.* Two wheels can mesh together only if they have the same base-circle pitch. The number of teeth on a wheel of given base-circle pitch is proportional to the base-circle diameter. The number of teeth is similarly proportional to the pitch-circle diameter, $2\times OP$ in Fig. 10.12. The pitch-circle circumference divided by the number of teeth is the *circular pitch.* The thickness of a tooth measured round the pitch circle is normally one half the circular pitch.

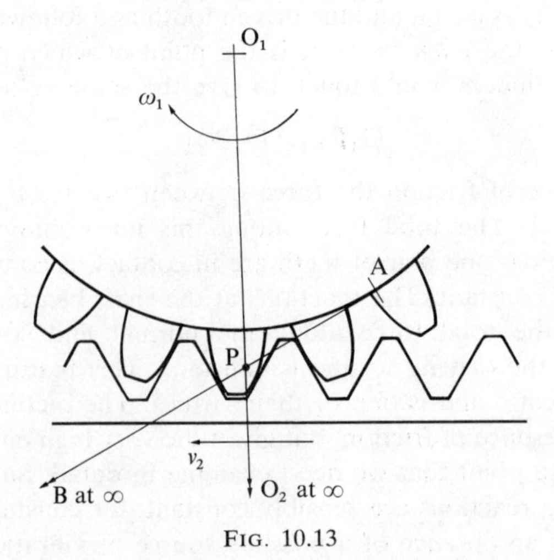

Fig. 10.13

The overall size of a tooth is determined by the pitch. Unfortunately neither the base-circle pitch nor the circular pitch can be expressed as an exact number because each is related to a circumference. For convenience of reference and in manufacture, the size of a tooth is measured by its *module,* which is the pitch-circle diameter divided by the total number of teeth in the wheel. As the module is directly proportional to the base circle and circular pitches, two gear wheels must have the same module if they are to mesh together.

The height of a tooth within its pitch circle is limited by the radial distance between the pitch and base circles. The active part of the tooth profile cannot extend within the base circle because an involute can exist only outside its base circle on account of its method of generation. In a tooth of standard proportions the height outside the pitch circle equals the module. This is normally sufficient to ensure continuity of action, that is to say, contact does not cease between one pair of teeth until it has been made between the next pair. The part of the tooth that lies outside the pitch circle is called the *addendum,* and that which lies inside it is called the *dedendum.* These features are illustrated in Fig. 10.14.

The angle ψ in Fig. 10.14, the angle between the normal to the tooth profiles and the tangent to the pitch circles at the pitch point is called the *pressure angle,* now commonly 20°. It relates to the pitch and base circle radii, r_p and r_b through $r_b = r_p \cos \psi$.

10.6. Gear trains

10.6.1. Simple gear trains

As two gears have the same speed at the pitch point their angular speeds are inversely proportional to their numbers of teeth:

$$\omega_2/\omega_1 = -N_1/N_2$$

where the minus sign indicates the difference in direction of rotation of the wheels, Fig. 10.15.

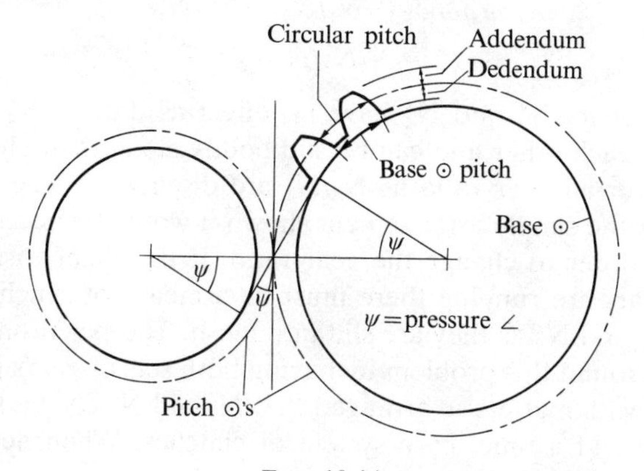

FIG. 10.14

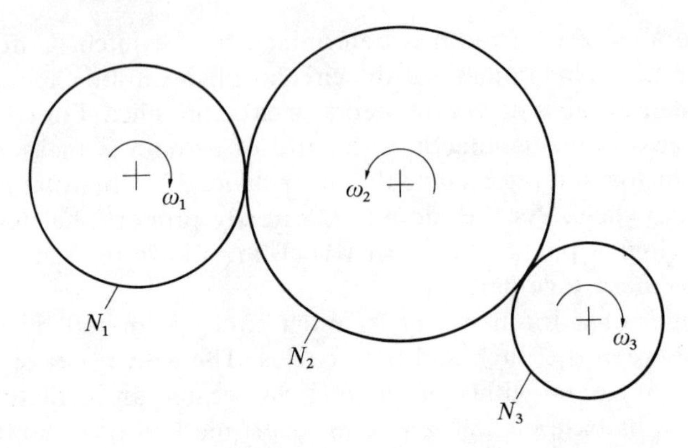

FIG. 10.15

If a third gear wheel is added to form a simple gear train then

$$\omega_3/\omega_2 = -N_2/N_3$$

so that

$$\begin{aligned}\omega_3/\omega_1 &= (-N_1/N_2)\times(-N_2/N_3)\\ &= N_1/N_3.\end{aligned}$$

The only effect of the intermediate wheel, or idler, is to change the direction of the output wheel. The overall speed ratio is unaffected by the size of the idler. It can be seen that the speed ratio depends only on the ratio of the sizes of the first and last wheel of a gear train and that any number of idlers can be introduced without any effect other than that of changing the sign of the speed ratio. If there is an odd number of idlers the overall gear ratio is positive; it is negative for an even number.

With the arrangement of Fig. 10.16 the intermediate gears N_2 and N_3 are fixed to a lay shaft so that $\omega_2 = \omega_3$ and are said to be compounded. The lay shaft rotates in fixed bearings. The overall gear ratio is now

$$\begin{aligned}\omega_4/\omega_1 &= (-N_1/N_2)(-N_3/N_4)\\ &= N_1N_3/N_2N_4.\end{aligned}$$

The extra gears N_3' and N_4' have no effect as drawn. N_4' and N_4 are integral with each other and can be slid bodily along their shaft to give a change of over-all gear ratio as N_3/N_4 are disengaged when N_3'/N_4' are engaged. For the simple arrangement shown it would be necessary to stop the gears in order to change the gear ratio. If the change is to be made when the gears are running there must be a means of synchronizing the speeds of N_3' and N_4' as they are slid into mesh. The synchromesh gearing of a car gets round this problem by having both sets of gears permanently engaged but with matters so arranged that N_4 and N_4' can be fixed to their shaft only one at a time, by a system of clutches. When neither gear is fixed to the output shaft the system is 'out of gear'.

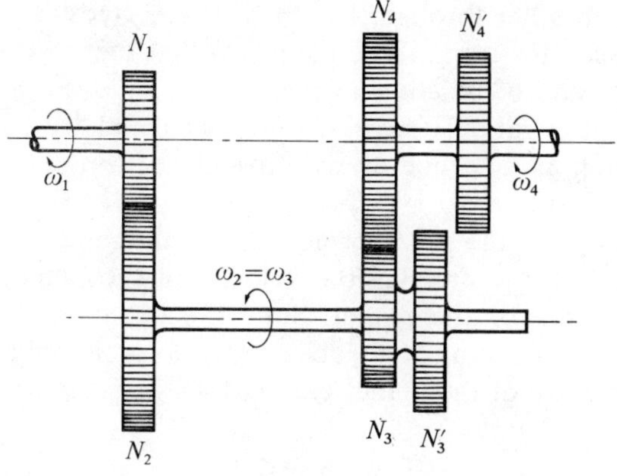

FIG. 10.16

10.6.2. Epicyclic gear trains

The gear train in Fig. 10.16 is simple but may be bulky. A more compact gear train is shown in Fig. 10.17. This is a form of epicyclic gear. It consists of two conventional spur gears (i.e. with teeth radiating outwards) and an internal gear, often referred to as an annulus. The geometry of the tooth profiles for the annulus is developed from that already given for involute gears and will not be given here.

One spur gear, the sun, is fixed to the input shaft. The other, the planet, meshes with the sun and turns freely on a spindle that is carried at the end of an arm fixed to the output shaft. As the sun wheel is turned the planet runs round inside the annulus and so drives the output. To keep the figure simple only one planet gear is shown in full. Frequently three planet gears are used at 120° to each other and they drive the output in

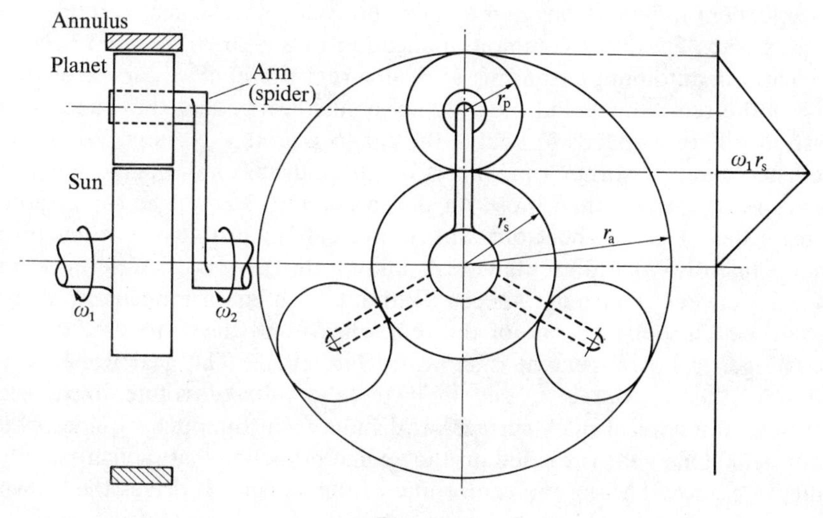

FIG. 10.17

parallel with each other through three arms connected together to form a spider. The reason for using three planets is that there is then automatic balance of mass and of reaction forces.

The overall gear ratio is most readily calculated by considering the speed distribution along a line drawn through the centres of the sun and planet.

At their pitch point the sun and planet have the same velocity $\omega_1 r_s$. At its pitch point with the annulus the velocity of the planet is zero. The variation in velocity of the planet along the diameter drawn between these two points, assuming it to be a rigid body, is linear so that the velocity at the centre of the planet is $\omega_1 r_s/2 = \omega_2 r_{arm}$. But $r_{arm} = (r_s + r_a)/2$, so that

$$\omega_1 r_s/2 = \omega_2(r_s + r_a)/2$$

$$\Rightarrow \qquad \omega_2/\omega_1 = r_s/(r_s + r_a)$$

$$= N_s/(N_s + N_a).$$

If the angular velocity ω_a of the annulus is not zero then

$$(\omega_1 r_s + \omega_a r_a)/2 = \omega_2 r_{arm}$$

and

$$\omega_2 = \omega_1 N_s/(N_s + N_a) + \omega_a N_a/(N_s + N_a).$$

When the output speed depends on two input speeds, ω_1 and ω_a in this case, the gear is said to be operating as a differential.

Compound epicyclic gears enable very substantial speed reductions to be achieved very compactly.

10.6.3. Differential gearing

A differential gear train has two degrees of freedom and accepts two independent inputs; it was noted above how an epicyclic gear can operate in this way. The most common application of a gear differential occurs, perhaps, in automobile transmissions. In a rear-wheel drive car the engine drives the rear wheels but must do so in such a way that the wheels can turn at different speeds to enable the car to go round corners. When the vehicle rounds a corner the wheels on the outside of the curve have to cover more ground than those on the inside and, having the same radii, must rotate faster. The front wheels present no problem as they turn independently. A differential gear allows the necessary difference in speed between the rear wheels. Without such an arrangement there would be slip between one of the rear wheels and the ground.

The general arrangement is shown in Fig. 10.18. The gear wheels are all bevel gears, that is to say their teeth of quasi-involute form, are formed on a conical pitch surface, as distinct from the pitch cylinder of a spur gear. One gear is carried on the engine propeller shaft which usually runs fore and aft along the centre line of the vehicle. It drives the crown wheel which is coaxial with the rear wheels, but is fixed to neither. Brackets

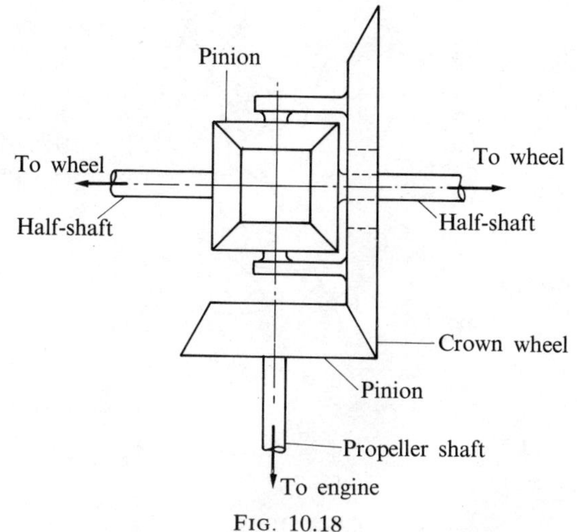

FIG. 10.18

on the crown wheel carry a pair of bevel pinions which mate with bevel gears at the ends of the shafts (half-shafts) to which the wheels are directly attached.

It can be seen that if the rear wheels are jacked clear of the ground and the propeller shaft is fixed, they will be able to turn at equal and opposite speeds $\pm\omega$. If these speeds are added to the common angular speed Ω due to the normal forward motion of the car then one rear wheel has speed $(\Omega+\omega)$ and the other $(\Omega-\omega)$. As ω is arbitrary it can freely assume the value needed to accommodate the motion of the car round a bend.

10.7. Friction in gearing

Proper lubrication of gears is essential, for without it heavy wear and premature failure are inevitable. Given proper lubrication power-loss due to friction is normally very small. Nevertheless, the effect of friction on the transmission of power through gearing is worthy of some consideration. In Fig. 10.19 a pair of teeth are in contact at C which is approaching the pitch point P. The two teeth have the same component of velocity along the common normal at C but have different components in the tangential direction. In the velocity diagram, oc_1 is perpendicular to O_1C and oc_2 is perpendicular to O_2C. So $ac_1=\omega_1 AC$ represents the tangential component of velocity of one tooth and $ac_2=\omega_2 BC$ represents the tangential component of velocity of the other tooth. The difference c_1c_2 is the slip velocity u,

$$\begin{aligned} u &= \omega_2 BC-\omega_1 AC \\ &= \omega_2(BP+PC)-\omega_1(AC-PC) \\ &= (\omega_1+\omega_2)PC, \end{aligned}$$

for $\omega_2 BP-\omega_1 AP=\omega_2 O_2P\sin\psi-\omega_1 O_1P\sin\psi=0$.

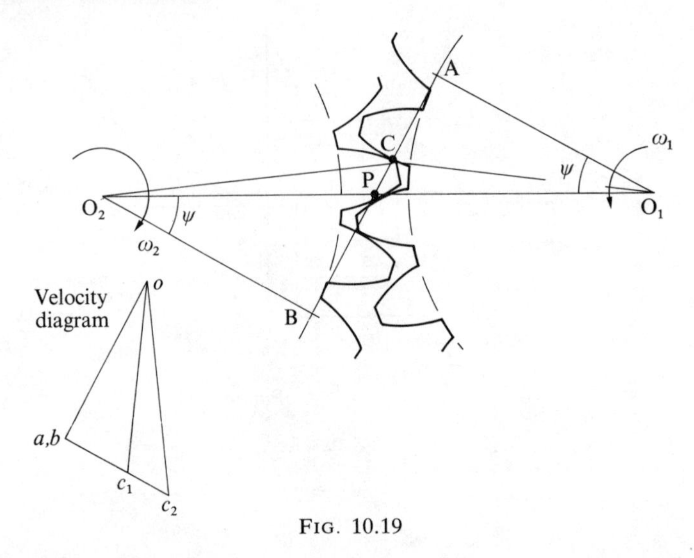

FIG. 10.19

The existence of the slip velocity means that there is a dissipation of energy except when contact is at the pitch point, where the sliding velocity is zero.

The normal force N between a pair of teeth can be assumed to be constant for as long as the same number of pairs of teeth are in contact. The power dissipation μNu at one contact varies with u. By calculating the total work dissipated for the period of time that it takes for Fig. 10.19 to repeat itself, it is possible to calculate an average efficiency. In general, the resulting expression is complicated, in spite of simplifying assumptions. Not only does u vary, but the number of pairs of teeth in contact also varies.

The situation is very much simplified if there are just two pairs of teeth in contact. The two points C must be on either side of P. If the total normal force is shared equally between the two pairs of teeth, say N on each, the two friction forces being in opposite directions constitute a couple. The distance between the two points C equals the base-circle pitch p_b and so is constant. For so long as there are just two pairs of teeth in contact the two gear wheels each experience a couple μNp_b opposing motion. The power dissipation is therefore $\mu Np_b(\omega_1+\omega_2)$.

10.8. Linkage mechanisms

As with gearing it is essential for linkage mechanisms to be properly lubricated if they are to function at high efficiency and without undue wear. The exception occurs when a mechanism is used as a latching device that relies on friction for its operation.

10.8.1. Bearing friction

We will confine our attention to mechanisms that employ plain journal bearings. The journal, or shaft, is circular and its diameter must be slightly less than that of the bearing that supports it. We will visualize that the axis of the shaft is horizontal. A vertical load acts through the centre of the shaft and moves with the shaft as it moves within the clearance in the bearing. A couple is applied to the shaft to overcome whatever resistance to motion there may be, so that the shaft turns with constant speed.

If the bearing is properly lubricated and not over-loaded the behaviour will be quite different from that which we will be considering in more detail below. We will note in passing a very simplified picture of that behaviour. In Fig. 10.20 the clearance between the bearing and the shaft is filled with lubricating fluid. In the figure the clearance is, of course, very much exaggerated. When the shaft turns anticlockwise the application of a light vertical load P causes it to move horizontally and to take up an eccentric equilibrium position as shown in the figure. The viscosity of the fluid causes it to be dragged round by the shaft. Any fluid that is transported across a typical section such as BB′ must also cross CC′ but, as the gap is narrower the mean velocity is higher. The increase in velocity is accompanied by a reduction in pressure so that the pressure at DD′ is less than that at AA′. This pressure differential means that the net vertical reaction on the shaft due to the pressure in the fluid is to the left of the applied force **P**. The resulting clockwise couple counterbalances that due to the viscous drag of the fluid.

If there is no lubricant, or if the load is so heavy that the fluid film is broken to permit direct contact between the shaft and the bearing, the behaviour is quite different.

Imagine that the shaft is stationary with the applied load **P** acting through the centre of the shaft as before. If the shaft is turned anti-clockwise it will start to roll to the left and, as it were, 'climb the hill'. It climbs until the hill becomes too steep and then it slips. From then on the shaft turns with the fixed point C, Fig. 10.21, as the point of contact. The movement of the shaft relative to the bearing is therefore in the opposite direction to what it is when the bearing is lubricated. The reaction **R** at C is at an angle ϕ to the normal OC, ϕ being the angle of friction. **R** is

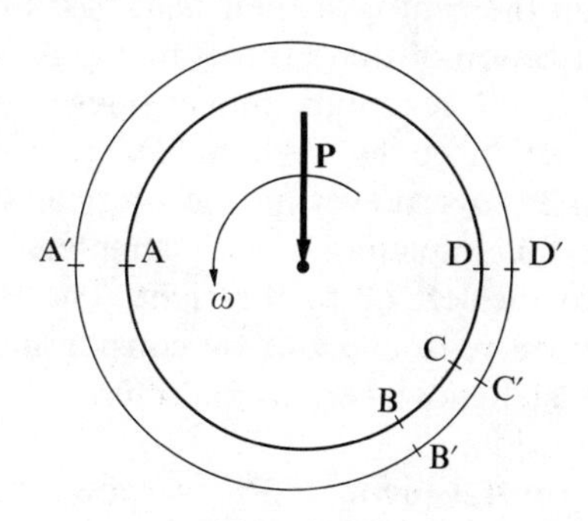

FIG. 10.20

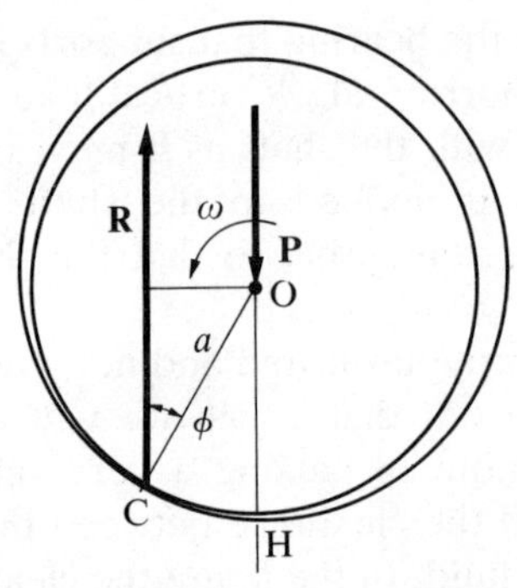

FIG. 10.21

equal in magnitude to **P** so that the magnitude of the couple that must be applied to maintain motion is $Pa \sin \phi$, where a is the radius of the shaft. If the coefficient of friction μ is small $\sin \phi \approx \tan \phi = \mu$. Hence the driving couple needed to overcome friction is μaP. If a smaller couple is applied the shaft simply rolls a little and rests in equilibrium without rotation and with the point of contact somewhere between H and C. A driving couple greater than μaP causes the shaft to accelerate.

If the applied load **P** turns the point of contact moves so as to maintain a constant distance between **P** and **R** thereby keeping the driving torque constant. It follows that whatever the direction of **P**, the reaction **R** is always tangential to a circle of radius concentric with the shaft. This circle whose radius is usually taken to be μa, is called the *friction circle.*

The concept of the friction circle is very helpful in explaining friction losses in mechanisms, and in accounting for any tendency to jam in certain configurations. Some simple examples will illustrate the use of the friction circle.

Example 10.1: A simple lever ABC is supported at its centre B. Pivots at A, B, and C are all of radius a. Determine the force $\mathbf{P}_1$ *that is necessary to overpower a force* $\mathbf{P}_2$ *at B if there is dry friction* μ *at each pivot.*

In the absence of friction $\mathbf{P}_1 = \mathbf{P}_2$ and the central reaction $\mathbf{R} = \mathbf{P}_1 + \mathbf{P}_2$. All forces act through the centre of their bearings, Fig. 10.22(a). If there is friction the lines of action of these forces move. As P_1' is increased from its original value P_1 the reactions move sideways until they are all tangential to their friction circles. When this state is reached the lever begins to move with $\mathbf{P}_1'$ overpowering the original $\mathbf{P}_2$. A glance at Fig. 10.22(b) shows that this explanation is incomplete, because each force could move either to the left or to the right. The problem is to decide which way. This can be reasoned out by considering the motion within each bearing. It is easier, however, to make use of the fact that friction invariably opposes motion.

Consider first the central pivot. If $\mathbf{P}_1'$ overpowers $\mathbf{P}_2$ the lever rotates anticlockwise. The friction at this pivot opposes this motion. It does so by

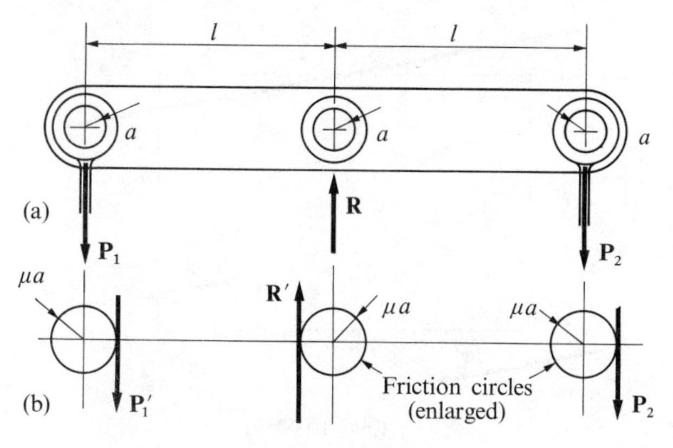

FIG. 10.22

causing the reaction $\mathbf{R}'$ to move to the left and so cause a clockwise torque to be applied to the lever. $\mathbf{P}_1'$ and $\mathbf{P}_2$ likewise move sideways so that they too apply clockwise moments to the lever.

The forces are eventually placed as shown in Fig. 9.22(b), where the friction circles have been drawn very much enlarged. Taking moments about a point on the line of action of the central reaction R' we find

$$P_1' = P_2(l+2\mu a)/(l-2\mu a)$$
$$\approx P_2(1+4\mu a/l).$$

Example 10.2: A parallel crank mechanism ABCD with AB = CD = a and BC = AD = b has bearings at A, B, C, and D of radius r. Determine the torque $\mathbf{Q}_A$ required at A to overcome a resisting torque $\mathbf{Q}_D$ at D when the mechanism is in the rectangular configuration if the coefficient of friction is everywhere μ.

If friction is negligible $Q_A = Q_D$ and the bearing reactions are equal to the thrust $\mathbf{P}$ in the coupler BC. With friction, the reactions must be tangential to the friction circles when motion is impending.

In Fig. 10.23 crank AB is in equilibrium under the action of the applied torque $\mathbf{Q}_A$, the bearing reaction $\mathbf{P}$ at A and an equal and opposite force $-\mathbf{P}$ applied to the crank at B by the coupler BC. A similar set of forces acts on crank DC.

The effect of $\mathbf{Q}_A$ is to turn AB clockwise. Accordingly, the bearing reaction at A moves its line of action so as to apply an anticlockwise moment to the crank. The coupler thrust at B also moves to give an anticlockwise moment, Fig. 10.23. Similar shifts occur at C and D. As the only forces which act on the couple are the forces $\pm\mathbf{P}$ at either end these two forces must be collinear. For equilibrium of the cranks, the reactions at A and D must both be parallel to the line of action of the coupler thrust, so giving the set of forces shown in Fig. 10.23.

It must be remembered that if the mechanism is in good condition the

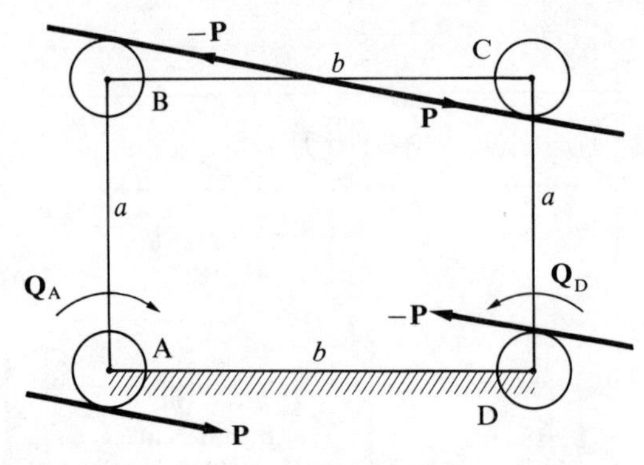

FIG. 10.23

radii of the friction circles will usually be very small compared with the main dimensions. Consequently

$$Q_A \approx P(a+2\mu r)$$

and

$$Q_D \approx P(a-2\mu r)$$

$$\Rightarrow \quad Q_A \approx Q_D(1+2\mu r/a)/(1-2\mu r/a)$$

$$\approx Q_D(1+4\mu r/a).$$

The similarity of this result to that of the previous example is purely coincidental.

As the mechanism in this last example rotates from the given position the driving torque needed to overcome a fixed resisting torque increases indefinitely and eventually the mechanism jams. This happens when the cranks have rotated to such an extent that the line of thrust of the coupler is collinear with the reaction at D, Fig. 10.24. Any further rotation would make it impossible for the reactions at C and D to be tangential to their friction circles. At one or other bearing the angle between the reaction and the normal at the point of contact would have to become less than the angle of friction. When this happens slipping must cease.

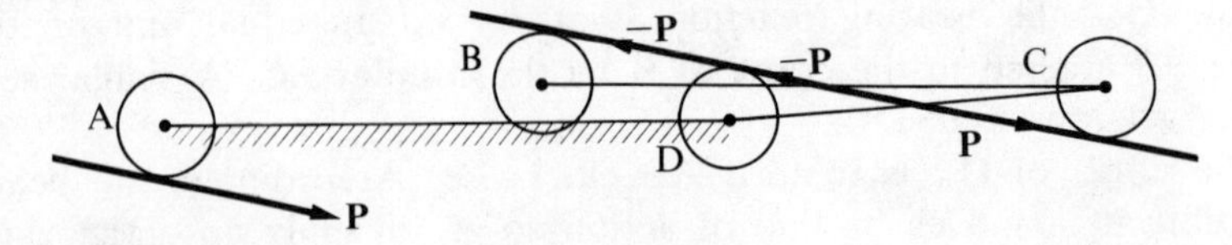

FIG. 10.24

10.9. Problems

1. A clutch of the single plate type, transmitting torque from both sides of the plate, is required to transmit 37·5 kW at 3000 rev min^{-1}. The coefficient of friction between the surfaces is 0·3 and the axial pressure is limited to 0·08 N mm^{-2} assumed to be uniform over the surfaces of the plate. If the external diameter is to be 1·5 times the internal diameter, determine the size of the plate.

2. Fig. 10.25 shows a cone clutch. Derive an expression for the maximum torque Q that it can transmit if the coefficient of friction is μ and the pressure between rubbing surfaces is assumed to be uniform.

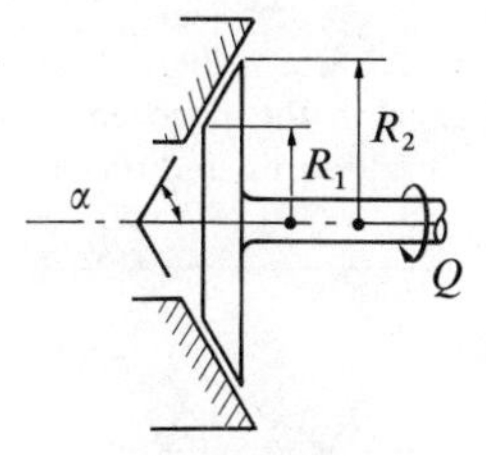

FIG. 10.25

3. A rope attached to a vessel passes three times round a fixed bollard on a dock, and a pull of 200 N is applied to the free end. The coefficient of friction between the rope and the bollard is 0·25. If the rope is on the point of slipping, find the pull being applied by the ship.

4. An internal expanding brake is shown in Fig. 10.26. The two brake shoes are freely pivoted to the vehicle at A and are subjected to equal forces P at their free ends. Taking the radius of the drum to be r and angle of lap of each shoe to be π, obtain expressions for the braking torque exerted on the drum

(a) assuming that the pressure distribution between the shoe and the drum is uniform, and
(b) assuming that the shoes are flexible and do not buckle.

Compare the results, using a reasonable value for the coefficient of friction.

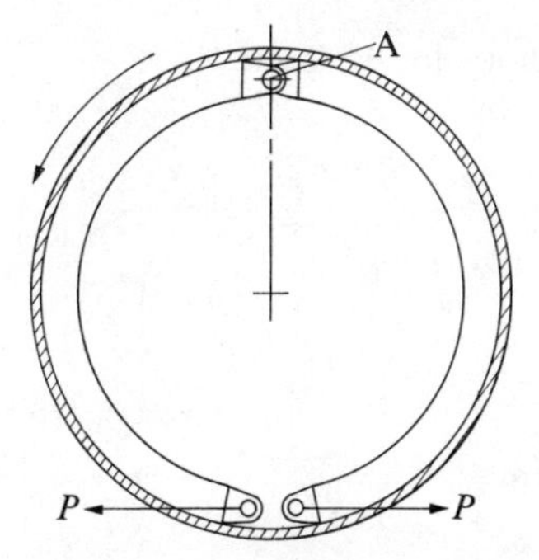

FIG. 10.26

5. A pulley mounted on a fixed horizontal axis has a light flexible band passed round 180° of its circumference, the ends being anchored as shown in Fig. 10.27. The springs have stiffnesses s_1 and s_2 respectively and initially each has a tension T. The coefficient of friction between the band and the pulley is μ.

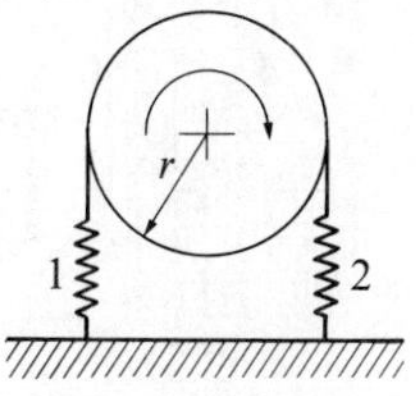

FIG. 10.27

The pulley is turned slowly in the direction shown until slipping occurs.

Obtain expressions for the angle through which it will have turned from the equilibrium position and the final torque.

6. A string passes through a ring which is attached to a weight W. If the string and the ring are in a state of limiting equilibrium in the position shown in Fig. 10.28, determine the coefficient of friction existing between the ring and the string, and the tensions in the string.

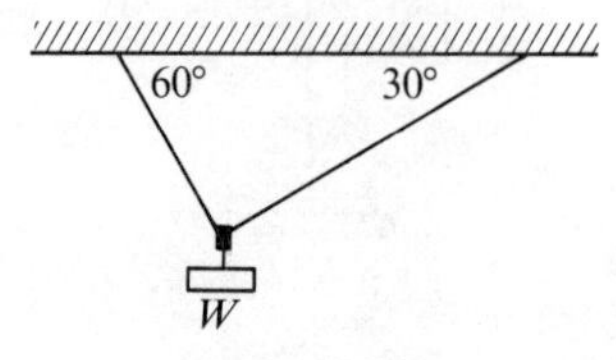

FIG. 10.28

7. A pair of identical spur gears each with 25 involute teeth and module 5 mm is designed to mesh without backlash at a pressure angle of 20°. By considering a pair of teeth mating at the pitch point, determine the tooth thickness and the gap between the teeth measured at the base circle, assuming that the involute profiles continue right down to the base circle.

8. Each of a pair of meshing spur gears has 70 involute teeth, and their pressure angle is 17°. The addendum equals the module m. Find the length of the path of contact in terms of m, and show that there are always at least two pairs of teeth in contact.

9. Fig. 10.29 shows an epicycle gear in which wheel P has 45 teeth and is geared with the internal gear Q through the planet R at the end of arm A. When P rotates at 63 rev min^{-1} clockwise, A rotates at 9 rev min^{-1}, also clockwise. Q is required to rotate at 21 rev min^{-1} anticlockwise.

Find the numbers of teeth needed on Q and R.

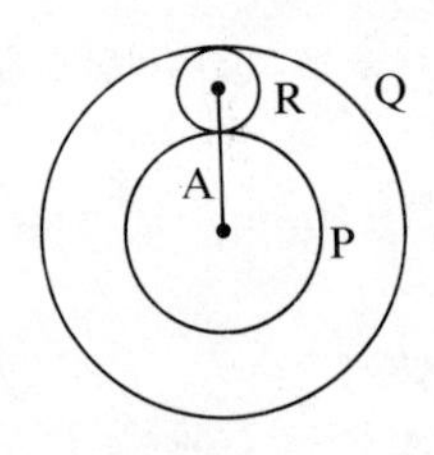

FIG. 10.29

10. Fig. 10.30 shows a compound epicyclic gear train in which gear A is fixed. Find:

(a) the speed ratio ω_0/ω_i,

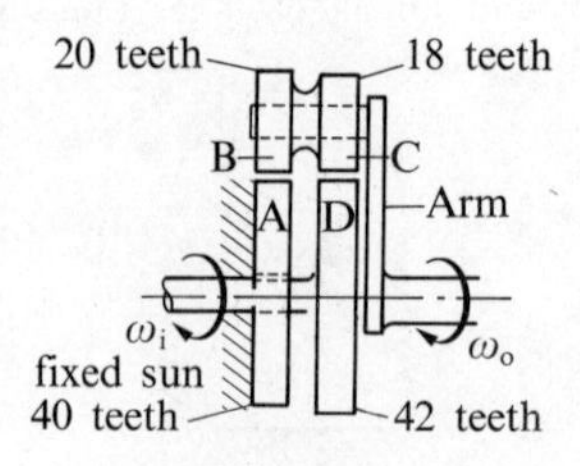

FIG. 10.30

(b) the fixing torque on gear A when the input torque is 1000 Nm, assuming an overall efficiency of 98 per cent.

11. Fig. 10.31 shows schematically a proposal for an epicyclic gear train. It comprises a sun wheel S with s teeth mounted on the input shaft X_1, an annular gear A with a teeth mounted on the concentric input shaft X_2 and the output shaft Z carrying a cage with planet gears P, each of p teeth. Find the ratio (s/a) for the output rotation z to be the mean of the two input rotations x_1 and x_2. Sketch any necessary modification to the design.

If power were now transmitted without loss through the gear train from shaft X_1 to shaft Z with shaft X_2 fixed and with a torque T on X_1, find the fixing torque on X_2, clearly indicating its sense relative to T.

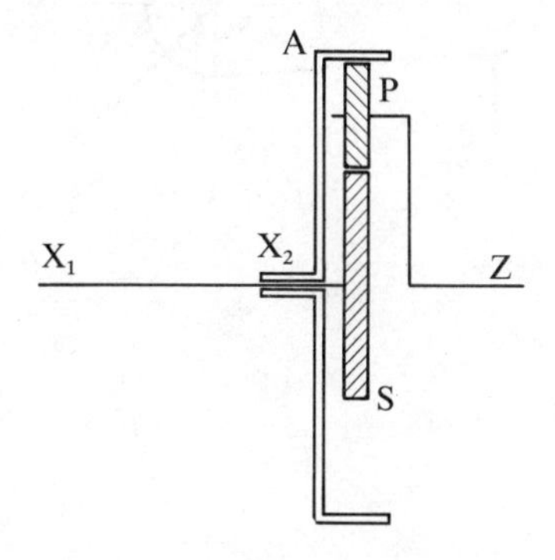

FIG. 10.31

12. In an epicyclic gear the sunwheel has 30 teeth and the planet wheels 15 teeth. Power is delivered to the gear through the shaft which carries the sunwheel, which rotates at 1000 rev min^{-1}. Find the magnitude and direction of the angular velocities of the shafts carrying the annulus and the spider on which the planet wheels are mounted, if equal power is delivered by the gear to these shafts and they rotate at different speeds.

13. Fig. 10.32 shows a gear box in which B is the input shaft and C the output shaft. The planet carriers X_1 and X_2 are integral with the output shaft. All wheels have the same module, and the numbers of teeth on the annuli and sun wheels are: $A_1 = A_2 = 80$, $S_1 = S_2 = 40$, $S_3 = 30$. Find the gear ratio (a) when A_1 is held, (b) when A_2 is held, (c) when S_3 is held.

For an input torque of 90 N m what torque is required to hold A_2 if the gear box efficiency is 95 per cent?

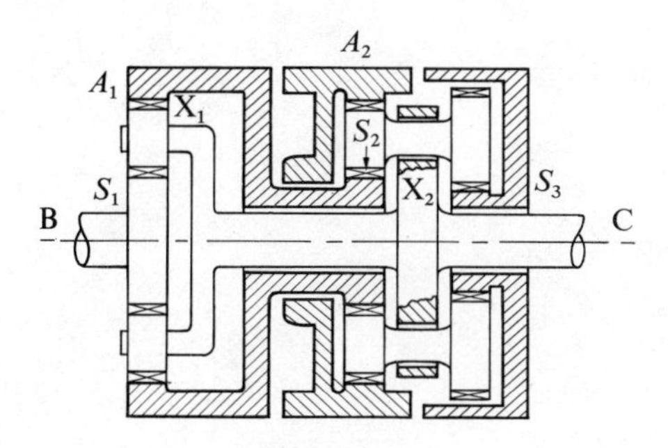

FIG. 10.32

14. In the bell-crank lever shown in Fig. 10.33 $a \ll l$. Determine the efficiency for **P′** overpowering **P** if the coefficient of friction at each of the pins is $\mu \ll 1$.

P′ and **P** are both fixed in direction but both can shift their lines of action laterally.

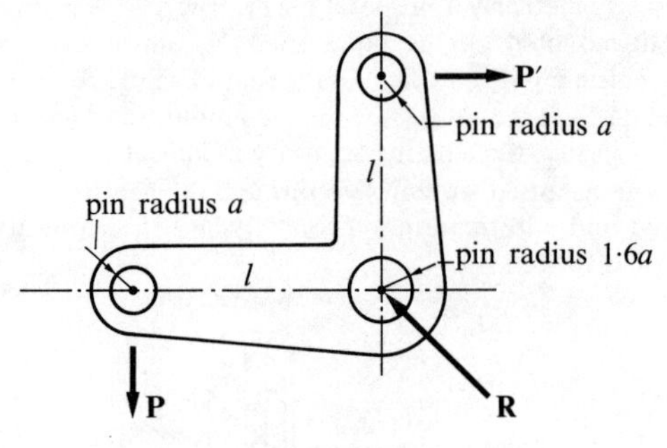

FIG. 10.33

Appendix 1. Some standard vector relationships

1.1. The distributive law for vector products

WE HAVE seen that

$$(\mathbf{a}\times\mathbf{b})\,.\,\mathbf{e}=(\mathbf{e}\times\mathbf{a})\,.\,\mathbf{b}.$$

Now let

$$\mathbf{b}=(\mathbf{c}+\mathbf{d})$$

so that

$$(\mathbf{a}\times\mathbf{b})\,.\,\mathbf{e}=(\mathbf{e}\times\mathbf{a})\,.\,(\mathbf{c}+\mathbf{d}).$$

Applying the distributive law for scalar products

$$\begin{aligned}(\mathbf{a}\times\mathbf{b})\,.\,\mathbf{e}&=(\mathbf{e}\times\mathbf{a})\,.\,\mathbf{c}+(\mathbf{e}\times\mathbf{a})\,.\,\mathbf{d}\\&=(\mathbf{a}\times\mathbf{c})\,.\,\mathbf{e}+(\mathbf{a}\times\mathbf{d})\,.\,\mathbf{e}\\&=(\mathbf{a}\times\mathbf{c}+\mathbf{a}\times\mathbf{d})\,.\,\mathbf{e}.\end{aligned}$$

But this relationship would hold for any vector **e**, so that

$$\mathbf{a}\times(\mathbf{c}+\mathbf{d})=\mathbf{a}\times\mathbf{c}+\mathbf{a}\times\mathbf{d},$$

and the distributative law for vector products is proven.

1.2. Evaluation of vector products and scalar triple-products

(i) Vector product as a determinant

Expressing two vectors **a** and **b** in terms of unit vectors **i**, **j**, **k**, we have

$$\begin{aligned}(\mathbf{a}\times\mathbf{b})&=(a_x\mathbf{i}+a_y\mathbf{j}+a_z\mathbf{k})\times(b_x\mathbf{i}+b_y\mathbf{j}+b_z\mathbf{k})\\&=a_xb_y(\mathbf{i}\times\mathbf{j})+a_xb_z(\mathbf{i}\times\mathbf{k})\\&\quad+a_yb_x(\mathbf{j}\times\mathbf{i})+a_yb_z(\mathbf{j}\times\mathbf{k})\\&\quad+a_zb_x(\mathbf{k}\times\mathbf{i})+a_zb_y(\mathbf{k}\times\mathbf{j})\\&=(a_xb_y-a_yb_x)(\mathbf{i}\times\mathbf{j})\\&\quad+(a_yb_z-a_zb_y)(\mathbf{j}\times\mathbf{k})\\&\quad+(a_zb_x-a_xb_z)(\mathbf{k}\times\mathbf{i})\\&=\begin{vmatrix}a_x&a_y\\b_x&b_y\end{vmatrix}\mathbf{k}+\begin{vmatrix}a_y&a_z\\b_y&b_z\end{vmatrix}\mathbf{i}-\begin{vmatrix}a_x&a_z\\b_x&b_z\end{vmatrix}\mathbf{j}\\&=\begin{vmatrix}\mathbf{i}&\mathbf{j}&\mathbf{k}\\a_x&a_y&a_z\\b_x&b_y&b_z\end{vmatrix}.\end{aligned}$$

(ii) Scalar triple-product as a determinant

$$(\mathbf{a}\times\mathbf{b})\,.\,\mathbf{c}=\begin{vmatrix}\mathbf{i} & \mathbf{j} & \mathbf{k}\\ a_x & a_y & a_z\\ b_x & b_y & b_z\end{vmatrix}\,.\,(c_x\mathbf{i}+c_y\mathbf{j}+c_z\mathbf{k}).$$

As $\mathbf{i}\,.\,\mathbf{j}=0=\mathbf{j}\,.\,\mathbf{k}=\mathbf{k}\,.\,\mathbf{i}$, we find on partially expanding the determinant

$$(\mathbf{a}\times\mathbf{b})\,.\,\mathbf{c}=\begin{vmatrix}a_x & a_y\\ b_x & b_y\end{vmatrix}c_z+\begin{vmatrix}a_y & a_z\\ b_y & b_z\end{vmatrix}c_x-\begin{vmatrix}a_x & a_z\\ b_x & b_z\end{vmatrix}c_y$$

$$=\begin{vmatrix}c_x & c_y & c_z\\ a_x & a_y & a_z\\ b_x & b_y & b_z\end{vmatrix}.$$

1.3. The triple-vector product

The compound product $\mathbf{a}\times(\mathbf{b}\times\mathbf{c})$ is known as the triple-vector product.

Consider first the special form $\mathbf{b}\times(\mathbf{b}\times\mathbf{c})$. As $(\mathbf{b}\times\mathbf{c})$ is perpendicular to both $\mathbf{b}$ and $\mathbf{c}$, and $\mathbf{b}\times(\mathbf{b}\times\mathbf{c})$ is perpendicular to $(\mathbf{b}\times\mathbf{c})$, it follows that $\mathbf{b}\times(\mathbf{b}\times\mathbf{c})$ is coplanar with $\mathbf{b}$ and $\mathbf{c}$, and perpendicular to $\mathbf{b}$. The magnitude is $\mathbf{b}^2\mathbf{c}\sin\theta$, where θ is the angle between $\mathbf{b}$ and $\mathbf{c}$.

$\mathbf{b}\times(\mathbf{b}\times\mathbf{c})$ can be resolved into two components, one parallel to $\mathbf{b}$, the other parallel to $\mathbf{c}$, see Fig. A1.1. The component in the direction of $\mathbf{b}$ has magnitude $(b^2c\sin\theta)\cot\theta$. The component in the direction of $\mathbf{c}$ has magnitude $(b^2c\sin\theta)\operatorname{cosec}\theta$. Hence

$$\mathbf{b}\times(\mathbf{b}\times\mathbf{c})=b^2c\cos\theta(\mathbf{b}/b)-b^2c(\mathbf{c}/c)$$
$$=\mathbf{b}(\mathbf{b}\,.\,\mathbf{c})-\mathbf{c}(\mathbf{b}\,.\,\mathbf{b}).$$

Similarly

$$\mathbf{c}\times(\mathbf{b}\times\mathbf{c})=\mathbf{b}(\mathbf{c}\,.\,\mathbf{c})-\mathbf{c}(\mathbf{b}\,.\,\mathbf{c}).$$

Any vector $\mathbf{a}$ can be resolved into three components in the directions $\mathbf{b}$, $\mathbf{c}$, and $\mathbf{b}\times\mathbf{c}$. This is provided that $\mathbf{b}$ and $\mathbf{c}$ are not parallel, in which case $\mathbf{a}\times(\mathbf{b}\times\mathbf{c})$ is zero for all $\mathbf{a}$. Let

$$\mathbf{a}=\beta\mathbf{b}+\gamma\mathbf{c}+\delta(\mathbf{b}\times\mathbf{c}).$$

$$\mathbf{a}\times(\mathbf{b}\times\mathbf{c})=(\beta\mathbf{b}+\gamma\mathbf{c})\times(\mathbf{b}\times\mathbf{c})$$
$$=\beta[\mathbf{b}(\mathbf{b}\,.\,\mathbf{c})-\mathbf{c}(\mathbf{b}\,.\,\mathbf{b})]+\gamma[\mathbf{b}(\mathbf{c}\,.\,\mathbf{c})-\mathbf{c}(\mathbf{b}\,.\,\mathbf{c})]$$
$$=\mathbf{b}[(\beta\mathbf{b}+\gamma\mathbf{c})\,.\,\mathbf{c}]-\mathbf{c}[(\beta\mathbf{b}+\gamma\mathbf{c})\,.\,\mathbf{b}]$$
$$=\mathbf{b}(\mathbf{a}\,.\,\mathbf{c})-\mathbf{c}(\mathbf{a}\,.\,\mathbf{b}).$$

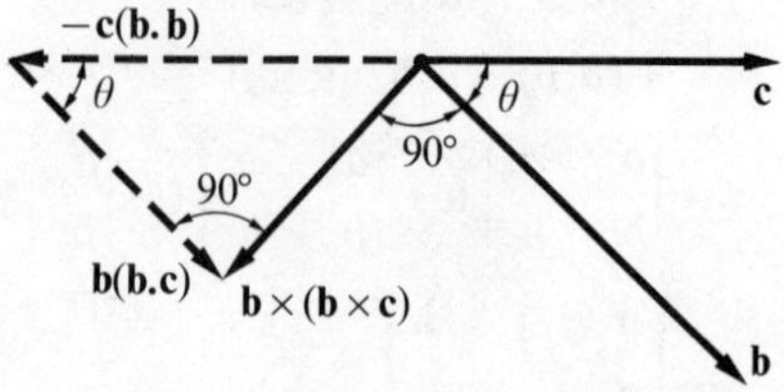

A1.1

Appendix 2. Relationships between the moments of inertia for laminae

2.1. Definitions

THE moment of inertia of a body about an axis is defined as $\sum n^2\,\delta m$, where n is the normal drawn from the element to the axis, and the summation extends over the total mass of the body.

The moments of inertia of a lamina about Cartesian axes Oxy drawn in its plane are

$$I_{xx} = \sum y^2\,\delta m$$

and

$$I_{yy} = \sum x^2\,\delta m.$$

The product of inertia with respect to Oxy is defined as

$$I_{xy} = \sum xy\,\delta m.$$

Oxy are principal axes if $I_{xy} = 0$, and I_{xx} and I_{yy} are then the principal moments of inertia.

I_{xy} is zero if either Ox or Oy is an axis of symmetry, and hence such an axis is automatically a principal axis.

2.2. The perpendicular-axis theorem

The moment of inertia of a lamina about an axis Oz normal to its plane is referred to as the polar moment of inertia. In conformity with the general definition above

$$I_{zz} = \sum r^2\,\delta m$$

where r is drawn from O to δm.

As $r^2 = x^2 + y^2$,

$$I_{zz} = I_{xx} + I_{yy}.$$

Thus the polar moment of inertia of a lamina about its normal through O is the sum of the moments of inertia about any pair of axes Oxy.

Two points are to be noted: (i) $r^2 = x^2 + y^2$ whatever the orientation of Oxy may be so that $(I_{xx} + I_{yy})$ is a constant $(= I_{zz})$. (ii) $I_{zx} = I_{zy} = 0$, because $z = 0$ for all δm, so I_{zz} is a principal axis.

2.3. The parallel-axis theorem

If O is at the centre of mass G of the lamina $\sum x\,\delta m = 0 = \sum y\,\delta m$.

The moments of inertia about axes O$'x'y'$ that are parallel to Gxy and with $O' \equiv (p, q)$, are related to I_{xx}, I_{yy}, and I_{xy} by the parallel axis theorem:

$$\begin{aligned} I_{x'x'} &= \sum (y-q)^2\,\delta m \\ &= \sum y^2\,\delta m - 2q \sum y\,\delta m + q^2 \sum \delta m \\ &= I_{xx} + Mq^2. \end{aligned}$$

Similarly

$$\begin{aligned} I_{y'y'} &= I_{yy} + Mp^2. \\ I_{x'y'} &= \sum (x-p)(y-q)\,\delta m \\ &= \sum xy\,\delta m - p \sum y\,\delta m - q \sum x\,\delta m + pq \sum m \\ &= I_{xy} + Mpq. \end{aligned}$$

If r_O is the distance from G to O, then by the perpendicular axis theorem

$$I_{z'z'} = I_{zz} + Mr_O^2.$$

The parallel axis theorem states that the moment of inertia of a lamina about any axis equals the moment of inertia about a parallel axis through G added to the product of the mass and the square of the distance between the two axes.

Care must be taken not to misinterpret this theorem: it does not apply as it stands to any pair of parallel axes, one must be through G. If neither axis passes through G the theorem must be applied twice over, first to find I about the parallel axis that does pass through G, remembering to subtract Mq^2, and then to move to the new axis.

2.4. Rotation of axes: Mohr's circle

The calculation of the moments of inertia for a lamina with reference to a given set of axes is most readily done when one axis, or even better both axes, is/are axes of symmetry. When a lamina has symmetry but the

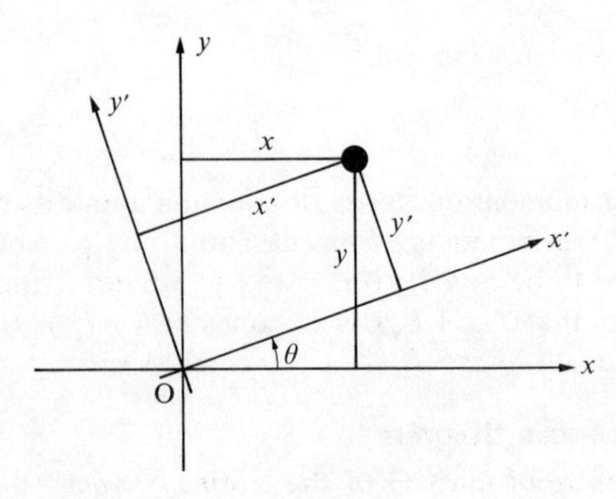

A2.1

moment of inertia is required about a non-principal axis it is usually best to determine first the principal moments of inertia and then to effect a change of co-ordinate system to give the moments of inertia about the required axes. To do this we need to know the relationships between moments of inertia for different sets of axes sharing the same origin.

Fig. A2.1 shows two sets of axes Oxy and O$x'y'$, the second being derived from the first by an anticlockwise rotation θ about O. The co-ordinates of an element δm in the O$x'y'$ system are related to those in the Oxy system by the equations

$$x' = x\cos\theta + y\sin\theta$$
$$y' = -x\sin\theta + y\cos\theta$$

so that

$$\begin{aligned} I_{x'x'} &= \sum y'^2\,\delta m \\ &= \sum(-x\sin\theta + y\cos\theta)^2\,\delta m \\ &= \sin^2\theta\sum x^2\,\delta m - 2\sin\theta\cos\theta\sum xy\,\delta m + \cos^2\theta\sum y^2\,\delta m \\ &= I_{xx}\cos^2\theta - I_{xy}\sin 2\theta + I_{yy}\sin^2\theta. \end{aligned}$$

Similarly

$$I_{y'y'} = I_{xx}\sin^2\theta + I_{xy}\sin 2\theta + I_{yy}\cos^2\theta$$

and

$$I_{x'y'} = \tfrac{1}{2}(I_{xx} - I_{yy})\sin 2\theta + I_{xy}\cos 2\theta.$$

These relationships can be used directly, and are easily evaluated on a small programmable calculator. They can also be evaluated graphically using the Mohr's circle construction. First, we rewrite the relationships in the form

$$\begin{aligned} I_{x'x'} &= \tfrac{1}{2}(I_{xx} + I_{yy}) + \tfrac{1}{2}(I_{xx} - I_{yy})\cos 2\theta - I_{xy}\sin 2\theta \\ I_{y'y'} &= \tfrac{1}{2}(I_{xx} + I_{yy}) - \tfrac{1}{2}(I_{xx} - I_{yy})\cos 2\theta + I_{xy}\sin 2\theta \\ I_{x'y'} &= \tfrac{1}{2}(I_{xx} - I_{yy})\sin 2\theta + I_{xy}\cos 2\theta. \end{aligned}$$

Now suppose that the principal axes OXY are derived by rotating from Oxy through an angle ϕ clockwise. Then we could write

$$\begin{aligned} I_{xx} &= \tfrac{1}{2}(I_{XX} + I_{YY}) + \tfrac{1}{2}(I_{XX} - I_{YY})\cos 2\phi \\ I_{yy} &= \tfrac{1}{2}(I_{XX} + I_{YY}) - \tfrac{1}{2}(I_{XX} - I_{YY})\cos 2\phi \\ I_{xy} &= \tfrac{1}{2}(I_{XX} - I_{YY})\sin 2\phi \end{aligned}$$

and $I_{x'x'}$, $I_{y'y'}$, and $I_{x'y'}$ would be obtained by writing $(\phi + \theta)$ in place of ϕ.

All these relationships are related by the geometrical construction in Fig. A2.2 and can be seen by inspection. The diagram assumes that $I_{xx} > I_{yy}$ and that I_{xy} is positive.

It is not necessary to know the principal axes and principal moments of inertia in order to draw the diagram. The circle has its centre at the point $\{-(I_{xx}+I_{yy}), O\}$ and it must pass through the points (I_{xx}, I_{xy}) and $(I_{yy}, -I_{xy})$.

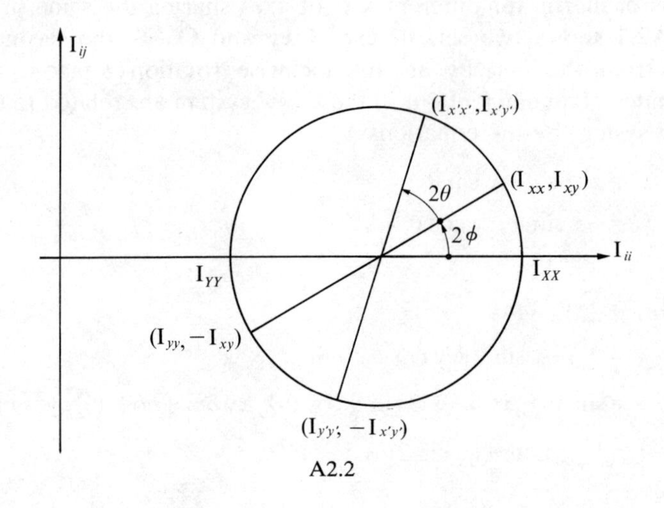

A2.2

Answers to problems

Chapter 1

1. $x = 0{\cdot}6714$, $y = 0{\cdot}9393$.
2. 15·91 km/hr, 45·3°W of N.
3. 36·87° ($\tan^{-1}\frac{3}{4}$) about (8,1), 143·13 about (2·167, 1·833).
4. (a) at the centre of the circle,
   (b) at the intersection of the normal to the walls at A and B,
   (c), (d) at the intersection of the normal to the wall at A and the normal to AB at the second contact.
6.

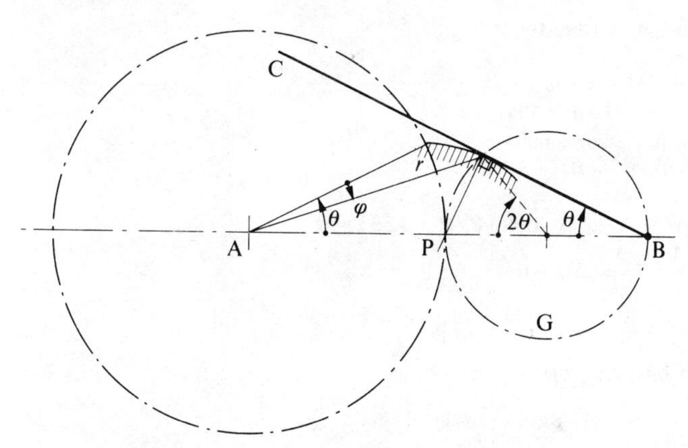

$r = 0{\cdot}25a(10 - 6\cos 2\theta)^{\frac{1}{2}}$, $\phi = \theta - \tan^{-1}\{\sin 2\theta/(3 - \cos 2\theta)\}$.
Note: The profile is an epicycloid generated by circle G rolling on the circle through P with centre at A.

8. I_{CBD} is at the intersection of the normals to the gliders at C and D. I_{BS} is at the intersection of I_{CBD}B and CS produced. $v = 10\sqrt{3}$ m/s.
9. 1115 rad/s, 26·5 m/s.
10. (a) 0·352 rad/s, (b) 1·00 m/s, (c) 3·26 m from bottom end.
11. (b)−6·01 rad/s, 7·92 rad/s, (c) 354 mm/s, (d) 69·7 mm/s.
12. 2·59 m/s.
13. 50 rad/s.
14. 2·28 rad/s.
15. 2·4 m/s.
16. 90°, 0·833 m/s; 270°, 7·50 m/s; (a) 2·18 m/s, (b) 0·84 m/s, (c) 0·29 m/s, (d) 72·8 rad/s.
17. $(\omega_4)_{min} = 0{\cdot}8\omega_2$ at $\theta = 0$; $(\omega_4)_{max} = 1{\cdot}33\omega_2$ at $\theta = 180°$; $\omega_4 = \omega_2$ at $\theta = 104{\cdot}5°$ and 255·5°; $75\omega_2$ mm/s; at the intersection of the perpendicular to BC from B and CA produced.
18. 26·4°, 86·4°.
19. (a) $F = 0$, (b) $F = 1$, (c) $F = 0$, (d) $F = 1$, (e) $F = 2$, (f) $F = 2$.

Chapter 2

1. 0·827 N at 57·03 to Ox.
4. $10\sqrt{(5/3)}$.

5. $F_{BE}=266\cdot7$ kN, $F_{DE}=-333\cdot3$ kN, $F_{BD}=-225\cdot8$ kN, $F_{AB}=336\cdot9$ kN, $F_{CB}=21\cdot0$ kN.
6. $2\sqrt{2}$ N cutting AB distance $2\cdot5a$ from A.
7. $a(3a^2+3b^2)^{\frac{1}{2}}/b$.
8. 0·604 N.
9. 709 N, 844 N, −597 N.
10. 10·92 kN.
11. $W \operatorname{cosec}\beta \sin\alpha \sin(\beta-\lambda)$.
12. 139 kN.
13. 0·93 W.
14. $W/\sqrt{3}$.
15. 17·7 N.
16. AC = 286, AD = 214, BC = 129, BD = 171, CE = 151, DF = 9, CD = 274, CF = 0.
17. $BC=5(3-\sqrt{3})/\sqrt{2}$, $CF=-CD=-DE=5(3-\sqrt{3})/2$, $AD=DF=-15/\sqrt{2}$.

Chapter 3

1. $1\cdot87\mathbf{n}-1.24\mathbf{t}$, $3\cdot10\mathbf{n}-0\cdot63\mathbf{t}$.
2. 1·502 km.
3. 0·72, 0·84, $5\mathbf{i}-4\mathbf{j}-3\mathbf{k}$.
4. 0·7017, $0\cdot5035\mathbf{i}+0\cdot7987\mathbf{k}$.
5. $\mathbf{\Omega}$, $\mathbf{r}$ and $\boldsymbol{\rho}$ are coplanar.
6. $-10/\sqrt{7}$, $1\cdot51\mathbf{i}-4\cdot04\mathbf{j}+3\cdot54\mathbf{k}$.
7. $-5\omega/3\sqrt{3}$, $2\cdot594\omega$.
9. $\mathbf{R}=-0\cdot161\mathbf{i}+0\cdot244\mathbf{j}-0\cdot469\mathbf{k}$, $R=0\cdot553$ Nm.
10. $\mathbf{R_A}=\left(\frac{11}{3}-\sqrt{3}\right)\mathbf{k}$; $\mathbf{R_B}=\left(-\frac{1}{2}-\frac{4}{\sqrt{3}}\right)\mathbf{i}+\frac{14}{3}\mathbf{k}$;
$\mathbf{R_C}=\left(-\frac{5}{2}+\frac{4}{\sqrt{3}}\right)\mathbf{i}-4\mathbf{j}+\left(\frac{11}{3}+\sqrt{3}\right)\mathbf{k}$.
11. 6·333, 2·981, 41·23 lbf.
12. $\frac{120}{\sqrt{70}}$, $\frac{-8}{\sqrt{14}}(15\mathbf{i}+4\mathbf{j}+8\mathbf{k})$; $\frac{40}{\sqrt{14}}(4\mathbf{i}-3\mathbf{j}-6\mathbf{k})$; $\frac{40}{\sqrt{7}}$.
13. $-2\cdot4\mathbf{j}+\mathbf{k}$; −3·23 kN, 6·52 kN.
16. $\frac{1}{3}, \frac{2}{3}, \frac{2}{3}$; 3, $-\frac{2}{3}$; $-\frac{2}{9}(\mathbf{i}+4\mathbf{j})$.

Chapter 4

1. 5·00 Nm.
2. −17·7 Nm.
3. 0·88 m/s, 0·363 m/s, 82·3 Nm.
4. $40/\sqrt{7}$.

Chapter 5

1. $\mathbf{v}=r\dot{\theta}\{(1-\cos\theta)\mathbf{i}+\sin\theta\mathbf{j}\}=r\dot{\theta}\mathbf{e}_1=2r\dot{\theta}\sin\frac{\theta}{2}\mathbf{t}$.
$\mathbf{a}=r\dot{\theta}^2\{\sin\theta\mathbf{i}+\cos\theta\mathbf{j}\}=-r\dot{\theta}^2\mathbf{e}_2=-r\dot{\theta}^2\left(\sin\frac{\theta}{2}\mathbf{n}-\cos\frac{\theta}{2}\mathbf{t}\right)$
2. 20·63 m/s, 122°13′E of N, -48×10^{-6} rad/s².
3. $-(v_A^2/l)\operatorname{cosec}^3\theta$, $-\frac{1}{2}(v_A^2/l)\operatorname{cosec}^3\theta$.
4. (i) 11·64 mm/s, 1·172 mm/s², (ii) 43·63 mm/s, 15·23 mm/s².
7. $625\sqrt{13}$ mm/s, $3125\sqrt{19}$ mm/s.
8. 260 rad/s², 11·7 m/s².
9. $5\cdot25a\omega^2$, 68·2 clockwise from $\overrightarrow{BA}$.
10. 10·33 rad/s², 14·8 m/s².

11. $80\ \mathrm{m/s^2}$.
12. $-190\ \mathrm{m/s^2}$.
13. $5{\cdot}70r\omega^2$.
14. $-6r\omega^2$.

Chapter 6

1. $2{\cdot}45\ \mathrm{m/s^2}$, 7·08 min./hr. gain.
2. 705 m.
3. 556 m.
4. 112 s, 1423 m.
6. (i) 55·6 m/s, (ii) 57·0 m/s, (iii) g/3 for 0·9 s every 18 s = uncomfortable.
7. 50 mm.
8. $\sqrt{(6mV/ea^2)}$.
9. 12·8 m.
10. $h(1+V/\sqrt{gh})$.
12. $\omega l\sqrt{31}$, $2\omega^2 l\sqrt{15}$.
13. 36 Mm.
15. $V-k(x^2+y^2)+\text{const.}$, $-2k\mathbf{r}$, S.H.M.
16. 17·2 N at 234·5° to Ox, 4·25 m/s.
19. (i) 370 MJ (392 MJ if g is assumed constant), (ii) 4·05 GJ.
20. $\frac{2}{5}mv_P$, impulse at $A=\frac{4}{35}mv_P$.
21. $(10-\Omega^2)^{-1}$ m, $(10-\Omega^2)^{1/2}/2\pi$ Hz.
24. 3·53 m/s.
25. $1{\cdot}52\sqrt{(ga)}$.

Chapter 7

1. $\mu g/\sqrt{2}$, $mv^2/l-0{\cdot}183\ \mu mg$.
2. 8·18 m.
4. v, $3v/4$; $3v/4$, $2v/3$.
5. 70 kN/m, 6·56 kJ.
6. 42·6 kN, 1·47 m/s, 0·155 m/s.

Chapter 8

1. Firing order A–C–B–D, or equivalent; $mar\omega^2\sqrt{2}$.
2. (a) $\frac{2}{5}Mr^2$, (b) $\frac{1}{12}M(a^2+b^2)$, (c) $0{\cdot}3Mr^2$, (d) $\frac{3}{80}M(h^2+4a^2)$, (k) $\frac{1}{24}Ma^2$, (f) $\frac{8}{9}M$, $\frac{1}{2}M$, 20·4° and 69·6° to side of length 4.
3. 236 s.
4. 6·72 m/s.
5. 0·7g.
6. AG = 0·906 m, $k=0{\cdot}608$ m.
7. $2\sqrt{(6g/7l)}$.
8. $\pi\sqrt{(2zr/gR)}$.
9. 1·16 m/s.
10. 2·17 rad/s.
11. $\{2af/(a^2+k^2)\}^{\frac{1}{2}}$.
13. 0·254 Hz.
14. $\frac{1}{6}ml^2\omega^2(7+2\sqrt{3})$.
15. $3mr^2\Omega\omega$.
16. 1·615 m/s.
17. $g/4\pi^2n^2$ below the support.
19. $(\sqrt{3}/4+\frac{1}{2})m\sqrt{gl}$.
20. 0·665 s.
21. $6a/5$.

22. 50·95 Nm.
23. (a) $a\omega\sqrt{3}$, (b) $a\omega^2/2\sqrt{3}$, (c) 0, (d) $R_A = R_D = ma\omega^2/6$.
24. $-\frac{5}{3}mr^2\omega^2$.
25. $\frac{2}{7}mg\sqrt{3}$.

Chapter 9

2. (a) 113 Nm, (b) 128 Nm; (a) 0·5, (b) 0·56.
3. 0·51 m.
4. (a) 0·1019 Nm, (c) 0·0364 Nm.
6. 22·84 kN, 6·18 kN, 148 kW, 285 kW.
7. $u(\log_e(M_0/M_1) + M_1/M_0 - 1)$;
   $[1 - M_1/M_0 - \log_e(M_0/M_1) + \frac{1}{2}(\log_e M_0/M_1)^2]u^2/g$.
8. 0·417 m/s^2.

Chapter 10

1. 238 mm outer diameter.
2. $(P/3 \sin\alpha)(R_2^3 - R_1^3)/(R_2^2 - R_1^2)$.
3. 22·3 kN.
4. (a) $8\pi\mu PR/(4 - \mu^2\pi^2) = 2{\cdot}42PR$ for $\mu = 0{\cdot}3$,
   (b) $2PR \sinh(\mu\pi) = 2{\cdot}18PR$ for $\mu = 0{\cdot}3$.
5. $\theta = (T/r)(e^{\mu\pi} - 1)/(s_2 e^{\mu\pi} + s_2)$; $(s_1 + s_2)r^2\theta$.
6. $W/2$, $W\sqrt{3}/2$.
7. 9·131 mm, 5·630 mm.
8. 6·045 m, contact ratio = 2·01.
9. 81, 18 teeth.
10. (a) 7:1, (b) 860 Nm.
11. $s/a = 1$, which is incompatible with the proposed design. Make A and S identical bevel gears, with P any convenient size, in the manner of Fig. 10.18.
12. −250 rev/min, 166·7 rev/min.
13. (a) $\frac{1}{3}$, (b) −1, (c) $\frac{2}{3}$; 166·5 Nm.
14. $(1 + 3{\cdot}36\mu a/l)P$.

Index